Wastewater Treatment Through Nature-Based Solutions

Naveen Dwivedi • Shubha Dwivedi
Ranjeet Kumar Mishra
Editors

Wastewater Treatment Through Nature-Based Solutions

Achieving Sustainable Development Goal 6

Springer

Editors
Naveen Dwivedi
Department of Biotechnology Engineering and Food Technology
University Institute of Engineering, Chandigarh University
Mohali, Punjab, India

Shubha Dwivedi
Department of Biotechnology
IIMT University
Meerut, Uttar Pradesh, India

Ranjeet Kumar Mishra
Department of Chemical Engineering
Manipal Institute of Technology, Academy of Higher Education
Manipal, Karnataka, India

ISBN 978-981-96-6698-0 ISBN 978-981-96-6699-7 (eBook)
https://doi.org/10.1007/978-981-96-6699-7

This Springer imprint is published by the registered company Springer Nature Singapore Pte Ltd.
The registered company address is: 152 Beach Road, #21-01/04 Gateway East, Singapore 189721, Singapore

Preface

The book's theme provides a comprehensive exploration of innovative, eco-friendly solutions to wastewater treatment, focusing on the role of nature-based approaches in addressing water and sanitation challenges. The book brings together interdisciplinary perspectives, offering an in-depth gaze at the potential of nature-based solutions (NBS) to support the realization of Sustainable Development Goal-6, ensuring access to water and sanitation for all and managing water resources sustainably. Nature-based solutions harness the innate assets of ecosystems; such as living microorganisms, forests, wetlands, and vegetation to address environmental and societal challenges. In wastewater treatment, these solutions exploit natural processes for the treatment of waste, water, mitigate pollution, and restore vital ecosystem services. By integrating these approaches into conventional infrastructure, we can enhance the resilience and sustainability of our water and wastewater treatment systems, reduce operational costs, and contribute to broader environmental and social goals.

Therefore, this book provides a broad impression of the latest technological innovations, practical case studies, and advanced research in fixing water pollution and contaminated environments through nature-based techniques and aims to highlight the potential of NBS in achieving SDG 6. The chapters explore a wide range of nature-based techniques such as vermifiltration, advanced integrated system, Urban Circularity Challenges via Nature-Based Solution, bio-hybrid system for scaling up treatment process, natural infiltration systems, and Bio-inspired/bio-nano materials for wastewater treatment to increase sustainability and resilience. Each solution is examined through the lens of its environmental sustainability, economic, and social benefits, providing a holistic perspective on their integration into water management strategies that align with climate resilience and the sustainable management of ecosystems.

The book is an essential resource for professionals involved in wastewater management, urban planning, environmental engineering, and sustainable development. It will also serve as a valuable textbook for students, research scholars in biotechnology engineering, environmental science, civil engineering, and related disciplines, offering both theoretical insights and practical applications.

We are thankful to the Springer publisher team, and all our contributors whose great efforts made this book a proper appearance.

Punjab, India Naveen Dwivedi

Contents

About the Editors and Contributors

About the Editors

Naveen Dwivedi is currently working as Professor and Head in the Department of Biotechnology Engineering and Food Technology, University Institute of Engineering, Chandigarh University, Mohali, Punjab, India. He obtained his M.Tech. degree in Biotechnology Engineering from Institute of Engineering and Technology, Lucknow. He has completed his M.Tech. dissertation work at Fermentation Division of Central Drug Research Institute (CSIR-CDRI), Lucknow, and Ph.D. research work at Bioenergy and Environmental Engineering Research Lab, Indian Institute of Technology Roorkee. He is broadly interested in the fields of Environmental Biotechnology, Bioprocess Engineering, Bioenergy, Environmental sustainability, Biomass conversion to materials and energy, and Waste valorization. He has more than 21 years of teaching and research experience. He has good number of research publications in peer-reviewed journals and conferences, edited books, and written around more than three dozen chapters in the books published by reputed international publishers.

Shubha Dwivedi is a researcher by profession. She is an Associate Professor and Head in the Department of Biotechnology, IIMT University, Meerut, India. She received her post-graduation degree in Biochemistry from Jiwaji University Gwalior and Ph.D. from Sustainable Processing and Water Treatment Research Lab, Indian Institute of Technology Roorkee. She was associated with Institute of Genomics and Integrative Biology (CSIR-IGIB), Delhi, during her master's. Her working area has been in the development of eco-friendly technology for the removal of hazardous contaminants from wastewater. She has more than 19 years of teaching and research experience. Her outstanding research work is a testament of a good number of research publications, books, book chapters, patents, etc.

Ranjeet Kumar Mishra is working as an Associate Professor, Department of Chemical Engineering, Manipal Institute of Technology, Manipal, Karnataka, India. He received his Ph.D. degree from Indian Institute of Technology Guwahati, Assam, in the year 2019. He completed his post-doctoral at University of Guelph Canada (2 years) and Indian Institute of Technology Madras (1 year). He has published more than 50 research articles in the academic press with an h-index of 30 and over

3732 citations. His current research interests are biomass conversion, pyrolysis, waste management, biochar, biofuel, hydrothermal liquefaction, water and wastewater treatment, etc. He is also associated with various SCI and Scopus International journals as an editor, associate editor, academic editor, and editorial board member.

Contributors

Gholamareza Abdi Department of Biotechnology, Persian Gulf Research Institute, Persian Gulf University, Bushehr, Iran

Oluwarinu A. Adepoju Department of Microbiology, Obafemi Awolowo University, Ile-Ife, Nigeria

Ibukun Modupe Adesiyan Department of Environmental and Occupational Health, School of Public Health, University of Medical Sciences, Ondo, Nigeria

Edward Terhemen Akange Department of Fisheries and Aquaculture, Joseph Sarwuan Tarka University, Makurdi, Nigeria

Austine Atokolo Department of Biological Sciences, College of Science and Technology, Covenant University, Ota, Nigeria

Shruti Awasthi Department of Life Sciences, Garden City University, Bangalore, Karnataka, India

Mounia Baidou Materials, Membranes and Nanotechnology Research Team, Moulay Ismail University of Meknès, Faculty of Sciences, Zitoune, BP: 11201, Meknès, MOROCCO

Latifa E.L. Bardiy Research Team "Data Science, Innovation and Development"; Affiliated with the Faculty of Legal, Economic and Social Sciences of Tangier, Tanger, Morocco

Mukul Machhindra Barwant Department Botany, Sanjivani Rural Education Society, Sanjivani Arts Commerce and Science College Kopargaon, Ahmednagar, Maharashtra, India

Department of Botany, YashavantraoChavan Institute of Science, Satara, Maharashtra, India

Maheswari Behera Department of Botany, College of Basic Science and Humanities, Odisha University of Agriculture and Technology, Bhubaneswar, Odisha, India

Abdelmjid Bouazizi Materials, Membranes and Nanotechnology Research Team, Moulay Ismail University of Meknès, Faculty of Sciences, Zitoune, BP: 11201, Meknès, MOROCCO

Interdisciplinary Laboratory for Modeling simulation and scientific innovation (LIMSIS), École Normale superieure of Meknès, Moulay Ismail University of Meknès, Meknès, Morocco

Laboratory of Materials, Membranes and Environment, Faculty of Sciences and Technologies of Mohammedia, Hassan II University of Casablanca, Casablanca, Morocco

Proma Chakraborty Department of Life Sciences, Garden City University, Bangalore, Karnataka, India

Chethana V. Chalapathy Department of Life Sciences, Garden City University, Bangalore, Karnataka, India

Bellam Darshitha Department of Life Sciences, Garden City University, Bangalore, Karnataka, India

Mohamed Douma Materials, Membranes and Nanotechnology Research Team, Moulay Ismail University of Meknès, Faculty of Sciences, Zitoune, BP: 11201, Meknès, MOROCCO

Zuwa Yoowa Duyann Department of Molecular Biology and Bioinformatics, Africa Centre of Excellence for Mycotoxins and Food Safety (ACEMFS), Federal University of Technology, Minna, Nigeria

Aarti Dwivedi Department of Chemistry, Amity School of Engineering & Technology, Amity University, Gwalior, Madhya Pradesh, India

Naveen Dwivedi Department of Biotechnology Engineering and Food Technology, University Institute of Engineering, Chandigarh University, Mohali, Punjab, India

Shubha Dwivedi Department of Biotechnology, IIMT University, Meerut, Uttar Pradesh, India

Vinay Dwivedi Amity Institute of Biotechnology, Amity University, Gwalior, Madhya Pradesh, India

Soukaina El Abbadi Materials, Membranes and Nanotechnology Research Team, Moulay Ismail University of Meknès, Faculty of Sciences, Zitoune, BP: 11201, Meknès, MOROCCO

Hajar El Moustansiri Materials, Membranes and Nanotechnology Research Team, Moulay Ismail University of Meknès, Faculty of Sciences, Zitoune, BP: 11201, Meknès, MOROCCO

Yetunde Mutiat Feruke-Bello Department of Microbiology, University of Ilesa, Ilesa, Osun State, Nigeria

Prajna Ghosh Department of Biotechnology Engineering and Food Technology, University Institute of Engineering, Chandigarh University, Mohali, Punjab, India

Lahar Khandelwal Department of Life Sciences, Garden City University, Bangalore, Karnataka, India

Prema Kulkarni Department of Life Sciences, Garden City University, Bangalore, Karnataka, India

Neeraj Kumar Department of Chemistry, Applied Science (NIET), NIMS University, Jaipur, Rajasthan, India

Riti Kumari Department of Biotechnology Engineering, University Institute of Engineering, Chandigarh University, Mohali, Punjab, India

Ayush Madan Department of Biotechnology, School of Research and Technology, People's University, Bhanpur, Bhopal, Madhya Pradesh, India

Institute of Tropical Aquaculture and Fisheries (AKUATROP), Universiti Malaysia Terengganu, Kuala Nerus, Terengganu, Malaysia

Gurmail Singh Malhi Department of Aerospace Engineering, Chandigarh University, Chandigarh, India

Emmanuel Ojochegbe Mameh Department of Biological Sciences, College of Science and Technology, Covenant University, Ota, Nigeria

Gbotemi Jerry Oni Department of Molecular Biology and Bioinformatics, Africa Centre of Excellence for Mycotoxins and Food Safety (ACEMFS), Federal University of Technology, Minna, Nigeria

Mohamed Ouammou Laboratory of Materials, Membranes and Environment, Faculty of Sciences and Technologies of Mohammedia, Hassan II University of Casablanca, Casablanca, Morocco

Mojisola Christiana Owoseni Department of Microbiology, Federal University of Lafia, Lafia, Nigeria

Samuel Adeniyi Oyegbade Department of Biological Sciences, College of Science and Technology, Covenant University, Ota, Nigeria

Piyus Kumar Pathak Department of Applied Science and Humanity, ITS Engineering College, Knowledge Park III, Greater Noida, Uttar Pradesh, India

Noor Hidayah Abd Rahman School of Chemical and Energy Engineering, Faculty of Engineering, UniversitiTeknologi Malaysia, Skudai, Johor, Malaysia

Moksh Rajpal Department of Biotechnology Engineering and Food Technology, University Institute of Engineering, Chandigarh University, Mohali, Punjab, India

Nurfadhila Nasya Ramlee School of Chemical and Energy Engineering, Faculty of Engineering, UniversitiTeknologi Malaysia, Skudai, Johor, Malaysia

Ramandeep Saini Chandigarh College of Technology, Chandigarh Group of Colleges, Landran, Greater Mohali, Punjab, India

Palak Sangal Department of Biological Sciences, Birla Institute of Technology & Science, Pilani, Rajasthan, India

Prashant Saxena Department of Biotechnology Engineering, University Institute of Engineering, Chandigarh University, Mohali, Punjab, India

Pooja Sharma Department of Biotechnology, IIMT University, Meerut, Uttar Pradesh, India

Rhydum Sharma Department of Biotechnology Engineering and Food Technology, UIE, Chandigarh University, Chandigarh, India

Anoushka Singh Department of Biotechnology Engineering and Food Technology, University Institute of Engineering, Chandigarh University, Mohali, Punjab, India

Lakshmi Singh Department of Botany, College of Basic Science and Humanities, Odisha University of Agriculture and Technology, Bhubaneswar, Odisha, India

Suhaily Suhaimi School of Chemical and Energy Engineering, Faculty of Engineering, UniversitiTeknologi Malaysia, Skudai, Johor, Malaysia

Queen Elizabeth Sule Department of Biological Sciences, College of Science and Technology, Covenant University, Ota, Nigeria

Najib Tijani Materials, Membranes and Nanotechnology Research Team, Moulay Ismail University of Meknès, Faculty of Sciences, Zitoune, BP: 11201, Meknès, MOROCCO

Jahnvi Vashishth NIMS Institute of Allied Medical Science and Technology, NIMS University, Jaipur, Rajasthan, India

Aline Viancelli Universidade Do Contestado, Concordia, SC, Brazil

Ahmed Wopa Wurie Department of Molecular Biology and Bioinformatics, Africa Centre of Excellence for Mycotoxins and Food Safety (ACEMFS), Federal University of Technology, Minna, Nigeria

Nuridayah Yusmaidi Department of Chemistry, Faculty of Science, UniversitiTeknologi Malaysia, Skudai, Johor, Malaysia

Advanced Integrated Nature-Based Solutions (NbS) for Fluoride Removal from Wastewater to Increase Sustainability and Resilience

1

Pooja Sharma and Shubha Dwivedi

Abstract

Nature-based solutions (NbS) mimic and integrate natural processes to endow with resilient and sustainable wastewater treatment methods. This study investigates complicated NbS frameworks that deal with wastewater issues by fusing modern technology with ecological engineering concepts. In order to maximise pollution exclusion while lowering energy and resource expenditure, the optional integrated systems make use of vegetated buffer strips, bioreactors supplemented with microbial consortia and built wetlands. Through an analysis of case studies and experimental projects, we exhibit how NbS may develop system resilience to climatic inconsistency, reduce pollution and rebuild ecosystems. This study emphasises NbS as a practicable way to endorse environmental sustainability in wastewater treatment and accomplish circular economy objectives. This chapter emphasises various methods employed by researchers in the area of fluoride removal from water and the mechanisms involved in fluoride removal.

Keywords

Sustainable technology · Nature-based solutions · Waste water treatment · modern technology

1.1 Introduction

The creative application of natural systems and processes to solve societal issues while providing environmental, social and economic advantages is known as advanced integrated nature-based solutions (NbS). These solutions increase their

P. Sharma · S. Dwivedi (✉)
Department of Biotechnology, IIMT University, Meerut, Uttar Pradesh, India

N. Dwivedi et al. (eds.), *Wastewater Treatment Through Nature-Based Solutions*,
https://doi.org/10.1007/978-981-96-6699-7_1

effectiveness, scalability and durability by combining cutting-edge scientific techniques, engineering and technology with conventional nature-based methodologies. Nature-based solutions (NbS) use natural processes to combat pollution, providing a sustainable and economical waste management option. Modern technologies are used with old ecological practices in advanced integrated NbS to increase its scalability and efficiency. For instance, microbial bioreactors and biofilters can be used with artificial wetlands to treat wastewater and promote biodiversity. Likewise, modern monitoring systems may be used in conjunction with phytoremediation techniques, which use plants to absorb or break down pollutants, to maximise performance in real time. These integrated techniques offer co-benefits including habitat development, carbon sequestration and community resilience in addition to reducing environmental impact. Advanced NbS offer a promising route to sustainable development and the objectives of the circular economy, as waste output is still rising due to urbanisation and industrialisation.

In the twenty-first century, Fluoride contagion of groundwater has become a major worldwide concern. Many people in nations like India, South Africa, and the East African valley suffer from an assortment of health issues as a consequence of consumption of groundwater tainted with fluoride. Numerous regulatory bodies, including the US EPA, WHO, CPCB and others, have set the maximum allowable limit for fluoride in drinking water at 1.5 mg/l. Minerals and rocks that are nearby and underneath the aquifer liquefy and discharge fluoride into the groundwater (Murray 1973). It gets into the soil through waste runoff, precipitation, or rock weathering. Additionally, a variety of industrial operations such as production of glass, bricks, nickel and copper, and burning of coal help raise the fluoride levels in water (Camargo (2003); Simons (1993); Pollution Prevention and Abatement Handbook Phosphate Fertilizer Plant (1998); State-wise Number of habitations (2016); Kabay et al. (2008); Hichour et al. (2000)).

Fluoride has been removed from water using a variety of physico-chemical processes, together with adsorption, ion exchange, electro-dialysis, coagulation/precipitation, reverse osmosis, nanofiltration, ultra filtration and others (Meenakshi et al. 2007; Haron et al. 2001; Sundaram et al. 2008; Chubar et al. 2005). Each of these methods has unique benefits and drawbacks, including lower efficiency, delicate working conditions and the creation of secondary sludge during use.

1.1.1 Analysis of the Challenges Associated with Fluoride in Water

In India, an estimated 40 million people are at danger of getting fluorosis, and 20 million have already had significant symptoms. Since the 1930s, when fluorosis was first discovered in India, the disease has been widespread in both urban and rural parts of different village blocks, and the number of individuals who have the condition has been steadily rising. There are around 62 million people with dental, skeletal, and non-skeletal fluorosis in India. Of them, six million are children under the age of fourteen according to The Foundation for Fluorosis Research and Rural Development (Susheela 2001) Fluorosis: Indian situation: A dissertation on Fluorosis. Twenty of the 36 states and union territories that constitute the Indian Republic have widespread fluorosis. The source of drinking water is responsible for the apparent variance in the prevalence of dental fluorosis. For the majority of India, groundwater is the most important source of drinking water. The primary issue influencing fluoride levels is groundwater stage; the greater the subterranean level the water comes from, the higher the fluoride level. The Ministry of Health & Family Welfare has recognised at least 132 areas across 19 states as harshly affected by extreme levels of fluoride in drinking water, a key contributor to fluorosis. According to a press release posted on Live Mint in July 2017, there are 75 districts in the state of UP, 51 in MP, 30 in Rajasthan and a few in Bihar, where the population is extremely vulnerable to fluorosis. According to Meenakshi et al. (2007), the amount of fluoride in groundwater varied across 17 Indian states. Prior reports indicated that fluoride was present in 230 districts across 19 states. According to the Ministry of Drinking Water and Sanitation's most recent data, 14,035 households in 17 states as of April 1, 2016, lacked access to safe drinking water. Furthermore, 11.53 million people live in areas where drinking water has excessive levels of fluoride, which puts them at risk (National Programme for Prevention & Control of Fluorosis (NPPCF) (2018); Fornasiero (2001)). The NPPCF survey and the presumed instances of dental and skeletal fluorosis were the main topics of Table 1.1. Figure 1.1 provides a percentage analysis based on the total number of individuals analysed in the community and at school using the data given in Table 1.1.

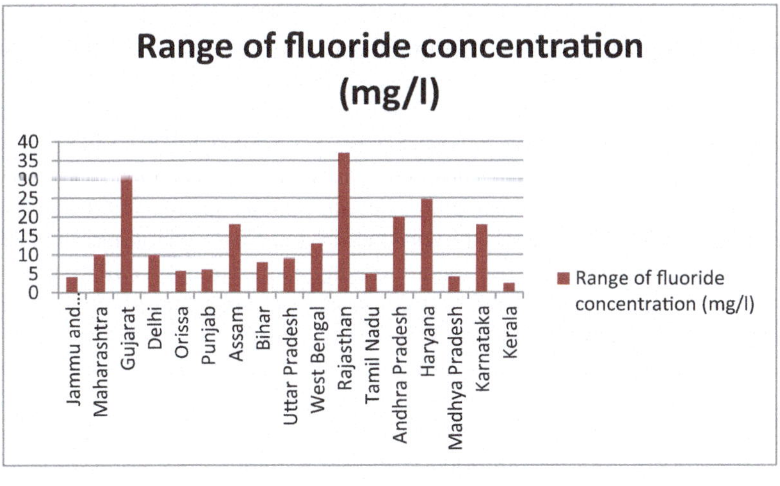

Fig. 1.1 The range of fluoride levels in several Indian states

Table 1.1 Fluoride levels in more than a few Indian states

S. No.	States	Array of fluoride absorption (mg/l)
1	Jammu and Kashmir	0.05–4.21
2	Maharashtra	0.11–10.2
3	Gujarat	1.58–31.0
4	Delhi	0.4–10.0
5	Orissa	0.6–5.7
6	Punjab	0.44–6.0
7	Assam	0.2–18.1
8	Bihar	0.6–8.0
9	West Bengal	1.6–13.0
10	Uttar Pradesh	1–10
11	Tamil Nadu	1.7–5.2
12	Rajasthan	0.3–38.0
13	Andhra Pradesh	0.11–20.0
14	Haryana	0.17–24.7
15	Madhya Pradesh	0.08–4.2
16	Karnataka	0.2–18.0
17	Kerala	0.2–2.5

1.1.2 Sources of Fluoride Contagion

The causes of fluoride pollution in water are both natural and anthropogenic (caused by humans) and each plays a major role in the ecosystem. Geological weathering of fluoride-rich minerals, which are frequently found in rocks and include fluorite (calcium fluoride), apatite (calcium phosphate fluoride) and cryolite (sodium aluminium fluoride), is the main source of fluoride. Increased fluoride levels result from the dissolution of these minerals by groundwater flowing through these rocks, particularly in places with significant geothermal activity, volcanic regions and alkaline soils. Gaseous fluoride released by volcanic eruptions has the ability to descend and settle on water. Further contributing to the natural fluoride cycle is the atmospheric deposition of sea spray in coastal areas, which can introduce fluoride into water systems.

Industrial and agricultural processes, human activity greatly increases fluoride pollution. Fluoride is released as a byproduct by industries including steel, ceramics, glass, phosphate and fertiliser manufacturing and aluminium smelting. Air pollutants or improper industrial effluent disposal can cause fluoride to build up in surrounding soil and water bodies. When phosphate-based fertilisers are widely used in agriculture, fluoride is also introduced into water sources through runoff. Burning coal releases compounds that include fluoride, especially in homes and power plants. These compounds end up in the environment and seep into waterways. Additionally, localised fluoride pollution is frequently caused by untreated effluents from the manufacture of pesticides and other chemicals. The various sources have been listed in Table 1.2.

Table 1.2 The natural and anthropogenic sources of fluoride contamination

Source type	Specific sources	Mechanism of contamination
Natural sources	Weathering of rocks that contain fluoride, such as fluorite, apatite and cryolite	Fluoride is dissolved from rocks and minerals by groundwater
	Deposition in the atmosphere (sea spray)	Sea salt aerosols deposit fluoride on adjacent land and water surfaces
	Activity of volcanoes	After being released into the atmosphere, fluoride gases and particles end up in soil and water
	Geothermal activity (hot springs, for example)	Geothermal activity (hot springs, for example)
	Soils with high fluoride content in arid or semi-arid areas	When groundwater is alkaline and evaporative, naturally occurring fluoride builds up
Anthropogenic sources	Burning coal (industrial and household)	Emissions from burning coal that include fluoride leak into the ground and water
	Industrial emissions (such as those from the steel, glass, phosphate fertiliser and aluminium smelting industries)	Inappropriate waste disposal and effluents discharge fluoride gases and particles into the air or water
	Fertilisers based on phosphate	Through agricultural runoff, fluoride from fertilisers finds its way into the land and water
	Production of chemicals and pesticides	Through untreated effluents, fluoride-rich waste from these companies can affect adjacent water bodies
	Urban waste (such as home chemicals, mouthwashes and fluoridated toothpaste)	Surface and groundwater become contaminated with fluoride due to improper sewage and disposal systems
	Ceramics industries and brick kilns	Through emissions and wastewater leakage, fluoride is released into the environment

1.2 Fluoride Within Consumption Water: Authoritarian Guidelines and Limitations

The ionic form of fluorine, fluoride, is the major electronegative element in the halogen family and does not originate in nature in its elemental state because of its high reactivity. The most potent oxidising agent known is fluorine, which is combustible and unpleasant in its elemental form. In nature, fluoride may be found in two different forms: inorganic and organic. Fluoride often comes from a variety of sources and is introduced into the surroundings. Sources of fluoride might be categorised into two groups: anthropogenic (including industrial operations) and natural (including fluorine originated in the natural world and in the Earth's crust, mostly as the

Table 1.3 Fluoride concentration in different food stuff

Food item	Fluoride concentration (mg/kg)
Mutton	3.2
Fishes	5.5
Beef	4.5
Rock salts	230.5
Tea	100
Pork	4

minerals fluorapatite and cryolite). Glassworks exhaust emissions, and the manufacturing of metals, bricks, ceramics and adhesives are further sources. The most widely accepted sources of fluoride found in groundwater are the mixture of rocks and soils or the weathering and atmospheric eruptions of volcanic particles. Both organic and inorganic forms of fluoride are present. When dissolved in water, fluoride has no colour, taste, or odour. As a result, it is difficult to ascertain via physical examination. The concentration of this ion may only be ascertained by chemical examination of the water samples. The colorimetric sodium 2-(parasulfophenylazo)-1,8-dihydroxy-3,6-naphthalene disulfonate scheme is the largest part extensively used technique for formative fluoride levels in groundwater samples. The level of fluoride in water may be measured in the field or in the lab using ion selective electrodes. Fluoride is one of the necessary micronutrients for human health and is necessary intended for strong bones and teeth. Approximately 95% of the fluoride in the human body is found in bones and teeth. The fluoride content of various foods is displayed in Table 1.3.

The following variables determine fluoride poisoning and the biological reaction that results in negative effects:

- Drinking water with excessive fluoride content
- Drinking water with high alkalinity and low calcium content
- Fluoride intake capacity (total/day), fluoride exposure duration and individual age especially the expectant and nursing mother are the most susceptible individuals in the populations because fluoride enters maternal milk and crosses the placenta without any barriers
- Hormonal profile disruption, either as a cause or as a result of fluoride toxicity, exacerbates the illness.

The toxicity of fluoride can be attributed to many processes. Fluoride that is consumed first affects the gut mucosa locally. It can cause gastrointestinal (GI) discomfort by forming hydrofluoric acid in the stomach. Additionally, fluoride affects how glucose is metabolised. Fluoride exposure has been linked to impaired glucose intolerance, according to the results. Because fluoride inhibits Na+/K + -ATPase, potassium may be released extracellularly, potentially resulting in hyperkalaemia. Dental fluorosis is caused by the secretary route, which is influenced by fluorides. The chemical mechanism of inorganic fluoride also affects a number of other cellular functions. Fluoride can change gene expression, cause cell death and produce oxidative stress, homeostasis and lipid peroxidation.

Redox status: Fluoride changes the confirmation of cell by changing its mitochondria count for example in Kidney cell. Fluoride hinders the process of cellular respiration by inhibiting the functions of mitochondria through fluoridesensitive enzymes of the Krebs cycle and glycolytic pathways, fluoride ions connect to the functional amino acid groups surrounding the active centre of an enzyme. Exposure to fluoride raises the production of anion superoxide, its concentration and its byproducts, including hydrogen peroxide. NO production is also increased by fluoride. The antioxidant activity is inhibited by fluoride. Apoptosis: When fluoride concentration is quite high, it causes mitochondrial malfunction and the activation of descending pathways, which are the main mechanisms of cell death. Fluoride also causes apoptosis by increasing oxidative stress-induced lipid peroxidation. Although fluoride ions can occasionally increase enzyme activity, fluoride itself functions as an enzyme inhibitor (Barbier et al. 2010).

1.3 Mechanisms of Fluoride Removal

Various technologies are available for defluoridation from water. One of the most popular methods is adsorption, in which fluoride ions connect to the outside of substances like bone char, activated alumina or biosorbents. By adding substances like alum or lime, precipitation techniques create insoluble fluoride compounds that may be extracted from water. Ion exchange methods use artificial resins to swap out fluoride ions for other anions. Reverse osmosis and nanofiltration are two membrane-based technologies that physically separate fluoride ions by pushing water through semipermeable membranes. Finally, electrochemical techniques encourage the deposition or removal of fluoride by use of electrical currents. The effectiveness of each technique is influenced by variables including fluoride content, water chemistry and operating circumstances. Three primary types of therapy processes are identified and explained below. These techniques include of chemical, biological and physical therapy.

1.3.1 Physical Treatment

Following physical methods used for the elimination of fluoride from ground/drinking water are divided into the following categories:

1.3.1.1 Adsorption

The removal of fluoride from contaminated water resources by using Adsorption process is a crucial step due to exceed of limit of fluoride concentration in the groundwater. Adsorption process is a surface based phenomenon where solid adsorbent are used to remove fluoride ions as the particles get adhere on the surface of solid adsorbent by using mechanism like electrostatic attraction, surface complexation and mainly ion exchange process. Examples of adsorbents used to get rid of a range of contaminants include bone charcoal, granular activated carbon, activated alumina and synthetic ion exchange resins. There are several factors like temperature, pH (slightly acidic to neutral pH), fluoride concentration, contact time and concentration of competing ions that affect the functionality of adsorption process (Wasewar et al. 2010; Feenstra et al. 2007). In this mechanism the solid bioadsorbent have active binding sites on its surface that comes in contact with fluoridecontaining water and binds the fluoride ions to active sites on the surface, reducing the fluoride concentration in the water. After saturation, the adsorbent can be restored by means of chemical solutions or can be replaced with other bioadsorbents (Alagumuthu et al. 2010; Popat et al. 1994; Gandhi et al. (2012); Dwivedi & Mondal (2018); Herrera et al. (2009)).

Benefits

- *Cost-Effectiveness:* When inexpensive materials like activated alumina or bio-based adsorbents are used, adsorption is frequently less expensive than more sophisticated techniques like reverse osmosis.
- *Simplicity:* The procedure is easy to execute and does not call for sophisticated equipment or a high level of operating knowledge.
- *Low Energy Consumption:* Adsorption typically takes less energy than membrane filtering systems.
- *Effectiveness at Low Concentrations:* Adsorption is a viable treatment option for regions with slightly polluted water since it can effectively treat low fluoride concentrations.
- *Eco-friendliness:* When used, many adsorbents, particularly natural ones like clay and charcoal, may be safely disposed away or regenerated.

1.3.1.2 Precipitation

Chemicals are added during precipitation procedures, and fluoride precipitates are created. These include calcium and aluminium salt-containing precipitations. According to Feenstra et al. (2007), a certain quantity of sludge is created daily through precipitation, and chemicals must be administered in stages.

Benefits:

- Easy to accomplish
- Inexpensive
- Requires some ability and practice but not a lot of competence
- Instruments are readily accessible; it is the oldest and the most extensively used

Limitations:

- Reactivity of the components to be titrated should be thoroughly studied as this may impact the end point
- Instruments must be calibrated correctly as this may affect the final findings

- Effective results need expertise and practice
- If done by hand, it takes a lot of time

1.3.1.3 Membrane Process

Filtration is the most widely used water treatment technique for eliminating particles from water. Coagulation, flocculation, and settling are employed to improve the efficiency of the filtration process. Because the coagulation and settling processes have become so effective, filtration may not always be necessary. However, where filtering has been neglected, slime accumulation in the mains has led to notable decreases in the water main carrying capacity.

A permeable membrane is used in these procedures to remove a variety of pollutants. In these processes, a permeable membrane is employed to eliminate a range of contaminants. Reverse osmosis (Sourirajan et al. 1972), ultra-filtration (Guo et al. 2001), nanofiltration, and electro-dialysis are membrane filtering methods that can be used to eliminate fluoride.This process takes place in opposition to the concentration gradient

Advantages:

- Chemicals such as coagulants, flocculants, disinfectants and pH adjusters are not required
- Constant quality of treated water in terms of chemical and microbial removal
- Easy automatisation
- Plant and process compactness
- Eliminate most of the dissolved non-ions and almost all contaminating ions

The thin film composite membrane and reverse osmosis membrane will be attacked and degraded by organic compounds and chlorine, which are captured by an activated carbon filter.

Limitations:

- Due to their low back pressure, household reverse osmosis devices consume a lot of water, with the remaining being released as waste water.
- In large-scale industrial/municipal systems, the recovery rate is usually between 75 and 80%, or even 90%, of the supply water.

1.3.1.4 Coagulation

The chemical process of coagulation is used to remove fluoride from water. Usually, coagulants, such as ferric chloride and aluminium sulphate, are added to the water to encourage the development of insoluble compounds that can capture and remove fluoride and other impurities. The following steps are involved in the process:

- Coagulant Addition: When coagulants are added to water, they create flocs, which are aggregated particles.

- Floc Formation: Fluoride ions and other waterborne particles gather to form bigger, easier-to-remove flocs as a result of the coagulants neutralising their charge.
- Floc Settling or Filtration: The flocs, which now contain fluoride and other impurities, either filter out of the water or settle out.

1.3.1.5 Biological Treatment

A newly developed method for treating water that makes use of widely accessible biomaterials is biological elimination and biosorption. For the elimination of fluoride, several biosorbents have been created. Chitin and chitosan-derivatives, neem bark, neem leaves, peepal leaves, etc., have gained a lot of attention as well-organised biosorbents owing to their low cost and high concentrations of amino and hydroxyl functional groups, which illustrate important adsorption potential for the elimination of a variety of aquatic pollutants (Sundaram et al. 2008; Maheshwari et al. 2015; Jamode et al. 2004). The appropriateness of chitin and chitosan as adsorbents for removing excess fluoride containing drinking water was evaluated Rangel et al. (2010).

The location of the metal extracted from the solution may also be used to categorise the biosorption:

1. Extracellular precipitation or accumulation
2. Sorption/precipitation on cell surfaces
3. Accumulation inside cells

1.3.1.6 Using Microorganisms for Treatment

More effective biological methods are replacing conventional and less effective physico-chemical methods. For example, The removal of fluoride from groundwater and soil by bacteria drives the development of ecologically feasible bioremediation methods (Chubar et al. 2008). In the bioremediation of iron and heavy metals, these bacteria were shown to be highly successful. Mandal et al. (2010) state that in the modern environment, bioremediation, bio-treatment and bio-removal of contaminants from waste water have become innovative methods for wastewater treatment. The proportion of fluoride removed by bio-adsorbent and microorganisms is displayed in Table 1.4 and Fig. 1.2 (Mondal et al. (2012); Naiya et al. (2009); Camargo (2003)).

1.4 Case Studies of Fluoride Removal Systems in High-Fluoride Regions

The case studies presented here are focused on the applicability of NbS in waste water treatment, especially for fluoride removal. In these case studies, various researchers opted the biophysical and bio-adsorption routes for defluoridation where most of the researchers used potential microorganisms for direct removal of fluoride using nature-based methods like phytoremediation and bio-adsorption.

Table 1.4 Percentage of fluoride removal using bio-adsorbent

S. No.	Bio-adsorbent	Removal	References
1	MnO2-coated Tamarind Fruit Shell (MTFS)	98	Sivasankar et al. (2010)
2	Neem & Kiker leaves	58	Kumar et al. (2008)
3	*Cordia Africana* seed	60	Abburi et al. (2012)
4	*Cynodon dactylon Activated carbon*	84	Alagumuthu et al. (2010)
5	*Spirogyra* IO1	64	Mohan et al. (2007)
6	*Trichoderma harzianum*	20	Koshle et al. (2016)
7	Rice husk	75	Deshmukh et al. (2009)
8	*Pleurotus ostreatus*1804	52	Ramanaiah et al. (2007)
9	*Pleurotuseryngii* ATCC 90888	97.03	Amin et al. (2015)
10	Neem (*Azadirachta indica*)	90	Goswami et al. (2015)
11	*Aspergillus nidulans*	29	Merugu et al. (2013)
12	*Ficus religiosa* leaves	74	Dwivedi et al. (2014a)
13	*Citrus limetta* peel	67.5	Dwivedi et al. (2014b)

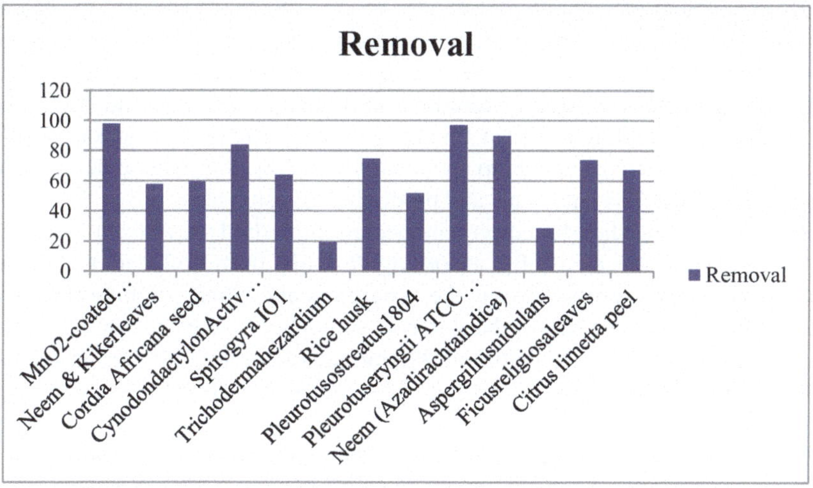

Fig. 1.2 The percentage of fluoride removed using bio-adsorbents

Named for the region where it was developed, the Nalgonda Method is an accepted and reasonably priced community-based fluoride elimination method in rural India. Lime and aluminium sulphate are added to the water, causing fluoride to precipitate as insoluble compounds. This approach works particularly well in places with low incomes and little access to cutting-edge technology. According to studies conducted in Andhra Pradesh and Telangana, the technique has lowered fluoride levels from above 5 mg/L to less than the 1.5 mg/L WHO-recommended limit. However, issues like operational knowledge and sludge removal continue to be major worries.

East African nations like Kenya and Tanzania, where groundwater frequently has fluoride levels above 10 mg/L, have adopted filtering systems based on activated alumina. Adsorption of fluoride onto the alumina surface is how these systems function. In collaboration with NGOs, community-scale units have been put in place, significantly improving the water quality in neighbourhoods and educational institutions. The systems are simple to operate and may reduce fluoride levels to less than 1 mg/L. Nonetheless, their helpfulness requires standard protection and resurgence of the activated alumina.

Bone char filtration is a long-standing and environment-friendly technique for removing fluoride from water in the Ethiopian Rift Valley. Bone char, a porous substance with a high fluoride adsorption capacity, is produced by heating animal bones without oxygen. Several high-fluoride localities have established community-based bone char plants, which have been effective in bringing fluoride concentrations down to acceptable drinking water levels from as high as 15 mg/L. The limited supply of raw resources and cultural reservations regarding the use of animal products are obstacles that must be overcome, notwithstanding their effectiveness.

Electrocoagulation technology has been used to alleviate excessive fluoride levels in drinking water in parts of Mexico, including San Luis Potosí. By running an electric current over iron or aluminium electrodes, this technique produces coagulants that eliminate fluoride. According to pilot studies, fluoride levels have decreased from 7 mg/L to less than 1.5 mg/L, making it a practical option for places with electrical power. Compared to chemical treatments, this method has the added advantage of using less chemicals and producing less sludge.

Fluoride-tolerant plant-based phytoremediation methods have been explored in China's high-fluoride areas. By absorbing fluoride from soil and water, these plants—like vetiver grass—lessen the environmental effect of fluoride. Henan Province field tests have shown that fluoride significantly reduces agricultural runoff, preventing groundwater pollution. Phytoremediation provides an inexpensive, eco-friendly option with the potential to scale, even if it is still in the experimental stage.

Ramanaiah et al. (2007) developed a biosorbent derived from the laccase fermentation process that uses the waste fungus *Pleurotus ostreatus* 1804 to remove fluoride from the aqueous phase. The pseudo-first-order rate equation was used in the batch sorption investigations. The Langmuir isotherm adsorption model was well-fitted by the adsorptive data that was isothermal data. Lower pH levels were shown to result in higher fluoride absorption and fluoride elimination in the aqueous phase. The chemical kind of interaction may be responsible for the fluoride sorption phenomenon on fungal biosorbent. Eight water samples taken from the region afflicted by fluorosis were used in batch tests to assess the effectiveness of these biosorbents for actual field use. The application of thermally activated carbon based on Cynodon dactylon for fluoride toxicity was examined by Alagumuthu et al. (2010). Batch adsorption techniques were used at neutral pH to ascertain the effects of co-anions, temperature, adsorbent dose, adsorbate concentration and contact duration.

Goswami et al. (2015) examined the possibility of defluoridating water using neem (Azadirachta indica) leaf powder. The fluoride ion sorption effectiveness is influenced by the adsorbents' particle size, pH and contact time. The treated leaf powder was tested with water solutions containing 10 mg F–/l at different pH values and contact times. According to the results, these inexpensive bio-adsorbents might be effectively employed to remove fluoride at a variety of concentrations. The treated bio-adsorbents were shown to be efficient in absorbing fluoride ions in the pH range of around 2 to 8. The size of the adsorbent particles and the period were shown to affect the percentage of fluoride removal at a particular initial solute concentration. Over time, the initial concentrations of the solute declined, but eventually increased. By altering the particle size, the treated biosorbents' adsorption ability was investigated. It was discovered that 50% of the fluoride ions were adsorbed at the maximum particle size of 1.4 mm. More active surface area is provided by smaller particle sizes, which leads to these outcomes.

Shubha Dwivedi et al. (2017) employed *Shewanella putrefaciens* bacteria in batch reactors in both bulk and immobilised phases. Studies have been conducted on bacterial cell development and acculturation under substrate stress and fluoride media. In the instant adsorption bioaccumulation method, peepal and mosambi leaves are employed as adsorbents in addition to the bacteria. The presentation under optimal circumstances has been contrasted with the process parameter optimisation for fluoride removal in the bulk phase. The simultaneous adsorption and bioaccumulation process, which is followed by bulk phase bioaccumulation and adsorption, has the highest fluoride removal capability (Fig. 1.2). In this case, 91.7% of the fluoride was eliminated throughout the bioaccumulation process. The initial fluoride content from 20 mg/L may be lowered by both SABA and bioaccumulation to less than or equal to 1.5 mg/L.

Aash Mohammad (2019) worked on the immobilised *Actinobacteria* species on the peel of a delicious lemon. Waste water is home to the water-living microorganism Actinobacteria. Actinobacteria, a microorganism, became immobile on the sweet lemon peel. Fluoride removal effectiveness rose approximately 60–90% under ideal conditions of 87 h (contact time), The removal effectiveness of the simple adsorption procedure was 59.59% under ideal circumstances of pH 4, dosage of 16 g/L and contact period of 60 min (Dwivedi et al. (2014a, b); National prevalence: Fluoride and fluorosis (n.d.); Picardal et al. (1993)).

1.5 Potential of NbS to Transform Wastewater Management and Fluoride Removal

A developing strategy for dealing with environmental issues, such as fluoride removal and wastewater management, is called nature-based solutions (NbS). As an alternative to traditional approaches, NbS addresses water quality concerns using designed ecosystems or natural processes. NbS might revolutionise fluoride removal and wastewater management in the following ways:

- Using natural processes including filtration, adsorption and bioremediation to enhance water quality is a common component of NbS wastewater treatment systems.
- Constructed Wetlands: These artificial ecosystems use plants, soils and microbes to filter and break down pollutants in wastewater, simulating natural wetlands. They work well to get rid of nutrients, infections and organic materials.
- Riparian Buffers and Vegetated Systems: By absorbing and filtering pollutants from wastewater before they reach natural water bodies, riparian buffers and swales' natural vegetation can enhance water quality and reduce pollution.
- Bioremediation is a low-cost, environment-friendly treatment option that uses plants, algae, or microbes to break down contaminants in wastewater. It is frequently used in conjunction with other NbS techniques.
- Fluoride Elimination Using NbS: Natural filtration and adsorption are promising methods and aligned with the approach of nature-based solutions for the removal of fluoride from ground water.
- Plants for Fluoride Elimination (Phytoremediation): Some plants may absorb and store fluoride from tainted water, lowering the level of fluoride in the water. Certain plant species, such *Salvinia* and *Pistia*, have been shown via research to be able to absorb fluoride, which is then eliminated by harvesting the plants.
- Natural materials such as zeolite, and charcoal have been studied as adsorbents for the elimination of fluoride from water. These materials are perfect for low-income or rural populations since they may be made from inexpensive, locally accessible resources.
- Bio-based Filtration Systems: When customised to local circumstances, natural filtration systems that use biosorbents—such as activated carbon made from algae or agricultural waste—can effectively remove fluoride from polluted water.

1.5.1 Benefits of NbS for Fluoride Removal and Wastewater Management

Cost-Effectiveness: Compared to high-tech, energy-intensive alternatives like reverse osmosis or sophisticated chemical treatments, NbS usually demand lower initial investment and operating expenses.

- Sustainability: By using natural processes, NbS create long-term environmental benefits and sustainability.
- Adaptability: Many NbS are adaptable to local settings, which makes them suitable for a range of climates, water types and community requirements.
- Community Involvement: NbS strategies frequently engage nearby communities, encouraging more participation in water management and a closer bond with the natural world.

1.6 Conclusion

A viable and sustainable strategy for improving wastewater treatment and tackling issues like fluoride pollution is nature-based solutions (NbS). NbS can provide efficient, economical and ecologically responsible substitutes for traditional treatment techniques by leveraging natural systems and processes. Structured wetlands, vegetated buffers and reed beds are examples of NbS that may be used in wastewater treatment to remove fluoride and other toxins, enhance water quality and promote biodiversity. By improving ecosystems' natural filtering capabilities, these technologies reduce the need for energy-intensive processes like chemical coagulation or reverse osmosis. NbS techniques including bio-sorption by fungus, algae and plants, as well as the use of materials like clay and charcoal, have shown promise in lowering fluoride levels in tainted water. Benefits of these natural materials include their low cost, renewable nature and little environmental effect. Co-benefits of NbS also include increased resistance to climate change, carbon sequestration and better ecosystem services. NbS use for fluoride removal and wastewater management, however, necessitates resolving issues including scalability, performance consistency and possible conflict with other land uses. NbS have the potential to supplement or even replace traditional water treatment technologies with more study and optimisation, helping create more robust and sustainable water systems globally.In summary, incorporating nature-based solutions into wastewater management plans for the removal of fluoride not only supports sustainability objectives but also provides creative ways to improve public health and environmental outcomes while reducing the ecological impact of conventional water treatment techniques.

References

Abburi MR, Belay K, Munagala DJ, Kaki S, Dubale A, Ahmed O (2012) Diminution of concentration of fluorides by bio adsorption using Cordia Africana in the waste waters around Aksum. J Chem Bio and Phy Sci 2:1067–1072

Alagumuthu G, Veeraputhiran V, Venkataraman R (2010) Adsorption isotherms on fluoride removal: batch techniques. Arch App Sci Res 2(4):70–185

Amin F, Talpur FN, Balouch A, Surhio MA, Bhutto MA (2015) Biosorption of fluoride from aqueous solution by whiterot fungus *Pleurotus eryngii* ATCC 90888. Environ Nanotech Mon & Mgt 3:30–37

Barbier O, Arreola-Mendoza L, Razo LMD (2010) Molecular mechanism of fluoride toxicity. Chem Biol Interact 188:319. https://doi.org/10.1016/j.cbi.2010.07.011

Camargo AJ (2003) Fluoride toxicity to aquatic organisms: a review. Chemosphere 50:251–264

Chubar NI, Samanidou VF, Kouts VS, Gallios GG, Kanibolotsky VA, Strelko VV, Zhuravlev IZ (2005) Adsorption of fluoride, chloride, bromide, and bromated ions on a novel ion exchanger. J Colloid Interface Sci 291:67–74

Chubar N, Behrends T, Behrends PV (2008) Biosorption of metals (Cu^{2+}, Zn^{2+}) and anions (F^-, $H2PO4^-$) by viable and autoclaved cells of the Gram-negative bacterium *Shewanella putrefaciens*. J Colloid Interface Sci 65:126–133

Deshmukh WS, Attar SJ, Waghmare MD (2009) Investigation on sorption of fluoride in water using rice husk as an adsorbent, nature environment and pollution technology. An Intern Quart Scientific J8:217–223

Dwivedi S, Mondal P, Majumder (2018) Role of microorganisms in the removal of fluoride. Publisher-CRC Press, Taylor & Francis Group. Chapter No. 11, ISBN: 9781771886673, E-Book ISBN: 9781351248914
Dwivedi S, Mondal P, Majumder CB (2014a) Bioadsorption of fluoride by *Ficus religiosa* (Peepal Leaf Powder): Optimization of process parameters and equilibrium study. Res J Chem Sci 4(7):52–60
Dwivedi S, Mondal P, Majumder CB (2014b) Removal of fluoride using *Citrus limetta* in batch reactor: kinetics and equilibrium studies. Res J Chem Sci 4:50–58
Dwivedi S, Mondal P, Majumder CB (2017) Bioremoval of fluoride from synthetic water using the Gram-negative bacteria *Shewanella putrefaciens*. J Hazard Toxic Radioact Waste 21:04016023-1–04016023-6. https://doi.org/10.1061/(ASCE)HZ.2153-5515.0000341
Feenstra L, Vasak LJ, Griffioen U (2007) Fluoride in groundwater: overview and evaluation of removal methods. Inter Groundwater Res Assessment Centre, pp 500–514
Fornasiero RB (2001) Phytotoxic effects of fluorides. Plant Sci 161(5):979–985
Gandhi N, Sirisha D, Shekar KBC, Asthana S (2012) Removal of fluoride from water and waste water by using low cost adsorbents. Inter J Chem Tech Res 4:1646–1653
Goswami R, & Kumar M (2015) Removal of fluoride from aqueous solution using nanoscale rice husk biochar. Groundwater for Sustainable Development, 1(1–2), 41–49
Guo L, Hunt BJ, Santsci PH (2001) Ultrafiltration behavior of major ions (Na, Ca, Mg, F, Cl, and SO4) in natural waters. Water Res 35(6):1500–1508
Haron MJ, Yunus WM (2001) Removal of fluoride ion from aqueous solution by a cerium–poly (hydroxamic acid). J Environ Sci Health A 36:727–734
Herrera O, Banihani V, Leon Q, Khatri G, Chandra F, James, Sierra-Alvarez R (2009) Toxicity of fluoride to microorganisms in biological wastewater treatment systems. Water Res 43:3177–3186
Hichour M, Persin F, Sandeaux J, Gavach C (2000) Fluoride removal from waters by Donnan dialysis. Sep Purif Technol 18:1–11
Jamode AV, Sapkal VS, Jamode VS (2004) Defluoridation of water using inexpensive adsorbents. J Indian Inst Sci 84:163–171
Kabay N, Arar O, Samatya S, Yuksel U, Yuksel M (2008) Separation of fluoride from aqueous solution by electrodialysis: effect of process parameters and other ionic species. J Hazard Mater 153:107–113
Koshle S, Mahesh S, Swami NS (2016) Isolation and identification of *Trichoderma harzianum* from ground water: an effective biosorbent for defluoridation of ground water. J Environ Bio 37:135–140
Kumar S, Gupta A, Yadav JP (2008) Removal of Fluoride by thermally activated carbon prepared from neem (*Azadirachta indica*) and kikar (*Acacia arabica*) leaves. J Environ Bio 29(2):227–232
Maheshwari U, Gupta S (2015) Removal of Cr (VI) from wastewater using activated neem bark in a fixed-bed column: interference of other ions and kinetic modeling studies. Desalin Water Treat 57(18):8514–8525
Mandal T, Maity S, Dasgupta D, Datta S (2010) Advanced oxidation process and biotreatment: their roles in combined industrial wastewater treatment. Desalination 250:87–94
Meenakshi S, Viswanathan N (2007) Identification of selective ion-exchange resin for fluoride sorption. J Colloid Interface Sci 308:438–450
Merugu R, Rao KK, Kumar G, Medi N (2013) Biodefluoridation of water using Aspergillus nidulans fungal bioadsorbent. Int J Environ Biolo 3(1):9–11
Mohammad A, Kumar S (2019) Adsorptive and bio removal of fluoride from synthetic waste water by using actinobacter immobilized on the surface of sweet lemon peel. Int J Eng Res Technol (IJERT) 8(06)
Mohan SV, Ramanaiah SV, Rajkumar B, Sarma PN (2007) Removal of fluoride from aqueous phase by biosorption onto algal biosorbent *Spirogyra* sp.-*IO2*: Sorption mechanism elucidation. J Hazard Mater 141(3):465–474

Mondal NK, Bhaumik R, Banerjee A, Datta JK, Baur TA (2012) Comparative study on the batch performance of fluoride adsorption by activated silica gel and activated rice husk ash. Inter J Env Sci 2(3):1643–1660

Murray JJ (1973) A history of water fluoridation. Br Dent J 134:250–350

Naiya TK, Bhattacharya AK, Mandal NS, Das SK (2009) the sorption of lead (II) ions on rice husk ash. J Hazard Mater 163:1254–1264

National prevalence: Fluoride and fluorosis (n.d.). Available from: http://www.fluorideandfluorosis.com/fluorosis/prevalence.html

National Programme for Prevention & Control of Fluorosis (NPPCF) National health profile (2018) Nutrition & IDD cell, Dte. GHS, Ministry of Health & Family Welfare

Picardal FW, Arnold RG, Couch H, Little AM, Smith ME (1993) Involvement of cytochromes in the anaerobic biotransformation of tetrachloromethane by *Shewanella putrefaciens* 200. Appl Environ Microbiol 59:3763–3770

Pollution Prevention and Abatement Handbook Phosphate Fertilizer Plant, World Bank Group (1998)

Popat KM, Anand PS, Dasare BD (1994) Selective removal of fluoride ions from water by the aluminum form of the aminomethylphosphonic acid-type ion exchanger. React Polym 23:23–32

Ramanaiah SV, Mohan SV, Sarma PN (2007) Adorptive removal of fluoride from aqueous phase using waste fungus (*Pleurotus ostreatus* 1804) biosorbent: Kinetics evaluation. Ecol Eng 31:47–56

Rangel CM, González MA, & López MA (2010) Fluoride levels in drinking water and prevalence of dental fluorosis in a rural community in Mexico. Revista de Salud Pública, 12(5), 730–738

Simons R (1993) Trace element removal from ash dam waters by nanofiltration and diffusion dialysis. Desalination 89:325–341

Sivasankar V, Ramachandra MT, Chandramohan A (2010) Fluoride removal from water using activated and MnO2-coated Tamarind Fruit (*Tamarindus indica*) shell: batch and column studies. J Hazard Mater 177:719–729

Sourirajan S, Matsurra T (1972) Studies on reverse osmosis for water pollution control. Water Res 6:1073–1086

State-wise Number of habitations affected by fluoride, arsenic, iron, salinity, nitrate and heavy metals such contaminations as reported by states in IMIS of the Ministry from April 2016 to April 2018 (From: Ministry of Drinking Water & Sanitation)

Sundaram CS, Viswanathan N, Meenakshi S (2008) Defluoridation chemistry of synthetic hydroxyapatite at nano scale: equilibrium and kinetic studies. J Hazard Mater 155:206–215

Susheela AK (2001) Fluorosis: Indian scenario: a treatise on fluorosis. Fluorosis Research and Rural Development Foundation, New Delhi

Wasewar KL, Kumar P, Chand S, Padmini BN, Teng TT (2010) Adsorption of cadmium ions from aqueous solution using granular activated carbon and activated clay. Clean Soil Air Water 38(7):649–656

Ecological Restoration Through Nature-Based Solutions and Minimizing Water Challenges

2

Shruti Awasthi, Proma Chakraborty, Prema Kulkarni, Chethana V. Chalapathy, Lahar Khandelwal, and Naveen Dwivedi

Abstract

In light of growing urbanization and environmental deterioration, ecosystem restoration and urban sustainability have emerged as crucial goals. This chapter investigates how nature-based solutions (NbS) might improve urban circularity, mitigate water-related issues, and manage ecosystem degradation. NbS provide multipurpose strategies for climate resilience, biodiversity preservation, and resource efficiency by using natural processes. This chapter starts out by outlining the main obstacles to attaining urban circularity, such as pollution, water scarcity, and ineffective resource management. It emphasizes how NbS can be integrated into sustainable approaches to removing these obstacles, such as urban forestry, wetland restoration, and green infrastructure. Further evaluation of NbS' contribution to ecosystem restoration highlights its capacity to restore ecosystem services such habitat formation, water purification, and carbon sequestration. In addition to addressing the issues of scalability, governance, and maintenance, this chapter discusses case studies of effective NbS installations that highlight the ecological, economic, and social advantages of this technology. This chapter also emphasizes how crucial it is to broaden the use of NbS in urban planning, make use of technology developments, and create appropriate legislative frameworks. This chapter is a call to action for adopting NbS as a fundamental component of sustainable urban development.

S. Awasthi · P. Chakraborty · P. Kulkarni · C. V. Chalapathy · L. Khandelwal
Department of Life Sciences, Garden City University, Bangalore, Karnataka, India

N. Dwivedi (✉)
Department of Biotechnology Engineering and Food Technology, University Institute of Engineering, Chandigarh University, Mohali, Punjab, India

N. Dwivedi et al. (eds.), *Wastewater Treatment Through Nature-Based Solutions*,
https://doi.org/10.1007/978-981-96-6699-7_2

Keywords

Nature-based solutions (NbS) · Ecosystem restoration · Urban circularity · Water management · Green infrastructure · Biodiversity conservation · Climate resilience

2.1 Introduction

In an era marked by rapid population growth, the relentless exploitation of natural resources has significantly disrupted our ecosystems. It is imperative that we prioritize ecological restoration to safeguard the intricate web of life that sustains our planet. By recognizing the interdependence of all living entities, we can work toward restoring balance and harmony within our environment. This collective effort is essential not only for the health of our ecosystems but also for the well-being of future generations. Ecological restoration means a proactive process of facilitating the recovery of ecosystems that are degraded, damaged, or destroyed. This process aims to expedite a natural ecological pathway, guiding the ecosystem along a positive trajectory over time. Through targeted actions and strategies, ecological restoration seeks to revive the health and functionality of these vital environments (Gann and Lamb 2006) (Fig. 2.1).

Within the last ten years, a number of other voices came into the picture and have been trying to put forth a fresh set of remedies aimed at the alleviation of the situation of global climate change caused by humans (Bauduceau et al. 2020). These options have been referred to as "nature-based solutions" (NbS) or "natural climate solutions" and include the protection, conservation, restoration, management, enhancement, or imitation of natural ecosystems and can incorporate, among many others, forest resource management, wetland restoration, biochar burial, and

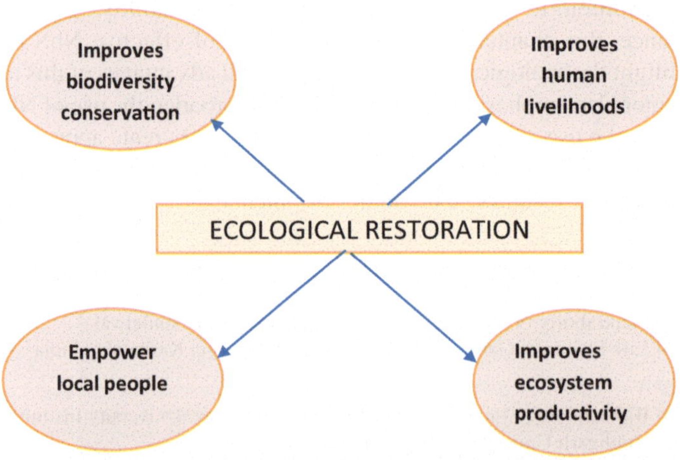

Fig. 2.1 Ecological restoration

Fig. 2.2 Restoration of the ecosystem using nature-based solutions

peatland conservation and restoration (Carrington 2019). This holistic approach not only alleviates environmental issues but also fosters social and economic benefits, aligning with worldwide sustainability goals (Fig. 2.2).

The world is struggling with harsh water challenges, including widespread scarcity, pollution, and insufficient access to clean drinking water (Hecht 2004), with over half of the global population facing water shortages for at least one month each year. According to the United Nations, approximately 785 million people lack access to basic drinking water services, resulting in significant health risks, including nearly 829,000 annual deaths from diarrheal diseases (Bonazzi 2023). This disaster is aggravated by climate change, population growth, and inexpert water management practices, particularly in waterless and densely populated regions, underscoring the urgent need for sustainable solutions to ensure water security for all.

2.2 Challenges in Achieving Urban Circularity

Urban circularity is a strategy involving sustainable practices, waste reduction, and resource efficiency (Romagnoli Sara et al. 2023). However, challenges like stakeholder resistance, political will, and finance hinder its implementation. Coordination across government and sectors is crucial.

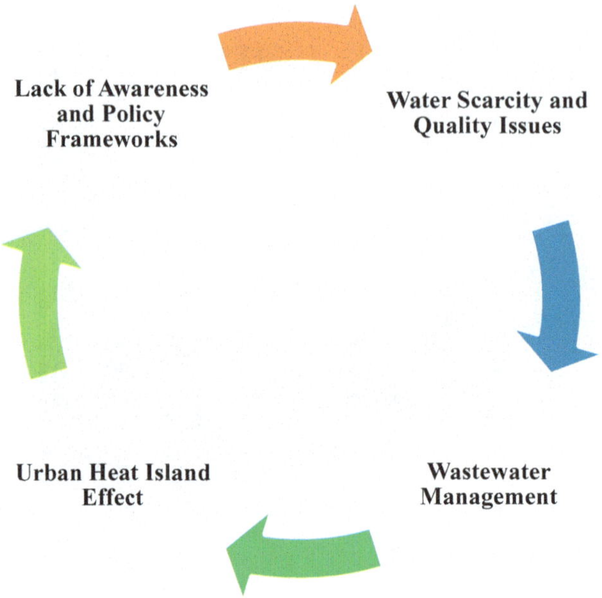

Fig. 2.3 Barriers hindering the implementation of urban circularity

Public-commercial cooperation and local community involvement are also essential. Cities can build resilient, ecologically friendly environments by establishing partnerships and raising awareness (Fig. 2.3).

However, several barriers hinder its implementation:

(a) *Water Scarcity and Quality Issues:* Water scarcity is a common problem in urban areas because of pollution and over-extraction (Ingrao et al. 2023). Water infrastructure is also often antiquated, which results in losses and inefficiencies.
(b) *Wastewater Management:* Ineffective wastewater treatment leads to water contamination, affecting ecosystems and public health (Rahman et al. 2023). To address this, financial and technological barriers in wastewater recycling must be overcome. Sustainable water management techniques, involving governments, corporations, and communities, are needed for future generations.
(c) *Urban Heat Island Effect:* Rapid urbanization worsens the urban heat island effect, increasing local temperatures and causing health issues (Qian et al. 2023). To reduce this, green infrastructure and urban planning techniques can be implemented, such as expanding green areas, using cool pavement and rooftops, and encouraging public transit. These strategies can save energy costs, improve air quality, and enhance the quality of life for citizens.
(d) *Lack of Awareness and Policy Frameworks:* The general public and institutions are not fully aware of circular systems, and legislative frameworks for urban circularity are often disjointed (Leow Abigail et al. 2022). This hinders the

widespread implementation of sustainable practices in cities. Lobbying and education campaigns are needed to highlight the benefits of circular systems and convince legislators to support them. A successful transition to a circular economy requires cooperation from all parties involved (Fig. 2.4).

These difficulties highlight how urgently creative solutions, such as NbS, are needed to strike a balance between environmental sustainability and urban expansion. Governments, corporations, and communities must work together to develop a unified strategy for putting circular systems into place in cities. Stakeholders can close the knowledge and policy gaps that are now impeding the transition to a more sustainable future by cooperating.

Public education and awareness-building regarding the advantages of a circular economy might also aid in gaining broad support for these projects. To overcome the challenges impeding the shift to a more sustainable urban environment, stakeholders must actively seek out chances for collaboration and engage in open communication.

Fig. 2.4 AI-imagined urban circularity using NbS

2.3 Implementing Nature-Based Solutions to Address Urban Circularity Challenges

Linear management of resources like water, food, materials, and energy in cities drives the global environmental issues of resource depletion, ecosystem degradation, biodiversity crisis, and natural capital loss. This translates into an increasing demand for resources on the one hand and waste production on the other—all under some banner of eternal growth. In this context, a circularity model (i.e., the (CE) Circular Economy) is suggested as an alternative to achieve economic growth without increasing new resource consumption and limiting environmental-detrimental impacts (Atanasova et al. 2021).

The three principles that form the basis of (CE)Circular Economy are:

(i) Regenerate natural capital
(ii) Keep resources in use
(iii) Design out waste externalities (Murray et al. 2017)

However, the current management of resources in cities largely relies on infrastructures that fail to accomplish these principles for a few reasons including resources that are imported and distributed to consumers, then consumed and treated and discharged from municipalities as residual (linear management); and existing infrastructure only containing single-functionality designed for waste water treatment (e.g., wastewater treatment or urban drainage). To adhere to the three principles of the CE: (i) preserve and enhance natural capital, (ii) optimize resource yields by closing technical and biological cycles, and (iii) promote system effectiveness, it requires significant modifications in infrastructure design and management through multisectoral and multidisciplinary approaches.

A nature-based solution (NbS) is a novel concept that is gaining great attention due to its enormous potential for sustainable solutions for various urban challenges (climate mitigation, air quality, water management participatory planning, and governing) (Raymond et al. 2017a, b). One of the most popular features of NbS is its co-benefits and effectiveness. Although used for a specific purpose (e.g., urban drainage), NbS can provide multiple co-benefits (e.g., treatment, evaporative cooling, and biodiversity). Using one Element Basins as an insulator for building waste water treatment plant (Raymond et al. 2017a, b; Oral et al. 2020; Kisser et al. 2020, Skar et al. 2020, Katsou et al. 2020)..

Multifunctionality is what makes NbS such an important and relevant resource management concept for cities in the context of the CE. Although the notion of NbS has been used for several decades, such as in the framework of related concepts like ecological engineering, green infrastructure, urban green (and blue) spaces, and ecosystem-based adaptation, there are many barriers to its systemic implementation: legislative gaps, lack of capacities and skills, lack of awareness, overly sectoral approach to resource management, etc. On a broad scale, while introducing NbS and alternative resource management may create quite a few dependability issues within the existing infrastructure—as it inherently changes the quantity and quality of resources flowing into such infrastructure—the impacts on availability must be carefully managed through innovation-led changes that alter rather than build new (hybrid) systems with a holistic contribution to all three CE principles (Katsou et al. 2020).

2.3.1 Urban Circularity Challenges for Shifting to Circular Management Resources and Nature-Based Solutions

Five challenges have been identified regarding urban circularity; the first is the difficulty of moving from linear resource management to circular management, and the second is the implementation of NbS. Municipalities face specific challenges having to do with linear, and therefore inefficient, economic models where waste is created. Overcoming these requires leveraging systems-level changes such as redesigning supply chains, building circular economies at the regional level, using renewable materials, and so on. Any cross-sectional networks that encompass the concept of green infrastructure and urban forests require a long-term commitment of resources, stakeholder cooperation, and public support for their long-term sustainability and people-centered implementation. Furthermore, the absence or inadequacy of policies, limited data on resource mobilization, and insufficient technological infrastructure remain significant impediments to the attainment of the stated objectives. Overcoming these barriers is crucial if the development of sustainable urban systems and increased resource utilization efficiency is to be achieved (Atanasova et al. 2021) (Fig. 2.5).

2.3.2 Nature-Based Solutions for Urban Circularity Challenges

Nature-based solutions (NbS) are an opportunity to achieve circularity in urban areas by creating opportunities for resource utilization through ecosystem support. These solutions integrate naturally occurring ecological processes into the urban built environment, facilitating the systematic reprocessing of resources and the recycling of materials. Urban trees help retain water and reduce runoff, green roofs mitigate water loss by slowing its release, and composting systems transform organic waste into nutrient-rich soil. Constructed wetlands are also used to manage the wastewater naturally at a lower cost than infrastructure intervention. So, by integrating NbS into urban plans of action, difficulties such as waste removal, water deficit, and climate resilience can be solved using systems that replicate and assimilate with nature to build closed-looping cycles that are in harmony with human actions (Pineda-Martos et al. 2024).

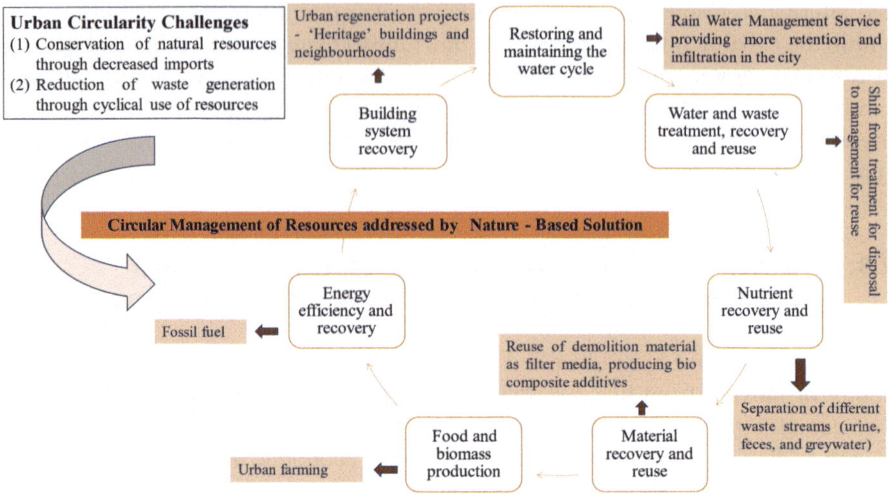

Fig. 2.5 Circular management of resources addressed by nature-based solution

2.3.3 Contribution of Nature-Based Solutions Toward Circularity

Nature-based solutions (NbS) are solutions that use natural processes and ecosystems to address problems affecting climatic, social, and economic domains. These solutions thus come close to emulating adaptive mechanisms derived from nature's systems, working with cycles that define a circular economy that seeks to optimize resource reuse. NbS can be used to identify solutions that will generate regenerative systems to address problems of environmental biophysical degradation, decrease depletion of natural resources, and increase biological vitality.

2.3.3.1 Enhancing Resource Efficiency Through NbS

Through the support of resource efficiency, NbS helps organizations become more circular. For example, green city systems, urban forests, green roofs, and wetlands act as natural filters of water, thus minimizing the need for energy-hungry water treatment structures. Similarly, agroforestry and regenerative agriculture enhance the retention of soil nutrients and water, thereby reducing dependence on synthetic fertilizers and irrigation. These practices promote circularity by recycling, balancing the ecosystem and prolonging the use of much-needed resources (Gemar et al. 2021).

2.3.3.2 Waste Management and Valorization

A key utility of NbS is waste management and valorization. Recycling nutrient-rich organic wastes into organic manure dowry is a natural process and reduces the pressure on the over-filled dumpsites while at the same time enriching the soils. Wetlands serve as natural systems for water treatment, while bioremediation techniques utilize plants and microorganisms to recycle wastewater and industrial effluents, enabling the recovery of water and valuable nutrients. These processes guarantee that derivatives are returned to the environment, thus closing the loop in the circular economy's agenda in turning waste into value-added products (Gemar et al. 2021).

2.3.3.3 Biodiversity and Ecosystem Resilience

Natural processes underpin the NbS framework and circularity by preserving the performance of system services. For instance, afforestation by re-establishing mangroves and coral reefs eradicates coasting erosion and improves fish production and people's income sources. In naturally diverse ecosystems, structures are naturally robust, and networks can restore themselves so that handling is minimal while it advances the natural loops as per the circular economy indicators (Åkerman 2016).

2.3.3.4 Climate Regulation and Carbon Sequestration

NbS helps enhance the circular economy in the contexts of climate change, carbon storage, and energy conservation. Carbon sequestration by afforestation and reforestation decreased the emission of CO_2 by using energy-efficient green building technologies. If combined, these approaches allow NbS to minimize the impact of anthropogenic actions on the environment and improve the stability of the communities and ecosystems. Solutions from the circular economy are useful as they reduce the effects of climate change and guarantee the stock of natural resources for future use (Cárdenas et al. 2021).

2.3.4 Economic and Social Benefits of NbS in Circular Systems

Implementing NbS in circular models develops the welfare of people and the environment by creating employment opportunities in the green economy. For instance, in construction and packaging purposes, bamboo is a far better solution than plastic and steel, which are limited resources. Environmentally, NbS positively affects human health by increasing the quality of air and water in addition to creating spaces that are healthy for people to be in. The principles of NbS promote a bottom-up approach for the participation of the community that reserves and enhances sustainable development within circularity. Nature-based solutions provide fresh and systemic strategies to make circularity by harmonizing human operations and natural processes. They improve resource utilization, allow for waste reconsideration as valuable resources, and support the stability of ecosystems as profitable and beneficial businesses and for the society. With global issues such as climate change and resource scarcity already becoming worse, mainstreaming NbS into the circular economy sector is a more sustainable and replenishing form of development (Cárdenas et al. 2021).

2.4 Assessing the Role of Nature-Based Solutions in Ecosystem Restoration

The foundation of attempts to repair damaged ecosystems is nature-based solutions (NbS), which use natural processes to restore functionality, biodiversity, and resilience. This section assesses how NbS can promote sustainable development while tackling the complex problems of ecosystem degradation. NbS can aid in restoring ecological balance and undoing the harm caused by human activity by using the power of nature.

NbS can significantly contribute to the restoration of ecosystems for the benefit of future generations by means of strategic implementation and cooperation with local communities. NbS is a realistic way to deal with environmental problems because it can also offer more affordable options than conventional restoration techniques. We can guarantee the long-term viability of ecosystem restoration initiatives by incorporating NbS into legislative frameworks and land management procedures (Diffie et al. 2022).

2.4.1 Restoring Ecosystem Services Through NbS

Degraded ecosystems often lose essential services like water filtration, carbon sequestration, and soil fertility. Natural biodiversity and processes can restore these services using NbS methods. Prioritizing NbS in restoration initiatives ensures ecosystems' long-term viability and future generations' essential services. Including NbS in land management practices improves climate change resilience and biodiversity, benefiting both the environment and society (Fig. 2.6).

(a) *Carbon Sequestration:* By absorbing atmospheric carbon dioxide, afforestation and reforestation initiatives improve soil stability and slow down climate change.

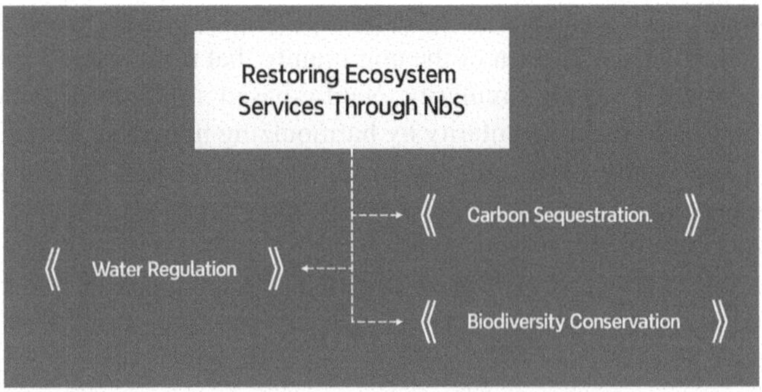

Fig. 2.6 Services that can be successfully restored via NbS

(b) *Water Regulation:* By filtering pollutants, lowering sedimentation, and preserving hydrological equilibrium, wetlands and riparian vegetation control water cycles.
(c) *Biodiversity Conservation:* By restoring natural vegetation and establishing wildlife corridors, NbS improves ecological connectedness and biodiversity by giving species a place to live (Brown et al. 2011).

2.4.2 Addressing Water Challenges in Ecosystem Restoration

Pollution and water scarcity are major contributors to ecosystem deterioration. By enhancing the quality and quantity of water, nature-based solutions (NbS) including riparian vegetation and wetland restoration can aid in addressing these water issues. Both people and wildlife gain from these behaviors, which also enhance the resilience and general health of the ecosystem. NbS can also lessen the effects of erosion and flooding by encouraging natural water filtration and lowering runoff. Restoring riparian vegetation and wetlands can also improve aquatic species' habitats, boosting ecosystem biodiversity (Kirezci et al. 2020). NbS provide sustainable approaches to manage and restore water resources (Fig. 2.7).

2.4.3 Case Studies of Successful NbS in Ecosystem Restoration

A. Mangrove Restoration in Coastal Areas

Initiatives to restore mangroves have increased fishing yields and carbon storage while successfully protecting coastlines from erosion and storm surges. Furthermore, by creating ecotourism opportunities, these projects have boosted local economies

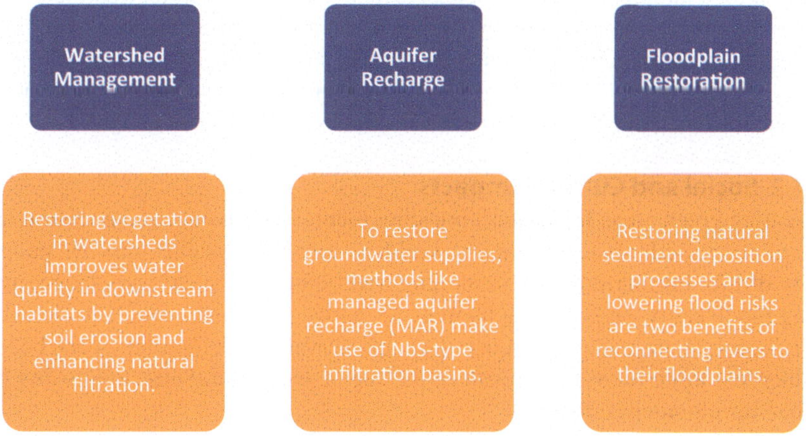

Fig. 2.7 Sustainable approaches to manage and restore water resources

and given home to a variety of species. Given the circumstances, mangrove restoration has shown itself to be an effective NbS strategy for ecosystem restoration (Casties et al. 2019).

In the Philippines, for example, mangrove restoration initiatives have significantly reduced the costs of storm damage while also improving local livelihoods. These initiatives have demonstrated the many advantages of restoring mangrove ecosystems while also making communities more resilient to the effects of climate change.

B. Urban Wetland Revitalization

Restoring urban wetlands has improved water quality, reduced flooding, and provided recreational opportunities for residents in places like Bengaluru, India. Cities can reduce the urban heat island effect and give wildlife vital habitat by reclaiming urban wetlands. Projects to revitalize urban wetlands have demonstrated the numerous advantages that investment in natural infrastructure can have for the environment and nearby populations.

For example, the restoration of Kaikondrahalli Lake in Bengaluru, India, turned a deteriorated area into a thriving environment by combining green infrastructure and community involvement. In addition to lowering flooding and improving the area's water quality, this project gave locals a place to enjoy the outdoors and wildlife. Other urban wetland restoration initiatives hoping to get comparable benefits should use Kaikondrahalli Lake's success as a model (Arrosyid et al. 2023).

2.4.4 Evaluating the Effectiveness of NbS

2.4.4.1 Economic Benefits

NbS frequently offers affordable substitutes for traditional infrastructure. Restoring wetlands, for example, is far less expensive than constructing water treatment facilities. Communities can save money, enhance water quality, and lower their risk of flooding by implementing NbS, such as wetland restoration. In addition to improving environmental sustainability, this economical strategy encourages economic efficiency in urban development initiatives (Beheshti Kathryn et al. 2023).

2.4.4.2 Social and Cultural Impacts

By creating green areas for leisure, boosting mental well-being, and encouraging a sense of connectedness to the natural world, restored ecosystems improve people's quality of life. Furthermore, by sustaining traditional ecological knowledge, maintaining biodiversity, and offering habitats for native plant and animal species, these restored ecosystems can also aid in cultural preservation. Communities can become healthier and more robust as a result of the social and cultural effects of using NbS in urban development initiatives.

2.4.4.3 Challenges in Implementation

NbS has several advantages, but they also have drawbacks, like:

- Limited political and financial backing
- Gaps in our understanding of their long-term performance
- Disagreements over land use in both rural and urban areas

NbS has shown itself to be invaluable in ecosystem restoration in spite of these obstacles, providing multipurpose solutions that consider social, economic, and environmental factors (Hao Xiao et al. 2023).

2.5 Evaluating the Contribution of Nature-Based Solutions to Urban Circularity

Because they increase resource efficiency, decrease waste, and support sustainable urban systems, nature-based solutions (NbS) are essential for urban circularity. They raise living standards and biodiversity while also improving water management, resource recovery, and energy efficiency. NbS can provide cost-effective solutions for problems with urban infrastructure and lessen the effects of climate change.

Reducing heat island impacts, improving air quality, and improving the well-being of inhabitants can all be achieved by implementing green spaces, roofs, and permeable surfaces. NbS can also control stormwater runoff, lessen flooding, and enhance water quality. Putting nature-based concepts into practice can improve property values, economic opportunities, and social cohesiveness (Kim Kwi-Gon et al. 2018).

2.5.1 Enhancing Water Circularity

Effective water management is essential to urban circularity, and NbS makes a substantial contribution in this area through creative design and natural processes (Fig. 2.8).

(a) *Rainwater Harvesting and Reuse:* Green roofs and urban wetlands help collect and store rainwater, providing an alternative water source for irrigation and cooling that is not drinkable. By lowering stormwater runoff, installing rainwater harvesting and reuse systems helps prevent flooding in addition to lessening

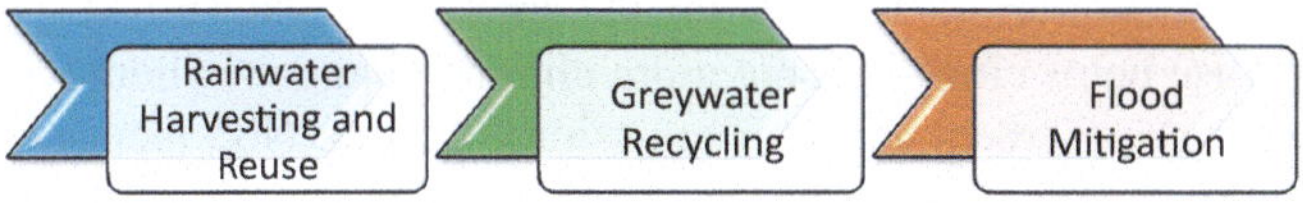

Fig. 2.8 Enhancing water circularity

the strain on conventional water supplies. These natural remedies can also enhance water quality and encourage biodiversity in urban settings.

(b) *Greywater Recycling:* Constructed wetlands can reduce wastewater discharge and the requirement for fresh water by processing greywater for reuse. Additionally, this sustainable approach can help buildings use less energy and pay less for utilities. Given the circumstances, greywater recycling makes the water management system more effective and sustainable (Seddon et al. 2020).

(c) *Flood Mitigation:* By restoring natural rivers and creating permeable surfaces, NbS refills aquifers, guarantees water availability, and lessens urban flooding. Green infrastructure, such rain gardens and bioswales, can help communities better control stormwater runoff and avoid flooding. Further fostering biodiversity in urban settings, these nature-based solutions also improve urban green spaces and offer wildlife habitat.

2.5.2 Promoting Resource Recovery

Urban circularity entails resource recovery and waste reduction. NbS offers effective, low-impact resource recovery solutions:

(a) *Composting of Organic Waste:* Urban green spaces can incorporate composting systems that convert organic waste into nutrient-rich soil amendments for local use. These technologies support the general sustainability of urban settings in addition to lowering the amount of waste that ends up in landfills. Composting can also promote regional food production efforts and enhance soil health (Mohajerani et al. 2017).

(b) *Biosolid Management:* Wetlands and other NbS can handle wastewater biosolids, converting them into resources for agriculture or landscaping. In addition to keeping garbage out of landfills, these solutions offer useful resources for environment-friendly operations. Cities may lessen their environmental effect and move toward a more circular economy by implementing these strategies in urban settings.

2.5.3 Energy Efficiency and Climate Resilience

By lowering cooling requirements and promoting renewable energy projects, NbS improves urban energy efficiency (Fig. 2.9).

(a) *Temperature Control:* Green roofs and urban forests reduce the need for energy-intensive air conditioning by reducing the urban heat island effect. This decrease in energy use helps cities become more climate change resilient while simultaneously lowering greenhouse gas emissions. Putting NbS tactics into practice can help create an urban environment that is more ecologically conscious and sustainable (Sheller et al. 2020).

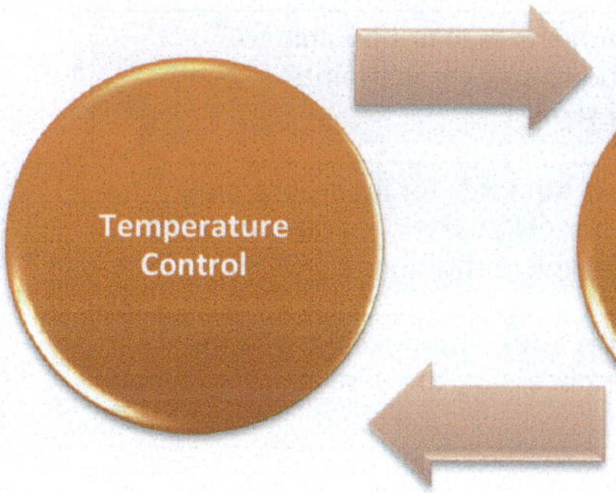

Fig. 2.9 NbS improves urban energy efficiency

(b) *Integration of Solar with Green Infrastructure:* Solar panels and green roofs can complement one another to provide sustainable energy and improve roof insulation. In addition to optimizing the advantages of both technologies, this integration assists cities in lowering their carbon footprint and transitioning to a more sustainable energy future. Cities can develop more eco-friendly and efficient urban environments by integrating solar electricity with green infrastructure.

2.5.4 Economic and Social Benefits of NbS in Urban Circularity

2.5.4.1 Cost Savings

NbS provides affordable alternatives to grey infrastructure, such as wetlands for water treatment and green roofs. These solutions are less expensive to construct and operate than traditional facilities and can save energy costs by providing natural insulation (Withers et al. 2016).

2.5.4.2 Improved Livability

Integrating Natural Resources (NbS) into urban development improves people's quality of life by creating green areas, reducing pollution, and enhancing air quality. This strategy creates healthier, livelier communities, attracts tourists and investment, and spurs metropolitan economies.

2.5.5 Limitations and Opportunities for Scaling NbS

NbS greatly promotes urban circularity; however, putting them into practice can be difficult (Fig. 2.10):

NbS presents opportunities for urban circularity and sustainable urban systems. As cities face increasing environmental demands, incorporating NbS into urban planning can improve quality of life and climate resistance. Prioritizing NbS

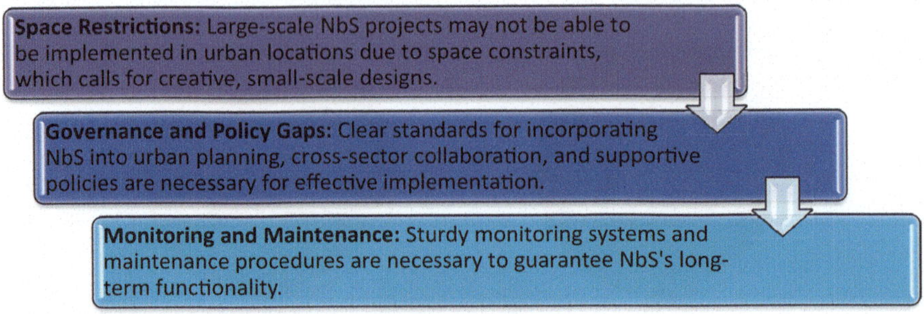

Fig. 2.10 Limitations and opportunities for scaling NbS

implementation leads to enhanced social and economic prospects, better environmental conditions, and improved environmental conditions (Wu et al. 2020).

2.6 Future Implications of Nature-Based Solutions in Urban Circular Systems

The global demand for water is surging due to factors such as population growth, economic development, and changing consumption patterns, particularly in developing and emerging economies. Climate change exacerbates this situation by altering the global water cycle, leading to wetter conditions in some regions and drier ones in others. Currently, approximately one billion people lack access to safe drinking water, and it is projected that by 2030, half of the world's population will reside in water-stressed areas. Urbanization compounds these challenges, as cities account for about 80% of global energy demand and face significant ecological footprints. Nature-based solutions (NbS) offer innovative approaches to address these issues by enhancing ecosystem management while providing social and economic benefits. Furthermore, the principles of circular economy (CE) are gaining traction in the water sector to promote sustainable resource management and balance urban water cycles. Addressing these challenges requires effective strategies for managing water resources to ensure a sustainable future for all (Amend et al. 2010).

2.6.1 Nature-Based Solutions for Urban Circularity Challenges

Nature-based solutions (NbS) represent a transformative framework that harnesses natural processes to enhance water availability, improve water quality, and mitigate risks associated with water crisis. By integrating NbS into water management strategies, immediate challenges can be addressed while promoting sustainable development, ecological resilience, and environmental equilibrium. Wastewater treatment ponds, soil infiltration systems, green roofs, and vertical gardens are key examples

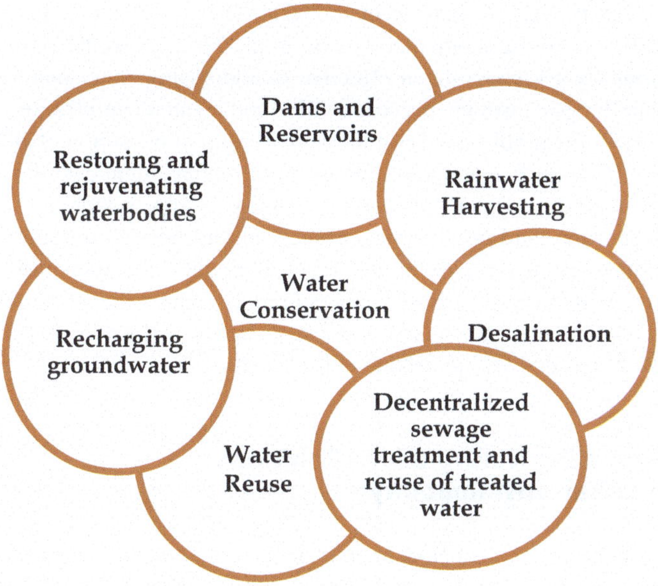

Fig. 2.11 Nature-based solution as a promising approach for water management

of natural infrastructure. NbS have steadily emerged as a paradigm of an integrated approach for sustainable and resilient urban development (Frantzeskaki et al. 2019; Sarabi et al. 2019). By combining natural elements with engineering approaches, NbS potential to provide ecosystem services can be maximized to support a sustainable and resilient urban water cycle (Fig. 2.11).

2.6.2 Nature-Based Solution Combined with Circular Economy

The circular economy promotes the efficient use of water resources by reducing water consumption and minimizing waste. Combining NbS with a circular economy approach helps connect different levels and improves cooperation between various urban resource systems and sectors (Kennedy et al. 2011). Circular NbS can create opportunities for sustainable food supply chains (Fader et al. 2018) while offering ecological, social, and economic co-benefits: water management, flood protection, carbon sequestration, biodiversity support, and climate change adaptation (MacKinnon et al. 2008; Raymond et al. 2017a, b). Implementing a circular economy in cities offers ecological, health, community, and economic benefits (Ramírez-Agudelo et al. 2021). Many urban strategies focus on "closing the loop" by promoting resource recovery and reuse (Petit-Boix and Leipold 2018; Williams 2019). However, these strategies often overlook water and nutrients, emphasizing waste management as the primary element of circularity plans. Sustainable water management is widely recognized as a key urban challenge (Almenar et al. 2021; Esmail and Suleiman 2020), particularly in advancing urban water circularity

(Langergraber et al. 2021; Atanasova et al. 2021). The urgent need for circular water systems in cities is well-documented (Arora et al. 2022). Specific types of nature-based solutions (NbS) have proven effective in addressing water management and circularity challenges through processes like conveyance, infiltration, detention, retention, evapotranspiration, and contaminant treatment (Castellar et al. 2022; Oral et al. 2020). NbS for urban water and wastewater management can deliver measurable benefits, including water purification, climate change adaptation and mitigation, natural resource and biodiversity protection, and improved quality of life for urban residents (Dumitru and Wendling 2021). To support this transition, it is essential to identify which types of NbS contribute to circular water reuse and resource recovery, understand their mechanisms and co-benefits, and determine how they can be integrated and scaled within urban environments.

2.6.3 Harnessing Nature-Based Solutions for Urban Sustainability

Strategically locating circular NbS in areas such as climate risk hotspots or underutilized urban spaces—whether public or private—enhances their efficacy (Saunders and Peirson 2013). Implementing these systems requires considering socio-ecological characteristics and climate sensitivity to ensure context-specific designs that align with local community needs and landscape identities. Circular NbS can be deployed across various spatial scales. At the household or building level, they can take the form of modular or hybrid systems such as green roofs, living walls, and internal water treatment installations, providing decentralized solutions for rainwater, stormwater, grey water, and wastewater collection, treatment, and reuse (Pérez Rubi and Hack 2021; Hunt et al. 2005). NbS can be integrated into road infrastructure, including sidewalks, pedestrian islands, roundabouts, and parking lots, optimizing storm water management (Bell 2015; Nichols and Lucke 2015; Coutts et al. 2013). Public green spaces and underutilized urban lands provide further opportunities for deployment, combining functionality with aesthetic and social benefits (Smith 2009). Climate and seasonality also play critical roles in system functionality, influencing plant selection, soil health, and the capacity to handle rainfall or drought events (Blecken et al. 2011; Yong et al. 2011; Li 2020). The design and performance of circular NbS must account for local climatic conditions and weather patterns, as these directly affect system maintenance, irrigation needs, and susceptibility to clogging (Saunders and Peirson 2013; Zinger et al. 2021). As urban populations continue to grow, climate change impacts intensify, and traditional water systems face increasing pressure, NbS can play a critical role in transforming urban water management in multiple dimensions, as presented in Fig. 2.12.

Nature-based solutions (NbS) in urban circular systems provide a transformative strategy for tackling water crises by integrating natural processes into city planning. As urban challenges like climate change, biodiversity loss, and economic instability escalate, NbS can enhance water management while delivering multiple benefits, including improved biodiversity and water quality. These solutions are

Fig. 2.12 Nature-based solutions for urban sustainability

cost-effective and sustainable, promoting resilience against extreme climate events and restoring degraded ecosystems. Investment in NbS is crucial for achieving broader environmental and social sustainability goals, ensuring water security for future generations. By emphasizing NbS, cities can strengthen their commitment to sustainable development while addressing pressing global challenges.

2.7 Conclusions and Discussion

Nature-based solutions (NbS) provide innovative approaches to addressing water management, urban circularity, and ecological degradation by integrating ecological principles into urban systems. NbS enhances resilience, restores ecosystems, and fosters sustainable resource cycles while offering economic benefits such as job creation, increased property values, and improved environmental sustainability. Techniques like rainfall harvesting, wetland restoration, and wastewater treatment help improve water quality, reduce pollution, mitigate water scarcity, and lower flood risks. Compared to traditional grey infrastructure, NbS delivers social, economic, and ecological advantages through multi-functionality, cost-effectiveness, and community engagement, promoting public involvement and climate resilience.

Moving forward, NbS presents a paradigm shift toward sustainability, addressing contemporary environmental challenges. Cross-sector collaboration among scientists, urban planners, policymakers, and communities is essential for tailoring NbS to local contexts and specific challenges. Education and advocacy efforts can build awareness and foster support for sustainable urban practices while integrating technologies like Internet of things (IoT), Geographic Information System (GIS), and remote sensing to enhance monitoring, evaluation, and adaptive management of NbS projects. These tools enable data-driven decision-making, helping cities achieve a balance between environmental health and human needs for a sustainable future.

References

Akerman E (2016) Development of circular economy Core indicators for natural resources. Master of Science Thesis, KTH Royal Institute of Technology Stockholm

Almenar JB, Elliot T, Rugani B, Philippe B, Gutierrez TN, Sonnemann G, Geneletti D (2021) Nexus between nature-based solutions, ecosystem services, and urban challenges. Land Use Pol 100:104898

Amend B, Barbeau B, Beyers S, Burns S, Eibing A, Fleischhauer B, Kus P, Poblete P (2010) A big foot on a small planet. GTZ and the Global Footprint Network

Arora M, Yeow LW, Cheah L, Derrible S (2022) Assessing water circularity in cities: methodological framework with a case study. Resour Conserv Recycl 178:106042

Arrosyid BH et al (2023) High-efficiency water filtration by electrospun expanded polystyrene waste nanofibers. ACS Omega 8

Atanasova N, Castellar JA, Pineda-Martos R, Nika CE, Katsou E, Istenič D, Pucher B, Andreucci MB, Langergraber G (2021) Nature-based solutions and circularity in cities. Circ Econ Sustain 1(1):319–332

Bauduceau N, Berry P, Cecchi C, Elmqvist T, Fernandez M, Hartig T, Krull W, Mayerhofer E, Sandra N, Noring L, Raskin-Delisle K (2020) Towards an EU research and innovation policy agenda for nature-based solutions & re-naturing cities. Final Report of the Horizon 2020 Expert Group on Nature-Based Solutions and Re-naturing Cities

Beheshti KM et al (2023) Large-scale field studies inform adaptive management of California wetland restoration. Restor Ecol 31

Bell S (2015) Renegotiating urban water. Prog Plann 96:1–28

Blecken GT, Marsalek J, Viklander M (2011) Laboratory study of stormwater biofiltration in low temperatures: total and dissolved metal removals and fates. Water Air Soil Pollut 219(1–4):303–317

Bonazzi D (2023) Imminent risk of a global water crisis, warns the UN World Water Development Report

Brown C (2011) Emergent permitting strategies for natural building systems in Ontario. J Green Build

Kennedy C, Pincetl S, Bunje P (2011) The study of urban metabolism and its applications to urban planning and design. Environ Pollut 159(8–9):1965–1973

Raymond CM, Frantzeskaki N, Kabisch N, Berry P, Breil M, Nita MR (2017a) A framework for assessing and implementing the co-benefits of nature-based solutions in urban areas. Environ Sci Pol 77:15–24

Cárdenas ML, Wilde V, Hagen-Zanker A, Seifert-Dähnn I, Hutchins MG, Loiselle S (2021) The circular benefits of participation in nature-based solutions. Sustain For 13(8):4344

Carrington D, Thunberg G (2019) We are ignoring natural climate solutions. The Guardian

Casties I, Briski E (2019) Life history traits of aquatic non-indigenous species: freshwater vs. marine habitats. Aquat Invasions

Castellar JAC, Torrens A, Buttiglieri G, Monclús H, Arias CA, Carvalho PN, Galvao A, & Comas J (2022) Nature-based solutions coupled with advanced technologies: An opportunity for decentralized water reuse in cities. J Clean Prod, 340, 130660. https://doi.org/10.1016/j.jclepro.2022.130660

Coutts AM, Tapper NJ, Beringer J, Loughnan M, Demuzere M (2013) Watering our cities: the capacity for water sensitive Urban Design to support urban cooling and improve human thermal comfort in the Australian context. Prog Phys Geogr 37(1):2–28
Diffie W, Hellman M (2022) Exhaustive cryptanalysis of the NBS data encryption standard. Democratizing Cryptography
Dumitru A, & Wendling L (2021) Evaluating the impact of nature-based solutions: A handbook for practitioners. European Commission. https://doi.org/10.2777/244577
Esmail BA, & Suleiman L (2020) Analyzing evidence of sustainable urban water management systems: A review through the lenses of sociotechnical transitions. Sustainability, 12(11), 4481. https://doi.org/10.3390/su12114481
Frantzeskaki T, McPhearson MJ, Collier D, Kendal H, Bulkeley A, Dumitru A (2019) Nature-based solutions for urban climate change adaptation: linking science, policy, and practice communities for evidence-based decision-making. Bioscience 69(6):455–466
Langergraber G, Castellar JA, Andersen TR, Andreucci MB, Baganz GF, Buttiglieri G (2021) Towards a cross-sectoral view of nature-based solutions for enabling circular cities. Water 13(17):2352
Gann GD, Lamb D (2006) Ecological restoration: a means of conserving biodiversity and sustaining livelihoods. Society for Ecological Restoration International, Arizona
Gemar G, Soler IP, Sánchez-Teba EM (2021) Waste management: valorization is the way. Food Secur 10(10):2373
Hao X et al (2023) How to improve environment, resources, and economic efficiency in the digital era? Resources Policy 80:103198
Hecht AD (2004) International efforts to improve access to water and sanitation in the developing world: a good start, but more is needed. Water Policy 6(1):67–85
Ingrao C et al (2023) Water scarcity in agriculture: an overview of causes, impacts, and approaches for reducing the risks. Heliyon 9:9
MacKinnon K, Sobrevila C, Hickey V (2008) Biodiversity, climate change, and adaptation: nature-based solutions from the World Bank Portfolio. The World Bank:1–112
Katsou E, Nika CE, Buehler D, Marić B, Megyesi B, Mino E, Babí Almenar J, Bas B, Bećirović D, Bokal S, Đolić M (2020) Transformation tools enabling the implementation of nature-based solutions for creating a resourceful circular city. Blue-Green Systems 2(1):188–213
Kim Kwi-Gon (2018) Planning models for climate resilient and low-carbon smart cities: an urban innovation for sustainability, efficiency, circularity, resiliency, and connectivity planning
Kirezci E et al (2020) Projections of global-scale extreme sea levels and resulting episodic coastal flooding over the 21st century. Sci Rep 10:11629
Kisser J, Wirth M, De Gusseme B, Van Eekert M, Zeeman G, Schoenborn A, Vinnerås B, Finger DC, Kolbl Repinc S, Bulc TG, Bani A (2020) A review of nature-based solutions for resource recovery in cities. Blue-Green Syst 2(1):138–172
Leow A (2022) Lack of awareness, vol 23, Early Years Educator, p 36
Li H (2020) Investigation of highway stormwater management pond capacity for flood detention and water quality treatment retention via remote sensing data and conventional topographic survey. Transport Res Rec 2674(7):514–527
Fader M, Cranmer C, Lawford R, Engel-Cox J (2018) Toward an understanding of synergies and trade-offs between water, energy, and food SDG targets. Front Environ Sci 6:112
Mohajerani A, Bakaric J, Jeffrey-Bailey T (2017) The urban heat Island effect, its causes, and mitigation, with reference to the thermal properties of asphalt concrete. J Environ Manag 197:522–538
Murray A, Skene K, Haynes K (2017) The circular economy: an interdisciplinary exploration of the concept and application in a global context. J Bus Ethics 140:369–380
Nichols PWB, Lucke T (2015) Local level stormwater harvesting and reuse: a practical solution to the water security challenges faced by urban trees. Sustain For 7(7):8635–8648
Oral HV, Carvalho P, Gajewska M, Ursino N, Masi F, Hullebusch ED, Kazak JK, Exposito A, Cipolletta G, Andersen TR, Finger DC (2020) A review of nature-based solutions for urban

water management in European circular cities: a critical assessment based on case studies and literature. Blue-Green Syst 2(1):112–136
Pérez Rubi M, Hack J (2021) Co-design of experimental nature-based solutions for decentralized dry-weather runoff treatment retrofitted in a densely urbanized area in Central America. Ambio 50:1498
Petit-Boix A, Leipold S (2018) Circular economy in cities: reviewing how environmental research aligns with local practices. J Clean Prod 195:1270–1281
Pineda-Martos R, Atanasova N, Calheiros CS, Junge R, Nickayin SS, Paço TA, Dominici L, Comino E, Andreucci MB, Theochari D (2024) Pucher B. Implementing nature-based solutions for a circular urban built environment. Creating a roadmap towards circularity in the built environment: 345–355
Qian W, LiX (2023) A cold Island connectivity and network perspective to mitigate the urban heat Island effect. Sustainable Cities and Society, 94, 104525
Rahman T et al (2023) The advancement in membrane bioreactor (MBR) technology toward sustainable industrial wastewater management. Membranes 13
Ramírez-Agudelo NA, de Pablo J, Roca E (2021) Exploring alternative practices in urban water management through the lens of circular economy–a case study in the Barcelona metropolitan area. J Clean Prod 329:129565
Raymond CM, Frantzeskaki N, Kabisch N, Berry P, Breil M, Nita MR, Geneletti D, Calfapietra C (2017b) A framework for assessing and implementing the co-benefits of nature-based solutions in urban areas. Environ Sci Pol 77:15–24
Romagnoli S et al (2023) The impact of digital technologies and sustainable practices on circular supply chain management. Logistics 7
Sarabi S, Han Q, Romme AGL, de Vries B, Wendling L (2019) Key enablers of and barriers to the uptake and implementation of nature-based solutions in urban settings: a review. Resources 8(3):121
Saunders NJ, Peirson WL (2013) Climate change adaptation of urban water management systems in the wet/dry tropics. Aust J Water Resour 17(2):180–192
Seddon N et al (2020) Understanding the value and limits of nature-based solutions to climate change and other global challenges. Philos Trans R Soc B 375:20190120
Sheller M (2020) Reconstructing tourism in the Caribbean: connecting pandemic recovery, climate resilience, and sustainable tourism through mobility justice. J Sustain Tour 29:1436–1449
Skar SL, Pineda-Martos R, Timpe A, Pölling B, Bohn K, Külvik M, Delgado C, Pedras CM, Paço TA, Ćujić M, Tzortzakis N (2020) Urban agriculture as a keystone contribution towards securing sustainable and healthy development for cities in the future. Blue-Green Systems 2(1):1–27
Smith BR (2009) Re-thinking wastewater landscapes: combining innovative strategies to address tomorrow's urban wastewater treatment challenges. Water Sci Technol 60(6):1465–1473
Williams J (2019) Circular cities. Urban Stud 56(13):2746–2762
Withers P, Flynn N, Warren G, Taylor M, Chambers B (2016) Sustainable management of biosolid phosphorus: a field study. Soil Use Manag 32:54
Wu H, Hao Y, Ren S (2020) How do environmental regulation and environmental decentralization affect green total factor energy efficiency: evidence from China. SSRN Electron J
Yong CF, Deletic A, Fletcher TD, Grace MR (2011) Hydraulic and treatment performance of pervious pavements under variable drying and wetting regimes. Water Sci Technol 64(8):1692–1699
Zinger Y, Prodanovic V, Zhang K, Fletcher TD, Deletic A (2021) The effect of intermittent drying and wetting stormwater cycles on the nutrient removal performances of two vegetated biofiltration designs. Chemosphere 267:129294

Vermifiltration in Wastewater Treatment and Waste Valorization

3

Rhydum Sharma, Naveen Dwivedi, and Gurmail Singh Malhi

Abstract

Increasing urbanization and industrialization have exacerbated the scarcity of freshwater resources. To address this challenge, authorities worldwide are compelled to consider the treatment and reuse of wastewater generated by industries and domestic sources. To date, vermifiltration has been successfully applied for treating domestic wastewater, and the limited results from industrial applications are also promising, demonstrating its potential for industrial wastewater remediation. Vermifiltration has garnered substantial academic and industrial interest in recent decades as a sustainable approach to the remediation of domestic and industrial wastewater. The chapter on vermifiltration examines the underlying mechanisms and the current status of this approach in remediating and reusing effluents from both domestic and industrial premises. Waste valorization is an essential strategy in sustainable waste management, focusing on converting waste materials into valuable products such as energy, fuels, and raw materials. Here, an attempt has been made to identify and thoroughly discuss the mechanisms involved in vermifiltration and waste valorization. This chapter examines the progress made in vermifiltration research during the previous decade, with a focus on the underlying mechanisms for pollutant removal, system performance characteristics, and the associated by-products and current technologies, challenges, and future directions in waste valorization, emphasizing its role in the circular economy. The review concludes with insights into advancing waste valorization for achieving environmental sustainability and economic viability.

R. Sharma (✉) · N. Dwivedi
Department of Biotechnology Engineering and Food Technology, University Institute of Engineering, Chandigarh University, Chandigarh, India

G. S. Malhi
Department of Aerospace Engineering, Chandigarh University, Chandigarh, India

N. Dwivedi et al. (eds.), *Wastewater Treatment Through Nature-Based Solutions*,
https://doi.org/10.1007/978-981-96-6699-7_3

Keywords

Vermifiltration · Wastewater treatment · Resource recovery · Circular economy · Waste valorization

3.1 Introduction

Addressing the challenges posed by inadequate wastewater management, including water scarcity, sanitation issues, health concerns, and environmental problems, wastewater treatment has emerged as a potential solution. Wastewater contains various valuable nutrients, such as nitrogen and phosphorus, which can serve as a resource for agriculture if appropriately treated. While intensive wastewater treatment systems, like activated sludge, can provide solutions, their high installation and operational costs often render them inaccessible for small communities. It is believed that the adoption of decentralized wastewater treatment technologies can introduce reliable and effective long-term solutions. The research on vermifiltration in wastewater treatment and waste valorization has highlighted several key aspects of this technology. Maggot cultivation, a related bioconversion technology, has been identified as a sustainable approach to organic waste management, with the potential to generate valuable resources such as animal feed and soil amendments (Hamilton et al. 2008). The broader concept of resource recovery and reuse in wastewater treatment facilities has also been explored, with the identification of various recovered resources that can contribute to a circular economy (Akyol et al. 2019). Vermicomposting of poultry litter, a specific application of vermifiltration, has been studied in depth, revealing the challenges associated with the physical and chemical properties of the waste, such as high ammonia content, low carbon content, and high bulk density. These studies highlight the importance of process optimization and the need for careful management of the vermifiltration system to ensure its successful implementation.

In the context of vermifiltration, the literature review suggests that this technology offers a promising solution for wastewater treatment and waste valorization, with the potential to generate valuable byproducts and contribute to the development of a circular economy. The decentralized valorization of organic waste streams, including kitchen waste and garden residues, has been proposed as an alternative to traditional urban waste management approaches. This decentralized approach offers the potential to recover valuable resources from these organic waste streams, such as nutrient-rich compost and energy, closer to the source of generation. This can help reduce the environmental impact and transportation costs associated with centralized waste management systems. Furthermore, decentralized valorization can empower local communities to take an active role in managing their own organic waste, fostering a more sustainable and circular economy. The vermifiltration technique, also referred to as lombrifiltration, can be applied for the remediation of wastewater generated from both domestic and industrial settings. Vermifiltration is a biofiltration process that uses earthworms to degrade organic matter and remove contaminants from wastewater (Yildiz 2012). The presence of earthworms in the filter bed enhances the microbial activity and improves the treatment efficiency. In the context of a circular economy, the management of wastewater has evolved from a focus on treatment and disposal to a broader perspective of water and resource recovery (Akyol et al. 2019). This transformation has led to the development of innovative technologies and approaches aimed at maximizing the recovery and valorization of valuable resources from municipal wastewater (Akyol et al. 2019; Rezaei et al. 2018), One promising technology that has gained attention is vermifiltration, a process that uses earthworms to filter and decompose organic matter in wastewater, while simultaneously producing valuable byproducts such as nutrient-rich vermicompost and biogas.

This chapter explores the potential of vermifiltration in the context of wastewater treatment and resource recovery, highlighting its environmental and economic benefits, as well as the challenges and opportunities associated with its implementation.

3.2 Types of Vermifiltration

Vermifiltration is a biofiltration process that combines the physical filtration capabilities of a porous media with the biological degradation activities of earthworms and associated microorganisms (Nascimento et al. 2018). The vermifiltration process integrates primary, secondary, and tertiary wastewater treatment stages within a single unit. The wastewater treatment system used in vermifiltration is the vermifilter, which comprises an active zone where earthworms reside and a filter bed that supports microbial growth. This integrated system is also referred to as a microbial-earthworm ecofilter. Various filter configurations have been explored and implemented in recent research to optimize the vermifiltration approach. The earthworms present in the filter bed play a crucial role in the treatment process by ingesting and breaking down organic matter, while also enhancing the microbial community (Sankaran et al. 2010).

The vermifiltration process can be divided into several key stages:

The vermifiltration process can be implemented through various filter configurations, including vertical subsurface flow vermifilters, horizontal subsurface flow vermifilters, and two-stage vermifilters. These configurations have been explored to optimize the treatment efficiency and resource recovery potential of the system.

- Pretreatment: This initial stage involves the removal of coarse solids, such as grit and debris, from the incoming wastewater.
- Physical filtration: The wastewater flows through the porous filter media, where physical filtration occurs, trapping suspended solids, bacteria, and other particulate matter.
- Biological degradation: The trapped organic matter is then broken down by the earthworms through their feeding and burrowing activities, which enhance microbial activity within the filter bed.

- Nutrient mineralization: The earthworms also contribute to the mineralization of organic nutrients, such as nitrogen and phosphorus, making them available for plant uptake.
- Effluent discharge: The treated wastewater, now with reduced organic load and nutrient content, is discharged from the vermifilter as the final effluent.

The vermifiltration process can be implemented through various filter configurations, including vertical subsurface flow vermifilters, horizontal subsurface flow vermifilters, and two-stage vermifilters. These configurations have been explored to optimize the treatment efficiency and resource recovery potential of the system (Odell 2016).

3.2.1 Vertical Subsurface Flow Vermifilters

Vertical Subsurface Flow Vermifilters (VSFVFs) are an eco-friendly wastewater treatment technology combining the principles of subsurface flow treatment systems with vermifiltration. By using earthworms, microorganisms, and filter media, these systems efficiently degrade organic pollutants, reduce pathogen levels, and improve water quality. Increasingly popular for their cost-effectiveness and low energy requirements, VSFVFs are suitable for domestic, agricultural, and industrial wastewater treatment.

3.2.1.1 Design and Components of VSFVFs

A typical VSFVF comprises the following key components:

1. *Filter Media:* Layers of gravel, sand, and organic material provide physical filtration and support microbial growth.
2. *Earthworms:* Earthworms such as *Eisenia fetida* (red wigglers) play a crucial role in degrading organic matter and aerating the filter bed.
3. *Drainage Layer:* The base contains coarse material to prevent clogging and ensure smooth water flow.
4. *Inlet and Outlet:* These ensure uniform wastewater distribution and collection of the treated effluent.
5. *Vegetation:* Optional integration of aquatic plants enhances nutrient uptake and aesthetics.

The system is configured vertically, enabling wastewater to flow downward through the layers, where filtration and biological treatment occur simultaneously.

3.2.1.2 Mechanism of Action

The treatment process in VSFVFs includes several stages:

1. *Physical Filtration:* As wastewater flows through the top layers, larger suspended solids are trapped.
2. *Biological Decomposition:* Earthworms and microorganisms break down organic pollutants into simpler, less harmful compounds. Earthworm activity enhances aeration, facilitating microbial action.

3. *Nutrient Removal:* Microbial action and optional plant uptake help remove nutrients such as nitrogen and phosphorus.
4. *Pathogen Removal:* The combined effects of microbial action, earthworm digestion, and physical filtration significantly reduce the pathogen load.

3.2.1.3 Advantages of VSFVFs

- *Eco-Friendliness:* The system promotes natural biological processes without harmful chemicals.
- *Low Energy and Maintenance Requirements:* Unlike conventional systems, VSFVFs operate with minimal energy input and low operational costs.
- *Improved Effluent Quality:* The treated water is suitable for irrigation or safe discharge.
- *Scalability:* Systems can be customized for small or large applications.

3.2.1.4 Applications

VSFVFs are used in:

- Domestic wastewater treatment in rural and urban areas.
- Agricultural runoff management.
- Industrial effluent treatment for low-load organic waste.

3.2.2 Horizontal Subsurface Flow Vermifilters

Horizontal Subsurface Flow Vermifilters (HSFVFs) are an innovative and sustainable approach to wastewater treatment, integrating horizontal subsurface flow systems with vermifiltration technology. This system relies on the combined action of earthworms, microorganisms, and filter media to treat wastewater by removing organic pollutants, pathogens, and nutrients. HSFVFs are particularly valued for their low maintenance, cost-efficiency, and ability to operate with minimal energy input.

3.2.2.1 Design and Components of HSFVFs

HSFVFs involve the following essential components:

1. *Filter Media:* Layers of gravel, sand, and organic materials serve as substrates for filtration and microbial growth.
2. *Earthworms:* Species such as *Eisenia fetida* (red wigglers) are introduced to the filter bed to aid organic matter decomposition and aeration.
3. *Vegetation:* Aquatic plants, such as reeds (*Phragmites australis*), may be planted on the surface to enhance nutrient uptake and improve water quality.
4. *Inlet Zone:* Wastewater enters the system horizontally, ensuring even distribution across the filter bed.
5. *Outlet Zone:* Treated effluent is collected and discharged or reused, depending on the application.

In this system, wastewater flows horizontally through the media layers beneath the surface, ensuring that it remains in contact with the filter media and the organisms facilitating treatment.

3.2.2.2 Mechanism of Action

The treatment process in HSFVFs involves the following key steps:

1. *Physical Filtration:* Large particles and suspended solids are trapped by the gravel and sand layers.
2. *Biological Degradation:* Earthworms and microorganisms break down organic matter, converting it into stable humus. The movement of earthworms enhances aeration, facilitating microbial activity.
3. *Nutrient and Pathogen Removal:* Microbial processes and plant roots absorb nutrients such as nitrogen and phosphorus, while pathogens are reduced through natural antagonism and filtration.
4. *Horizontal Flow:* The flow pattern ensures prolonged contact time between wastewater, microorganisms, and filter media, optimizing treatment efficiency.

3.2.2.3 Advantages of HSFVFs

- *Sustainability:* HSFVFs use natural processes without chemical additives.
- *Low Energy Consumption:* The horizontal flow design minimizes the need for pumps, reducing energy requirements.
- *High Treatment Efficiency:* The combination of physical, biological, and vegetative processes ensures effective pollutant removal.
- *Minimal Odor and Mosquito Breeding:* Subsurface operation prevents the release of odors and minimizes mosquito breeding.
- *Versatility:* These systems can treat a wide range of wastewaters, including domestic sewage, agricultural runoff, and some industrial effluents.

3.2.2.4 Applications

HSFVFs are widely used for:

- Treating domestic wastewater in decentralized systems.
- Managing agricultural runoff and livestock wastewater.
- Providing pre-treatment for industrial effluents with low to moderate organic loads.

3.2.2.5 Limitations

- *Land Requirement:* Horizontal systems may require more space compared to vertical configurations.
- *Climate Sensitivity:* Performance can vary with temperature, as microbial and earthworm activity decreases in colder conditions.

3.2.3 Two-Stage Vermifilters

Two-stage vermifilters are an advanced configuration of vermifiltration technology designed for enhanced wastewater treatment. They integrate two sequential vermifiltration units to optimize the removal of organic pollutants, pathogens, and nutrients. By combining physical filtration, biological decomposition, and microbial activity, this system provides a robust solution for treating a variety of wastewater types, including domestic, agricultural, and industrial effluents.

This dual-stage design enhances treatment efficiency and ensures higher-quality effluent, making it a preferred choice for locations with stringent wastewater discharge standards.

3.2.3.1 Design and Components of Two-Stage Vermifilters

A two-stage vermifilter typically consists of the following components:

1. *Primary Vermifilter (Stage 1):*
   - *Filter Media:* Coarse gravel and sand layers trap larger suspended solids and provide a substrate for microorganisms.
   - *Earthworms: Eisenia fetida* or similar species aid organic matter degradation and enhance aeration.
   - *Drainage Layer:* Facilitates the outflow of pre-treated water.
   - This stage focuses on removing larger organic particles and reducing the initial pollutant load.
2. *Secondary Vermifilter (Stage 2):*
   - *Fine Filter Media:* Sand and finer gravel are used to provide additional filtration.
   - *Earthworms and Microbial Community:* Further reduce organic matter, pathogens, and nutrients in the effluent.
   - *Vegetation (Optional):* Plants like reeds or canna may be incorporated to aid nutrient uptake and improve aesthetics.

The treated water from the primary stage flows into the secondary stage for additional polishing and removal of residual contaminants.

3.2.3.2 Mechanism of Action

The treatment process in two-stage vermifilters involves the following steps:

1. *Stage 1—Primary Treatment:*
   - *Physical Filtration:* Suspended solids are trapped in the coarse filter media.
   - *Biological Decomposition:* Earthworms and microorganisms degrade organic matter into simpler compounds.
   - *Pathogen Reduction:* Initial reduction in pathogens occurs due to microbial activity and earthworm digestion.

2. *Stage 2—Secondary Treatment:*
   - *Fine Filtration:* Residual suspended particles and finer solids are removed.
   - *Advanced Nutrient Removal:* Microbial activity and plant uptake (if vegetation is used) reduce nitrogen and phosphorus concentrations.

- *Final Effluent Polishing:* Effluent quality is improved to meet reuse or discharge standards.

3.2.3.3 Advantages of Two-Stage Vermifilters

- *Enhanced Treatment Efficiency:* The sequential process ensures comprehensive removal of contaminants.
- *Flexibility:* Suitable for a variety of wastewater types with varying pollutant loads.
- *Low Energy Requirements:* Operates passively without the need for complex machinery.
- *Sustainability:* Uses natural processes without reliance on chemicals.
- *Scalability:* Can be adapted for small or large applications.

3.2.3.4 Applications

- *Domestic Wastewater Treatment:* Ideal for decentralized systems in residential areas.
- *Industrial Effluent Treatment:* Effective for industries with low to moderate organic pollutant loads.
- *Agricultural Wastewater Management:* Handles nutrient-rich runoff efficiently.

3.2.3.5 Limitations

- *Space Requirements:* Requires sufficient area to install two stages.
- *Maintenance:* Regular monitoring is necessary to ensure optimal earthworm activity and avoid clogging.

3.3 Mechanisms of Vermifiltration

The vermifiltration process is driven by a combination of physical, biological, and biochemical mechanisms that work synergistically to improve wastewater treatment and enable resource recovery (Neralla et al. 2000; Bertanza et al. 2010; Kirkpatrick and Asano 1986). The earthworms in the vermifilter establish a symbiotic relationship with the diverse microbial communities present in the filter media (Fig. 3.1). The earthworms contribute to the degradation of organic pollutants through their feeding and burrowing activities, which enhance microbial activity within the filter bed (Nascimento et al. 2018). Specifically, the earthworms break down the larger organic matter particles into smaller, more readily available substrates for the microorganisms to further decompose. The microorganisms, on the other hand, play a crucial role in the biodegradation of organic pollutants, converting them into simpler, less harmful compounds. In addition to the synergistic relationship between earthworms and microorganisms, the presence of earthworms also contributes to the overall microbial diversity and activity within the filter bed.

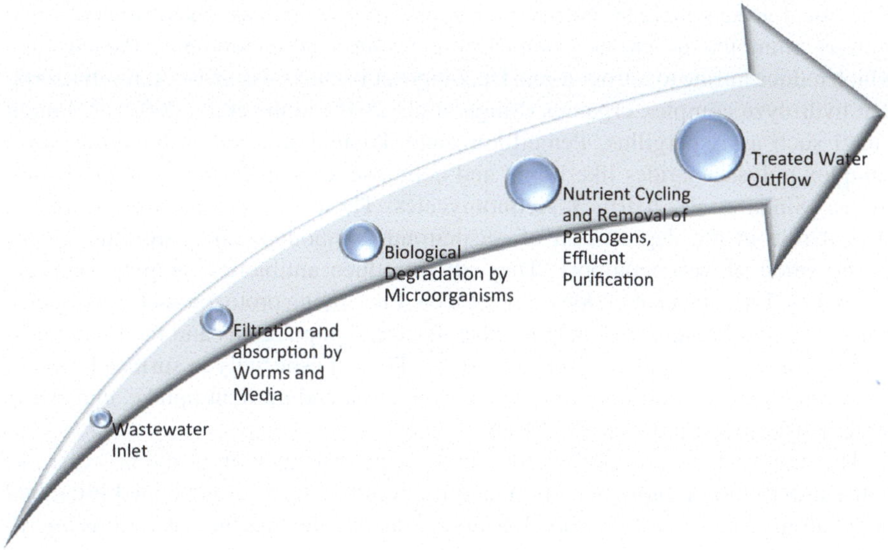

Fig. 3.1 Flowchart for mechanism of vermifiltration

The vermifiltration process facilitates the transformation and recovery of valuable nutrients, such as nitrogen and phosphorus, from the wastewater. The earthworms and microorganisms within the vermifilter work in concert to convert organic forms of nitrogen and phosphorus into more readily available inorganic forms. Specifically, the earthworms excrete nitrogenous compounds in the form of ammonium, which is then oxidized by nitrifying bacteria to nitrite and nitrate. The earthworms' burrowing activities and the porous structure of the filter media also enhance the movement and distribution of nutrients within the vermifilter, promoting their efficient uptake by plants or their removal through denitrification processes.

The vermifiltration process also contributes to the removal of pathogenic microorganisms, heavy metals, and other emerging contaminants from the wastewater (Fig. 3.1). The earthworms and microorganisms present in the vermifilter can immobilize or transform these pollutants, rendering them less harmful or facilitating their removal from the treated effluent.

3.4 Types of Microbes in the Vermifilter

The diverse microbial communities present in the vermifilter play a crucial role in the wastewater treatment and resource recovery processes. These microbial communities are typically dominated by bacteria, with the presence of fungi, archaea, and protozoa as well. Some of the key microbial groups found in vermifiltration systems include:1. Bacteria: Dominant in vermifilters, bacteria are responsible for the degradation of organic matter, nutrient cycling, and pathogen removal.

Functional groups include: Nitrifying bacteria (Nitrosomonas, Nitrobacter), which convert ammonia to nitrate; Denitrifying bacteria (Pseudomonas, Paracoccus), which reduce nitrate to nitrogen gas; Decomposer bacteria (Bacillus, Actinobacteria) that hydrolyze complex organics (Singh et al., 2018; Sinha et al., 2008). 2. Fungi: Fungi such as Aspergillus, Penicillium, and Trichoderma aid in breaking down complex polysaccharides like lignin and cellulose, especially in organic-rich substrates (Sinha et al., 2010). 3. Actinomycetes: These filamentous bacteria play a critical role in the degradation of recalcitrant compounds and contribute to the earthy smell of vermicompost. They also produce antibiotics, helping suppress pathogens (Lazcano et al., 2008). 4. Protozoa: Free-living protozoa such as Amoeba, Vorticella, and Paramecium help regulate bacterial populations and facilitate nutrient mineralization (Li et al., 2009). 5. Algae: Present especially in surface layers of vermifilters, algae contribute to oxygen production and nutrient uptake, enhancing system efficiency (Yadav et al., 2013).

The digestive tract of earthworms serves a pivotal function in the processes of nutrient dissolution, bioremediation, and the reinforcement of beneficial microbial populations. Analyses of microbial diversity uncover the presence of various microorganisms in vermicast, such as *Azotobacter*, *Rhizobium*, *Nitrobacter*, nitrogen-fixing bacteria, phosphate-solubilizing bacteria, as well as *Basidiomycetes* and *Ascomycetes*.

Earthworms have been observed to significantly impact the soil microbial community composition and diversity. Studies have demonstrated that the passage of organic material through the earthworm gut can selectively enrich certain microorganisms, such as cellulose-decomposing bacteria, *actinomycetes*, and fungi. This selective enrichment of microorganisms within the vermifilter's ecosystem can enhance the overall efficiency of the wastewater treatment process. The effectiveness of vermifiltration in wastewater treatment and resource recovery has been extensively studied, with numerous research efforts exploring the various aspects of this technology.

Examination of the bacterial composition in the system indicates a higher prevalence of aerobic bacteria compared to facultative and anaerobic species. Furthermore, Arora et al. identified 26 bacterial species in the vermifilter, with Bacillus as the dominant genus, in contrast to only 11 species in the filter without earthworms. Bacillus is known to produce metabolites with antimicrobial, cytotoxic, or antifungal properties. Moreover, biofilm analyses revealed that the feeding and excretion activities of the earthworms altered the bacterial population within the vermifilter. Additionally, the use of Automatic Testing Bacteriology expression and sequencing data demonstrates that the presence of earthworms leads to an increase in microbial diversity and metabolic activities in the biofilm.

The presence of earthworms in the vermifilter system significantly enhances the bacterial population. Research by Wang et al. on the biofilms revealed a substantially higher pore number in the vermifilter compared to a filter without earthworms or a non-vermifilter. Additionally, the total organic carbon was higher in the VF. Dehydrogenase activity and adenosine triphosphate, indicators of microbial activity, were better in the VF than in the NVF, with an improvement of more than

16%. The burrowing activities of the earthworms can also increase parameters such as hydraulic conductivity, specific surface of the substrate, nutrient exchange rate, and aeration of the system. Furthermore, earthworms play a role in the vermifilter by consuming the bacterial population that competes with the microorganisms responsible for pollutant degradation. Earthworms secrete a viscous fluid that traps these harmful bacteria, while the remaining microbes compete with pathogens for nutrients. Additionally, earthworms have been shown to accumulate heavy metals in their gut, leading to a substantial decrease in pollution levels in the final treatment products.

Various studies have demonstrated the potential of vermifiltration in the treatment and valorization of different types of wastewater, including municipal sewage, agricultural runoff, and industrial effluents. The vermifiltration process has shown promising results in reducing organic matter, suspended solids, nutrients, and pathogens from wastewater while also facilitating the recovery of valuable resources such as plant nutrients (Bhadauria and Saxena 2009; Weibao and Yang 2008; Nadana et al. 2020; Ray 2018).

Vermifiltration is a biological process that employs earthworms to remove organic matter, suspended solids, and nutrients from wastewater (Cornejo et al. 2018). The process involves passing the wastewater through a bed of soil or organic media, where the earthworms break down the organic matter and extract nutrients for their own growth and reproduction. The resulting vermicompost can be used as a nutrient-rich fertilizer, while the biogas produced during the decomposition process can be captured and used for energy generation. The application of vermifiltration in wastewater treatment has several advantages over conventional methods, such as improved organic matter and nutrient removal, reduced sludge production, and the potential for resource recovery Jiménez-Benítez et al. (2020). Furthermore, vermifiltration can be implemented at the household or community level, making it a decentralized and scalable solution for wastewater management (Kraker et al. 2019).

Research has shown that vermifiltration can achieve high removal rates for organic matter, suspended solids, and nutrients, with studies reporting removal efficiencies of up to 90% for biochemical oxygen demand, 80% for total suspended solids, and 70% for total nitrogen and phosphorus (Akyol et al. 2019; Rulkens 2007). Additionally, vermifiltration can reduce sludge production by up to 50% compared to conventional activated sludge processes, thereby reducing the burden and cost of sludge management (Nascimento et al. 2018, Qin and He 2017).

3.5 Implementation of a Vermifilter and Factors to Consider

The design and implementation of a vermifilter system involve several key considerations. The application of vermifiltration has expanded beyond its initial focus on domestic wastewater, demonstrating its potential for treating a diverse range of wastewater sources. Studies have reported the successful implementation of this technology in treating effluents from various industries, including dairies, palm oil processing plants, textile operations, piggeries, feedlots, and even hospital wastewater. This broad spectrum of applications highlights the versatility of vermifiltration and its suitability for handling different types of wastewaters. Furthermore, the system is capable of accommodating a wide range of organic loading rates, from relatively low levels around 100–200 mg/L of chemical oxygen demand to higher concentrations. Vermifiltration has been demonstrated as an effective treatment process for highly polluted water. Studies conducted by Manyuchi et al. have shown successful application of this technology in treating distillery wastewater with chemical oxygen demand exceeding 92,000 mg/L and biochemical oxygen demand over 25,000 mg/L, achieving removal rates of approximately 89% and 91%, respectively. Similar promising results have been reported by other researchers. Despite fluctuations in organic loading at the inlet, vermifilters have exhibited the capability to operate effectively over extended periods. However, the performance of vermifilters can be influenced by the varying characteristics of different wastewaters. For instance, industrial effluents may contain toxic compounds that can inhibit earthworm activity, thereby reducing the overall treatment efficiency. Notably, earthworms possess a remarkable ability to withstand a wide range of contaminants, including herbicides, heavy metals, and organic pollutants commonly found in industrial, factory, and hospital wastewater, as well as sewage sludge. This resilience allows earthworms to accumulate and reduce the concentrations and toxicity of these wastewater contaminants within their body tissues.

The hydraulic loading rate (HLR), which represents the volume of water passing through the filter per unit area and time, is a crucial parameter in water filtration. It has been observed that both the treatment performance and the susceptibility to clogging in a vermifilter are closely linked to the applied HLR. The natural activities of the earthworms, such as feeding and burrowing, help enhance the porosity of the filter bed, which is beneficial for the filtration process. However, when the HLR is increased, the earthworm activity tends to decrease, leading to the accumulation of clogging materials in the initial section of the filter bed, which can result in severe clogging problems. According to Suhaib and Bhunia, the instability of the earthworm casts within the filter bed at higher HLRs can further diminish the bed's permeability. Furthermore, their studies have revealed that a filter bed operating at an HLR of 2.34 m3/m2/day experiences an increase in organic and suspended solid loads compared to a HLR of 0.66 m3/m2/day, resulting in a significant accumulation of clogging materials, such as biofilm biomass and suspended solids, at the inlet end of the filter bed.

The earthworms used in vermifiltration systems are typically epigeic, also known as surface-dwelling earthworms. These species inhabit the upper soil layers and produce wormcasts on the surface, playing a crucial role in the breakdown of organic matter. Commonly employed species include Lumbricus rubellus, Eisenia fetida, and Eudrilus eugeniae. When implementing a vermifiltration system, the bedding materials and active layer are essential parameters to consider, as they influence the wastewater's residence time in the filter due to their varying porosities, permeability, and other characteristics. Macrophyte Vertical Filters have emerged as a promising technique for removing organic and mineral matter from wastewater.

Studies indicate that planted VFs demonstrate superior average removal rates to simple VFs, with the effluent from planted VFs showing lower BOD5 concentrations, around 15 mg/L, compared to around 30 mg/L in simple VFs. Earthworms are sensitive to temperature fluctuations, and factors such as fecundity, cocoon production, and maturation can be affected by temperature changes. However, under optimal temperature conditions, earthworms possess the remarkable ability to biologically transform a wide range of organic compounds, including environmental pollutants.

Vermifiltration has been demonstrated as a promising approach for the valorization of various types of organic waste, including agricultural residues, animal manure, and municipal solid waste.

The application of vermifiltration in wastewater treatment not only improves the quality of the effluent but also offers opportunities for resource recovery. The vermicompost produced during the process is a nutrient-rich soil amendment that can be used in agriculture, landscaping, and other applications, providing a valuable resource that can offset the costs of wastewater treatment (Qin and He 2017; Rulkens 2007).

In addition to the vermicompost, the biogas generated during the decomposition of organic matter can be captured and used for energy generation, further enhancing the sustainability and cost-effectiveness of the vermifiltration process.

3.6 Benefits and Applications of Vermifiltration

Vermifiltration can achieve high removal rates for organic matter, suspended solids, and nutrients in wastewater (Jadhav et al. 2021; Sankaran et al. 2010; Rulkens 2007). Removal efficiencies of up to 90% for biochemical oxygen demand, 80% for total suspended solids, and 70% for total nitrogen and phosphorus have been reported (Sankaran et al. 2010). The process can reduce sludge production by up to 50% compared to conventional activated sludge processes. Vermifiltration can be implemented at the household or community level, making it a decentralized and scalable solution for wastewater management. Vermifiltration produces a nutrient-rich vermicompost that can be used as a soil amendment in agriculture, landscaping, and other applications.

While vermifiltration offers numerous benefits, there are also several challenges and considerations that need to be addressed for its widespread adoption. One of the key challenges is the need for a stable and optimal environment for the earthworms, which requires careful control of factors such as temperature, moisture, and organic loading rate. Additionally, the scaling up of vermifiltration systems to handle larger volumes of wastewater can be challenging, and the long-term performance and maintenance of these systems need to be thoroughly evaluated.

Another consideration is the potential for the accumulation of heavy metals and other contaminants in the vermicompost, which may limit its use as a soil amendment. Careful monitoring and pre-treatment of the wastewater may be necessary to ensure the safety and quality of the vermicompost.

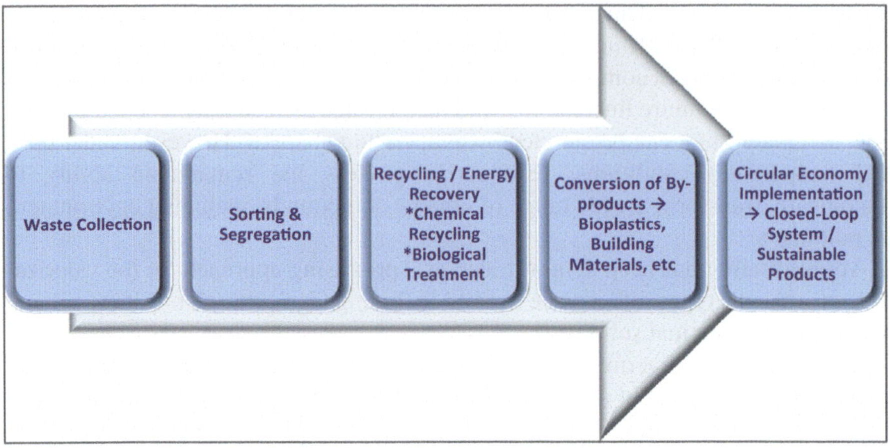

Fig. 3.2 Waste valorization process flowchart

3.7 Waste Valorization

Waste management remains one of the most pressing environmental challenges of the twenty-first century. Global waste generation is projected to reach 3.4 billion tons annually by 2050, necessitating innovative strategies to mitigate its impacts (Kaza et al. 2018). Waste valorization offers a sustainable alternative to traditional disposal methods, such as landfilling and incineration, by transforming waste into valuable resources. This paper reviews the current landscape of waste valorization, focusing on its processes (Fig. 3.2), applications, challenges, and future potential.

3.7.1 Processes in Waste Valorization (Fig. 3.2)

3.7.1.1 Material Valorization

Material recovery is the most widely practiced form of waste valorization, targeting recyclable waste streams such as plastics, metals, and glass. Mechanical recycling is predominant, involving the collection, sorting, and reprocessing of materials into new products. For biodegradable waste, composting is a key process, producing nutrient-rich compost for agricultural use.

Innovative approaches, such as chemical recycling, are gaining attention for their ability to break down complex polymers into monomers, enabling the production of high-quality recycled materials (Geyer et al. 2017).

3.7.1.2 Chemical Valorization

Chemical valorization converts waste into valuable chemicals or fuels. Key processes include:

- *Pyrolysis:* Converts plastics and biomass into oils, syngas, and char at high temperatures in the absence of oxygen.

- *Gasification:* Produces syngas from organic waste, which can be used to synthesize chemicals or generate electricity.
- *Fermentation:* Uses microorganisms to convert organic waste into bioethanol, bioplastics, or other biochemicals.

These processes are particularly effective for non-recyclable and mixed waste streams.

3.7.1.3 Energy Valorization

Energy valorization focuses on using waste as a source of renewable energy. Technologies such as anaerobic digestion and waste-to-energy incineration play significant roles.

- *Anaerobic Digestion:* Decomposes organic matter in the absence of oxygen to produce biogas and digestate.
- *Refuse-Derived Fuels (RDF):* Converts high-calorific waste into fuel for industrial applications.

Energy valorization provides dual benefits of waste reduction and renewable energy generation, aligning with global efforts to transition to cleaner energy sources.

3.7.2 Applications of Waste Valorization

3.7.2.1 Industrial Applications

- *Material Recovery in Manufacturing*

Recycling waste materials, such as metals, plastics, and glass, reduces the demand for virgin raw materials in manufacturing industries. For instance, electronic waste (e-waste) recycling enables the recovery of precious metals like gold, silver, and palladium, which are crucial for the electronics sector (Cucchiella et al. 2015).

- *Alternative Fuels for Industry*

Refuse-derived fuels (RDF) and solid recovered fuels (SRF) are used as substitutes for fossil fuels in energy-intensive industries such as cement and steel manufacturing. These fuels provide a sustainable energy source and help industries reduce carbon emissions (Arena 2012).

- *Chemical Production*

Chemical valorization processes, such as pyrolysis and gasification, convert plastics and other waste into valuable chemicals like hydrogen, syngas, and oils, which are used as raw materials for industrial chemical production (Meys et al. 2020).

3.7.2.2 Agricultural Applications

- *Composting for Soil Fertility*

Biodegradable waste, such as food scraps and agricultural residues, is composted to produce organic fertilizer. This compost enriches soil fertility, supports sustainable farming practices, and reduces the need for chemical fertilizers (Pérez-Piqueras et al. 2020).

- *Biochar Production*
  Biochar, produced from agricultural and forestry waste through pyrolysis, is used to enhance soil structure, improve water retention, and sequester carbon, thereby contributing to sustainable agricultural practices (Lehmann and Joseph 2015).
- *Biogas for Farm Energy*
  Animal manure and crop residues undergo anaerobic digestion to produce biogas, a renewable energy source that can be used for heating, electricity generation, or as vehicle fuel in rural areas (Holm-Nielsen et al. 2009).

3.7.2.3 Energy Applications

- *Waste-to-Energy (WTE)*
  Non-recyclable waste is incinerated in WTE plants to generate electricity and heat. This process reduces landfill dependency while recovering energy from waste, contributing to renewable energy goals (Panepinto et al. 2015).
- *Biogas Production*
  Organic waste, such as food and agricultural residues, is processed in anaerobic digesters to produce biogas (methane and carbon dioxide). This renewable energy source can be used in power plants or as an alternative to natural gas in homes and industries (Weiland 2010).
- *Hydrogen Generation*
  Gasification of organic waste produces syngas, which can be further processed to extract hydrogen, a key energy carrier in the transition to low-carbon energy systems (Materazzi et al. 2019).

3.7.2.4 Environmental Remediation

- *Phytoremediation of Polluted Soils*
  Plants grown on organic waste composts can aid the remediation of polluted soils by absorbing heavy metals and other contaminants, thus restoring soil quality (Vangronsveld et al. 2009).
- *Mitigation of Greenhouse Gas Emissions*
  Valorizing organic waste through composting or anaerobic digestion prevents methane emissions from landfills, significantly reducing the carbon footprint of waste management (Boldrin et al. 2009).

3.7.2.5 Consumer and Retail Applications

- *Production of Bioplastics*
  Organic waste can be used to produce biodegradable plastics through microbial fermentation, offering an eco-friendly alternative to conventional plastics (Chen 2010).

- *Textiles and Packaging*
  Waste valorization has applications in creating textiles from recycled materials and producing sustainable packaging solutions, reducing environmental impacts of traditional manufacturing (Hawkins et al. 2013).

3.7.2.6 Circular Economy Integration

Waste valorization plays a central role in enabling a circular economy by closing material loops. Products derived from waste streams are reintroduced into the supply chain, promoting resource efficiency and minimizing waste generation (Geissdoerfer et al. 2017).

3.7.3 Challenges in Waste Valorization

Despite its potential, waste valorization faces several challenges:

- *Economic Viability:* High initial costs of advanced technologies and fluctuating markets for recovered materials limit widespread adoption.
- *Technological Barriers:* Inefficiencies in waste sorting and treatment technologies hinder scalability.
- *Regulatory Issues:* Inconsistent policies and inadequate enforcement reduce the effectiveness of waste valorization programs.
- *Public Perception:* Resistance to waste-to-energy plants due to concerns over emissions and environmental impacts remains a significant obstacle.

3.7.4 Future Directions of Waste Valorization

As global waste generation continues to rise, waste valorization presents significant opportunities to create value while addressing environmental, economic, and societal challenges. Future directions in waste valorization focus on advancing technologies, fostering policy innovation, and integrating these systems into broader sustainability frameworks.

3.7.4.1 Technological Advancements

- *Development of Advanced Recycling Technologies*
  Emerging technologies, such as chemical recycling and enzymatic depolymerization, are expected to overcome the limitations of mechanical recycling. These processes can efficiently handle mixed and contaminated waste streams, converting them into high-quality raw materials (Meys et al. 2020).
- *Integration of Artificial Intelligence (AI)*
  AI and machine learning are poised to revolutionize waste sorting and processing by improving the accuracy of material separation, thereby enhancing the efficiency of recycling and valorization processes (Patsavellas et al. 2021).

- *Biotechnological Innovations*
  Advances in biotechnology, such as engineered microbes and enzymes, are enabling the production of bio-based chemicals and fuels from organic waste. These innovations can optimize processes like fermentation and anaerobic digestion, making them more cost-effective and scalable (Chen et al. 2020).
- *Energy Valorization Optimization.*
  Emerging energy recovery technologies, such as plasma gasification and supercritical water oxidation, offer promising pathways for converting waste into cleaner energy forms with higher efficiency and lower emissions (Arena 2012).

3.7.4.2 Policy and Regulatory Frameworks

- *Extended Producer Responsibility (EPR)*
  Governments are increasingly adopting EPR policies to hold producers accountable for the end-of-life management of their products, incentivizing design for recyclability and promoting waste valorization (Geissdoerfer et al. 2017).
- *Carbon Pricing and Incentives*
  Introducing carbon credits or taxes linked to waste valorization activities can encourage industries to adopt environment-friendly practices and reduce landfill disposal (Lehmann et al. 2020).
- *Global Standardization.*
  Harmonizing waste management standards and certifications across countries will facilitate international collaboration and create markets for waste-derived products (Velis et al. 2012).

3.7.4.3 Integration with Circular Economy

- *Closed-Loop Systems in Industries*
  Industries are increasingly focusing on closed-loop systems where waste from one process serves as an input for another. For example, industrial symbiosis can link multiple industries to share waste streams, optimizing resource utilization (Chertow 2000).
- *Product Redesign for Valorization*
  Future products will likely incorporate modular designs and biodegradable materials, making them easier to disassemble and valorize at the end of their life cycle (Stahel 2016).

3.7.4.4 Waste-to-Value Innovations

- *Bioenergy and Biofuels*
  The transition to renewable energy sources is driving innovations in converting organic waste into advanced biofuels, such as biodiesel and hydrogen. These fuels could replace fossil fuels in transportation and industrial applications (Weiland 2010).

- *High-Value Products from Waste*
  Innovative processes are being developed to extract high-value compounds, such as proteins, vitamins, and fibers, from organic waste, opening new markets in food, pharmaceuticals, and cosmetics (Koutinas et al. 2014).

3.7.4.5 Urban Waste Valorization Strategies

- *Smart Waste Management Systems*
  Smart cities are adopting Internet of Things (IoT) technologies to optimize waste collection and valorization. Sensors and data analytics enable real-time monitoring, reducing inefficiencies in waste management (Foschi et al. 2018).
- *Decentralized Waste Valorization Units*
  Small-scale, community-based waste valorization units, such as micro-composters and biogas plants, are being developed for urban and rural areas to reduce transportation costs and improve waste recovery rates (Holm-Nielsen et al. 2009).

3.7.4.6 Education and Public Awareness

- *Behavioral Change Campaigns*
  Raising public awareness about the benefits of waste valorization can lead to better waste segregation at the source, improving the quality of feedstocks for valorization processes (Pérez-Piqueras et al. 2020).
- *Skill Development*
  Training programs focused on waste valorization technologies can build a skilled workforce, promoting the adoption of advanced systems in industries and municipalities (Lehmann and Joseph 2015).

3.8 Conclusion

Vermifiltration is a promising technology that offers a sustainable approach to wastewater treatment and resource recovery. By leveraging the natural abilities of earthworms to decompose organic matter and extract nutrients, vermifiltration can improve the quality of wastewater effluent while simultaneously producing valuable byproducts such as vermicompost and biogas. The implementation of vermifiltration can contribute to the development of a circular economy by reducing the environmental impact of wastewater management and creating new opportunities for resource recovery and reuse. As research and development in this field continue, the potential of vermifiltration to transform the way we manage wastewater and recover resources will be further explored and realized. Similarly, the future of waste valorization is promising, with opportunities for technological, economic, and environmental growth. By fostering innovation, strengthening policies, and integrating waste valorization into the circular economy, societies can transition toward more sustainable and resilient systems of waste management.

References

Akyol Ç, Foglia A, Özbayram EG, Frison N, Katsou E, Eusebi AL, Fatone F (2019) Validated innovative approaches for energy-efficient resource recovery and re-use from municipal wastewater: From anaerobic treatment systems to a biorefinery concept. In: Akyol Ç, Foglia A, Özbayram EG, Frison N, Katsou E, Eusebi AL, Fatone F (eds) Critical reviews in environmental science and technology, vol 50(9). Taylor & Francis, p 869. https://doi.org/10.1080/10643389.2019.1634456

Arena U (2012) Process and technological aspects of municipal solid waste gasification. A review. Waste Manag 32(4):625–639

Bertanza G, Collivignarelli MC, Crotti BM, Pedrazzani R (2010) Integration between chemical oxidation and membrane thermophilic biological process. In: Bertanza G, Collivignarelli MC, Crotti BM, Pedrazzani R (eds) Water science & technology, vol 61(1). Pergamon Press, p 227. https://doi.org/10.2166/wst.2010.793

Bhadauria T, Saxena KG (2009) Role of earthworms in soil fertility maintenance through the production of biogenic structures. In: Bhadauria T, Saxena KG (eds) Applied and environmental soil science, vol 2010. Hindawi Publishing Corporation, p 1. https://doi.org/10.1155/2010/816073

Boldrin A, Andersen JK, Møller J, Christensen TH, Favoino E (2009) Composting and compost utilization: accounting of greenhouse gases and global warming contributions. Waste Management & Research 27(8):800–812

Chen GQ (2010) Scarcity and sustainability issues in the supply of strategic metals: A review. Resources, Conservation and Recycling 54(12):1073–1080

Chen GQ, Patel MK, Pavanan KC (2020) Bioplastics and bio-based polymers: a sustainable alternative to plastics. Chem Rev 120(7):3846–3879

Cornejo PK, Becker JG, Pagilla K, Mo W, Zhang Q, Mihelcic JR, Chandran K, Sturm B, Yeh DH, Rosso D (2018) Sustainability metrics for assessing water resource recovery facilities of the future. In: Cornejo PK, Becker JG, Pagilla K, Mo W, Zhang Q, Mihelcic JR, Chandran K, Sturm B, Yeh DH, Rosso D (eds) Water environment research, vol 91(1). Wiley, p 45. https://doi.org/10.2175/106143017x15131012187980

Chertow MR (2000) Industrial symbiosis: literature and taxonomy. Annu Rev Energy Environ 25(1):313–337

Cucchiella F, D'Adamo I, Rosa P (2015) Waste management and valorization: A technological and economic overview. Waste Management 45:1–15

do Nascimento T, Fdz-Polanco F, Peña M (2018) Membrane-based technologies for the up-concentration of municipal wastewater: a review of pretreatment intensification [review of membrane-based technologies for the up-concentration of municipal wastewater: a review of pretreatment intensification]. In: Separation and purification reviews, vol 49(1). Taylor & Francis, p 1. https://doi.org/10.1080/15422119.2018.1481089

Foschi E, Bonoli A (2018) The commitment of packaging industry in the framework of the circular economy: The case of compostable bioplastics. Sustainable Production and Consumption 17:162–174

Geissdoerfer M, Savaget P, Bocken NMP, Hultink EJ (2017) The circular economy—a new sustainability paradigm? J Clean Prod 143:757–768

Geyer R, Jambeck JR, Law KL (2017) Production, use, and fate of all plastics ever made. Sci Adv 3(7):e1700782

Hamilton DW, Murie ME, Khan A, Ndegwa PM (2008) In: Hamilton DW, Murie ME, Khan A, Ndegwa PM (eds) Vermicomposting of poultry litter: process optimization. Providence, Rhode Island, June 29—July 2, 2008. https://doi.org/10.13031/2013.24712

Hawkins TR, Singh B, Majeau-Bettez G, Strømman AH (2013) Comparative environmental life cycle assessment of conventional and electric vehicles. Journal of Industrial Ecology 17(1):53–64

Holm-Nielsen JB, Al Seadi T, Oleskowicz-Popiel P (2009) The future of anaerobic digestion and biogas utilization. Bioresource Technology 100(22):5478–5484

Jadhav AL, Saraf RV, Dakhore AN (2021) Energy recovery from waste water treatment plant sludge. In: In AL, Jadhav RV, Saraf, Dakhore AN (eds) Materials today proceedings, vol 42. Elsevier BV, p 1224. https://doi.org/10.1016/j.matpr.2020.12.871

Jiménez-Benítez A, Ferrer FJ, Greses S, Ruiz-Martínez A, Fatone F, Eusebi AL, Mondéjar N, Ferrer J, Seco A (2020) AnMBR, reclaimed water and fertigation: two case studies in Italy and Spain to assess economic and technological feasibility and CO2 emissions within the EU innovation Deal initiative. In: Jiménez-Benítez IA, Ferrer FJ, Greses S, Ruiz-Martínez A, Fatone F, Eusebi AL, Mondéjar N, Ferrer J, A. (eds) Seco, journal of cleaner production, vol 270. Elsevier BV, p 122398. https://doi.org/10.1016/j.jclepro.2020.122398

Kaza S, Yao L, Bhada-Tata P, Van Woerden F (2018) What a waste 2.0: A global snapshot of solid waste management to 2050. World Bank Publications

Koutinas AA, Roy I, Keshavarz T (2014) Advances in microbial biotechnology for biofuels and high-value products. Crit Rev Biotechnol 34(4):309–320

Kirkpatrick WR, Asano T (1986) Evaluation of tertiary treatment systems for wastewater reclamation and reuse. In: Kirkpatrick WR, Asano T (eds) Water science & technology, vol 18(10). Pergamon Press, p 83. https://doi.org/10.2166/wst.1986.0115

de Kraker J, Kujawa-Roeleveld K, Villena M, Pereira CPP (2019) Decentralized valorization of residual flows as an alternative to the traditional urban waste management system: the case of Peñalolén in Santiago de Chile. In: de Kraker J, Kujawa-Roeleveld K, Villena M, Pereira CPP (eds) Sustainability, vol 11(22). Multidisciplinary Digital Publishing Institute, p 6206. https://doi.org/10.3390/su11226206

Lazcano C, Gómez-Brandón M, Domínguez J (2008) Comparison of the effectiveness of composting and vermicomposting for the biological stabilization of cattle manure. Chemosphere 72(7):1013–1019

Lehmann J, Joseph S (2015) Biochar for environmental management: science, technology, and implementation. Routledge

Lehmann H, Ludwig C, Scherer A (2020) Transformational products and production: Circular economy and eco-design. In: Sustainable Materials and Technologies 25:e00188

Li Z, Lu H, Ren L, He L (2009) Earthworm species and their interactions with microorganisms in vermicomposting: A review. Bioresource Technology 102(24):11167–11172

Materazzi M, Lettieri P, Taylor R, Chapman C (2019) Performance analysis of RDF gasification in a twostage fluidized bed–plasma process. Waste Management 59:385–399

Meys R, Kätelhön A, Bachmann M, Winter B, Zibunas C, Suh S, Bardow A (2020) Achieving net-zero greenhouse gas emission plastics by a circular carbon economy. Science 370(6519):1299–1304

Nadana GRV, Kunjiappan S, Pandian S, Palanichelvam K (2020) Coelomic fluid of earthworms extruded by cold stress method has commercially significant compounds and trigger seed germination in Vigna radiata L. In: Nadana GRV, Kunjiappan S, Pandian S, Palanichelvam K (eds) Environmental technology & innovation, vol 19. Elsevier BV, p 100814. https://doi.org/10.1016/j.eti.2020.100814

Neralla S, Weaver RW, Lesikar BJ, Persyn RA (2000) Improvement of domestic wastewater quality by subsurface flow constructed wetlands. In: Neralla S, Weaver RW, Lesikar BJ, Persyn RA (eds) Bioresource technology, vol 75(1). Elsevier BV, p 19. https://doi.org/10.1016/s0960-8524(00)00039-0

Odell LH (2016) Groundwater treatment options abound. In: Odell LH (ed) Opflow, vol 43(1). Wiley, p 18. https://doi.org/10.5991/opf.2017.43.0003

Panepinto D, Fiore S, Zappone M, Genon G, Meucci L (2015) Evaluation of the environmental sustainability of a small wastewater treatment plant. Environmental Engineering and Management Journal 14(7):1571–1580

Patsavellas J, Dermatas D, Stefanakos I (2021) The role of AI in waste management: predictive models and sustainable practices. Resour Conserv Recycl 170:105607

Pérez-Piqueras J, Martínez-Blanco J, Aldaco R (2020) Environmental assessment of valorization routes for the management of municipal solid waste. Science of the Total Environment 717:137281

Qin M, He Z (2017) Resource recovery by osmotic bioelectrochemical systems towards sustainable wastewater treatment. In: Qin M, He Z (eds) Environmental Science water research & technology, vol 3(4). Royal Society of Chemistry, p 583. https://doi.org/10.1039/c7ew00110j
Ray S (2018) In: Ray S (ed) Introductory chapter: earthworms—the ecological engineers of soil. InTech eBooks. https://doi.org/10.5772/intechopen.78264
Rezaei N, Diaz-Elsayed N, Mohebbi S, Xie X, Zhang Q (2018) A multi-criteria sustainability assessment of water reuse applications: a case study in Lakeland, Florida. In: Rezaei N, Diaz-Elsayed N, Mohebbi S, Xie X, Zhang Q (eds) Environmental science water research & technology, vol 5(1). Royal Society of Chemistry, p 102. https://doi.org/10.1039/c8ew00336j
Rulkens WH (2007) Sewage sludge as a biomass resource for the production of energy: overview and assessment of the various options. In: Rulkens WH (ed) Energy & fuels, vol 22(1). American Chemical Society, p 9. https://doi.org/10.1021/ef700267m
Sankaran S, Khanal SK, Jasti N, Jin B, Pometto AL, van Leeuwen J (2010) Use of filamentous fungi for wastewater treatment and production of high value fungal byproducts: a review [review of use of filamentous fungi for wastewater treatment and production of high value fungal byproducts: a review]. In: Critical Reviews in Environmental Science and Technology, vol 40(5). Taylor & Francis, p 400. https://doi.org/10.1080/10643380802278943
Singh S, Srivastava M, Bhunia H (2018) Performance evaluation of a vermifiltration system for wastewater treatment and characterization of microbial communities. Environmental Technology & Innovation 10:108–121
Sinha RK, Bharambe G, Chaudhari U (2008) Sewage treatment by vermifiltration with synchronoustreatment of sludge by earthworms: A low-cost sustainable technology over conventional systems with potential for decentralization. Environmentalist 28(4):409–420
Sinha RK, Herat S, Bharambe G, Chaudhari U (2010) Vermistabilization of sewage sludge (biosolids) by earthworms: A novel sustainable technology for sanitation and nutrient recovery. American Journal of Environmental Sciences 6(3):259–268
Stahel WR (2016) Circular economy. Nature News 531(7595):435
Vangronsveld J, Herzig R, Weyens N, Boulet J, Adriaensen K, Ruttens A, Mench M (2009) Phytoremediation of contaminated soils and groundwater: lessons from the field. Environmental Science and Pollution Research 16(7):765–794
Velis CA, Wagland S, Longhurst PJ, Robson B, Sinfield K, Cheeseman CR (2012) Solid recovered fuel: Influence of waste stream composition and processing on fuel properties. Waste Management 32(12):2316–2325
Velis C, Brunner PH (2013) Recycling and resource efficiency: it is time for a change from quantity to quality. Waste Manag Res 31(6):539–540
Weibao S, Yang H (2008) Effects of earthworm and microbe on soil nutrients and heavy metals. In: Weibao S, Yang H (eds) Agricultural sciences in China, vol 7(5). Elsevier BV, p 599. https://doi.org/10.1016/s1671-2927(08)60058-9
Weiland P (2010) Biogas production: current state and perspectives. Appl Microbiol Biotechnol 85(4):849–860
Yadav KD, Tare V, Ahammed MM (2013) Vermicomposting of source-separated human faeces for nutrient recycling. Waste Management 33(2):521–529
Yildiz BS (2012) Water and wastewater treatment: biological processes. In: Yildiz BS (ed) Elsevier eBooks. Elsevier BV, p 406. https://doi.org/10.1533/9780857096463.3.406

Advances in Microbial Wastewater Treatment Systems

4

Ibukun Modupe Adesiyan, Yetunde Mutiat Feruke-Bello, Mojisola Christiana Owoseni, and Oluwarinu A. Adepoju

Abstract

The increasing population and demand for resources have led to rapid urbanization globally. The aftermath of this is the arbitrary use of resources and services, with consequent enormous waste generation which has put the environment under undue and mounting stress. Improper management of various wastes can lead to disease transmission if not critically tackled. Wastewater from households and industries usually comprises diverse noxious substances that are not only dangerous to our environment but detrimental to the health of humans. Therefore, the treatment of wastewater before disposal is an important issue for safeguarding human health and in achieving environmental security.

Several treatment methods for wastewater are adopted among which the biological treatment process is more proffered. Further research into the biological treatment method has validated the good use of microbes as the most sustainable and eco-friendly in the remediation of diverse (organic and inorganic) pollutants in wastewater systems. This present chapter therefore discusses the suitability of the microbial wastewater treatment method as the most effective and sustainable treatment method. We further reviewed the various advances in the microbial

I. M. Adesiyan (✉)
Department of Environmental and Occupational Health, School of Public Health, University of Medical Sciences, Ondo, Nigeria

Y. M. Feruke-Bello
Department of Microbiology, University of Ilesa, Ilesa, Osun State, Nigeria

M. C. Owoseni
Department of Microbiology, Federal University of Lafia, Lafia, Nigeria

O. A. Adepoju
Department of Microbiology, Obafemi Awolowo University, Ile-Ife, Nigeria

N. Dwivedi et al. (eds.), *Wastewater Treatment Through Nature-Based Solutions*,
https://doi.org/10.1007/978-981-96-6699-7_4

wastewater treatment process that confer the proven effectiveness and favor its continuous adoption.

4.1 Introduction

One of the biggest issues faced globally in recent times is the provision and availability of safe and clean water. The continuous increase in human population, industrial development, and urban expansion has exerted pressure on freshwater resources, thereby necessitating more endurable and efficient management systems (Adesiyan et al. 2018).

The growing pressure on water resources has led to a critical gap in accessibility. The number of people that have access to safe and clean water is about 2.6 billion globally, with approximately 96% benefiting in the urban areas and only 4% in the rural areas (Mishra 2023). However, estimates show that 663 million of the populaces still do not have access to treated and clean water, most of whom reside in Africa and South Asia.

Waste generated by households, agriculture, and industries are significant sources of harmful pollutants that can impair both human and environmental health (Karri et al. 2021). Therefore, if wastewater from these sources is not properly treated before discharge into the environment, it could pose a threat to biodiversity, aquatic and terrestrial ecosystems, and more importantly human health due to the potential to cause waterborne diseases (Chowdhary et al. 2020).

Several methods of conventional wastewater treatment exist and many of them have proven to be effective but not without some limitations (Shah et al. 2020; Saravanan et al. 2021). The activated sludge system is known to be energy demanding and ineffective in the complete removal of some pollutants including heavy metals, while the use of chemicals for waste treatment also leaves some of the by-products in the treated waste. These methods have preferences of the waste they treat, are very expensive to sustain, and are mostly not environmentally sustainable (Shah et al. 2020; Anekwe et al. 2022). For instance, in many developing nations, there is a lack of effective infrastructure for the increasing waste being generated due to the associated cost implication of conventional treatment plants. Therefore, there is need for eco-friendly and affordable waste treatment methods. Microbial treatment of wastewater has become the emerging solution. The use of microorganisms for the treatment of waste water lies in the intrinsic capabilities of many microorganisms to degrade organic substances, consume pollutants as nutrients, and detoxify hazardous constituents (Haripriyan et al. 2022; Raghav et al. 2022).

Using the biological processes exhibited in microbial communities, microbial wastewater systems provide a more eco-friendly, cost effective and energy efficient solution compared to the conventional methods. The microbial consortium in microbial wastewater includes fungi, bacteria, algae, and archaea, which are well-adapted and live in different environmental settings, making them very useful agents for wastewater treatment and the bioremediation process (Raghav et al. 2022).

Therefore, substantial progress has been accomplished in developing and optimizing microbial wastewater treatment methods over the last few decades. These developments are revolutionizing wastewater management and creating new opportunities for pollutant removal, resource recycling, and energy recovery. Specifically, the performance, scalability, and reliability of microbial treatment have improved with the advent of bioaugmentation techniques, microbial fuel cells (MFCs), anaerobic treatment procedures, and genetic engineering of microbes (Ghoshal et al. 2021; Rana and Gupta 2022).

However, one of the main issues with conventional treatment systems that MFCs may be able to solve is the high energy consumption. MFCs can lower the overall energy consumption of treatment facilities for wastewater by transforming the chemical energy inside wastewater to electrical energy (Rana and Gupta 2022). Furthermore, current research endeavors to enhance the scalability and cost-effectiveness of MFCs, rendering them more feasible for extensive deployment.

Anaerobic membrane bioreactors (AnMBRs) are now more durable and efficient, with problems like membrane fouling and operating expenses being minimized because to recent developments in membrane materials and design. AnMBRs are becoming a more appealing choice for businesses wishing to implement environment-friendly wastewater treatment techniques as a result of these advancements.

This chapter reviews some of the different strategies and technologies currently being adopted in the treatment of wastewater using microbial enhancement systems. We discussed the various wastewater treatment methods currently practiced and evaluated the suitability of microbial wastewater treatment systems as the most sustainable treatment method. We further reviewed the various advances in the microbial wastewater treatment process that confer the proven effectiveness and favor its continuous adoption.

4.1.1 Overview of Wastewater Treatment Methods

Wastewater refers to a type of water whose quality has been negatively impacted by human activities, including surface water runoff, sewage, industrial, household, and agricultural sources. Numerous contaminants found in this waste effluent are released into rivers, lakes, streams, and oceans (Zhang et al. 2015), endangering drinking water and creating serious environmental problems (Zhou et al. 2004).

Water is a necessary component for all living beings. The scarcity of potable water is caused by erratic weather patterns and environmental pollution that lowers the quality of most water sources. In most developing countries, while urban water supplies are better, drinkable water is lacking. In addition to improving public health, having access to clean water is essential for food production, poverty alleviation, and sustainable growth. Due to rising population pressure on available water supplies, the amount of water consumed globally has practically increased (Van Vliet et al. 2017). The rising trepidations over water pollution have resulted in a broad investigation and improvement in water management methods to encourage the recycling of used water and increase the quality of available water for the populace. Recycling wastewater requires treating it and applying the appropriate wastewater management techniques. Wastewater management plan is an environmentally critical field that includes a detailed analysis of several dynamics, including chemical structures, biological impacts, and waste analysis.

The biological and chemical processes used in the different treatment methods for wastewater usually result in the positive treatment of water to some extent. The specific ways of treating wastewater are determined by the colors, components of the sewage, cost of the processes, and additional pollutants present in the wastewater (Nishat et al. 2022).

4.1.1.1 Physical and Mechanical Methods

In the management of wastewater, mechanical techniques such as skimming and screening are employed to remove solid contaminants. Screening is an old treatment technique that eliminates several pollutants from the waste stream, prevent the damage of downstream equipment, plant operational disruptions, and safeguard the entry of suspended materials into the primary settling tanks.

The physical technique is employed because of its simplicity, great efficacy, and capacity to recycle the pollutant (Foroutan et al. 2019; Wong et al. 2019). Compared to other remedies, physical treatment is incredibly dependable because it does not rely on living things (Samsami et al. 2020). To remove particles from wastewater, all available physical forces are applied, including aeration, adsorption, filtration, and sedimentation; no chemicals are introduced. While aeration entails passing air (O2) through the wastewater, sedimentation is a technique used to suspend insoluble particles from the wastewater. Conversely, filtration uses filters to remove impurities from wastewater; the grease that is present on the wastewater's surface can also be removed through filtration. The filters used in wastewater management are specifically designed and the most frequently used filter is the sand filter (Harleman and Murcott 1999; Abdiyev et al., 2023). Currently, adsorption is being used by investigators for its high efficacy and low operational cost (Foroutan et al. 2019).

4.1.1.2 Chemical Methods

This is the treatment method in which chemicals are applied for the cleaning of water. The oxidizing chemical, chlorine, is commonly used in killing bacteria that contaminate and decompose water. Another wastewater oxidizing agent commonly used for purification is ozone. The neutralization technique is used to adjust the pH of water to its natural pH (pH 7) through the addition of acid or base. The addition of chemicals inhibits the growth of bacteria, preventing them from reproducing in the water, hence the production of pure water (Ødegaard and Karlsson 1994, Mathew et al., 2022).

4.1.1.3 Biological Methods

The biological material found in wastewater, including detergent and human and animal waste, is broken down using a variety of biotic pathways. By breaking down organic material in the presence of oxygen, bacteria in wastewater use an aerobic method to produce the carbon dioxide that plants need. To aid in the treatment of the wastewater, they can also employ the composting method, which involves mixing sawdust and other carbon sources with the wastewater and fermenting it at a specific temperature (Wei et al. 2003; Arun, 2011).

Biological treatment processes are still undeniably vital choices for the biodegradation of most pollutants in wastewater after pre-treatment steps because these are economical and sustainable compared to the physico-chemical management method (Guimarães et al. 2018). Therefore, it is imperative to migrate to green, sustainable technologies like microbial treatment of wastewater which is a long-lasting alternative to the traditional method of wastewater treatment.

4.2 Microbial Wastewater Treatment

Certain bacteriological populations are globally recognized for thriving exclusively on the impurities present in wastewater (Silva-Bedoya et al. 2016).

Water bodies naturally contain microorganisms that can break down organic wastes. However, high levels of biological wastes in the aquatic environment cause microorganisms to require more oxygen, which lowers the amount of total dissolved oxygen in water and has detrimental effects on marine life and results in poor quality of water (Abdel-Raouf et al. 2012). The usage of microorganisms (bacteria, fungi, algae, and protozoa) in wastewater management for the removal of organic contaminants is termed "microbial remediation."

Wastewater treatment often takes place in centralized and decentralized treatment systems through biological, physical, and chemical removal of organic and inorganic pollutants (Bajpai et al. 2019). In most centralized wastewater plants, there are four main stages of treatment: preliminary, primary, secondary, and tertiary treatments.

The preliminary stage, also known as the mechanical stage, removes bulky solid matter from the influent. Next is the primary treatment stage which removes inorganic and organic solids. Using filtration and sedimentation processes. Biochemical oxygen demand (BOD) and about 90% of the suspended material are removed by secondary or biological treatment, the stage where natural self-purification processes take place. The biological treatment process includes the activated sludge treatment process which uses microorganisms for the removal of inorganic components of the wastewater and enable the nitrogen cycle processes.

The tertiary or disinfection stage further purifies the wastewater by killing microorganisms and reducing their population to an acceptable level before effluent discharge. Disinfectants can be in physical or chemical forms such as ozone, ultraviolet (UV) light, hydrogen peroxide, pure chlorine, chlorine dioxide, chlorine gas, and chlorine compounds such as calcium and sodium hypochlorites (Owoseni and Okoh 2022).

Chemical treatment/disinfectants are commonly used in the treatment of wastewater. In decentralized wastewater treatment processes, the three stages may not be fully incorporated but still achieve a considerable level of treatment, leaving the wastewater devoid of contaminants.

Microbial wastewater treatment can be conducted in oxygen-rich (aerobic) or oxygen-free or poor (anaerobic) conditions. The process of wastewater treatment involves the breaking down of organic substances, and nitrification and denitrification processes. The procedure may be natural (sewage pond) or artificial (trickling filter and activated sludge), or through decentralized wastewater treatment systems, which include constructed wetlands, membrane bioreactors, and anaerobic digestion systems (Capodaglio et al. 2017).

Aerobic Biological Treatment

Aerobic and facultative microorganisms oxidize most of the organic matter in the waste in the presence of dissolved oxygen in the water. The time taken by aerobic biological treatment is usually short and in suitable situations, 80–90% BOD5 can be removed, yielding water devoid of smell.

Anaerobic digestion or fermentation is the process by which anaerobic and facultative anaerobic bacteria break down organic materials in trash. It is mostly used for managing organic waste at higher concentrations and in organic slurry. The process takes place in an oxygen-free (sealed) environment, enabling bio-energy production, particularly methane gas.

Anaerobic digestion systems are energy-efficient and straightforward bioreactors designed for wastewater treatment. These systems effectively break down organic matter by forming a dense sludge blanket at the reactor's base. The upward flow of wastewater promotes bacterial aggregation and ensures the retention of active anaerobic sludge, enhancing treatment efficiency. A notable example of this technology is the Up-flow Anaerobic Sludge Blanket (UASB) reactor. A UASB optimally reduces COD up to 9800 mg/L at 80% efficiency from concentrated black water at ambient temperature while generating methane gas at 1.8 m^3 CH_4/m^3 (de Graaff et al. 2010).

The microbial treatment of wastewater is the secondary stage in the effluent management scheme and it involves the use of activated sludge systems, biofilm method, stabilization pool, aerated lagoons, and trickling filters.

(i) *Stabilization/Oxidation Ponds*

These ponds contain partially refined effluent and are left to enable microorganisms such as bacteria and algae that have the potential to degrade the waste to grow. Wastewater treatment ponds are classified based on their position within the treatment system and the type of effluent they receive. These include raw sewage oxidation ponds, stabilization ponds, and polishing ponds. Additionally, they can be categorized according to the dominant natural processes occurring within them. Facultative ponds operate under both aerobic and anaerobic conditions, aerobic ponds rely on oxygen for treatment, and anaerobic ponds function in the absence of oxygen. The pools are simple, constructed to adapt to ample turns in movement, and assist in the control of waste in a regular scheme at a very low cost.

Solids in the effluent are preserved or degraded in the septic tank before the waste is released into the laguna. The wastewater is released below the surface of the water system to the middle of the pond to allow for equal distribution and more efficient management. The stabilization pool allows for the degradation of organic matter, which acts as nutrient for the development of algae; the algae will further break down the organic matter, leading to the production of oxygen. One of the main factors to consider in the use of stabilization ponds is the temperature because biological reactions are highly sensitive to temperatures between 0 °C and 40 °C. Other factors are location (unrestricted access to sunlight or air motion), construction of a dike around the pool, and lining with clay soil to prevent leakage. The oxidation ponds are a productive management system with minimum danger to the environment. However, it is inclined to fill up quickly due slow degradation and accumulation of microbial cells during the degradation of sewage.

(ii) *Trickling Filters*

Trickling filter is a biological treatment process under the secondary wastewater treatment stage comprising a filter bed that contains gravel or plastic material that serves as a sieve. A circulating sprinkler suspended over a layer of permeable material distributes sewage in an aerobic process.

A trickling filter contains strata of rocks enclosed with microorganisms. The spread of the effluent throughout the stratum promotes the growth of microorganisms. On the stratum, the microbial development manifests as a slime. Hence, a slime layer of microorganisms is found on the surface of the gravel bed forming a biofilm over it. The effluents tardily seep through the porous stratum and gather to the underside. As organic nutrients in the effluent trickle down the porous stratum, the microbial community uses it as a home, absorbing and breaking them down (Tchobanoglous et al. 2003). Trickling filters are categorized as high-rate and low-rate systems with a removal rate of 80–85% BOD.

The maneuver of air across the stratum supplies aeration to the effluent; the effluents is redistributed severally through the filter to lessen the BOD.

High-rate filters are designed to process greater hydraulic and organic loads, while low-rate filters operate at a slower pace, providing more stable and consistent effluent quality. In high-rate filter systems, effluent recycling is employed, whereas it is not used in low-rate filters. Recirculating wastewater enhances contact between waste and microorganisms, ensuring more effective treatment (van Haandel and van der Lubbe 2007). Despite the advantages of this method, a surplus of nutrients results in an inordinate quantity of slime on the stratum and reduction in aeration. In addition, part of the slime may break off from the stratum and must be totally removed through settling tank. The trickling filter treatment system is commonly used because it is inexpensive and relatively simple to use.

(iii) *Activated Sludge*

Activated sludge is a commonly used aerobic system for sewage and wastewater treatment. This process relies on a massive concentration of microorganisms, mostly bacteria, fungi, and protozoa, which form a loose mass of clumped fine particles.

These microorganisms are kept suspended in the wastewater through continuous stirring to facilitate the breakdown and elimination of organic matter (Templeton and Butler 2011).

Among the various wastewater treatment systems, the activated sludge process is considered one of the most efficient and versatile. Inside the aeration tank, microbial biomass actively degrades organic pollutants, substantially improving water quality. The microbial structure of activated sludge consists of nearly 70–90% organic matter and 10–30% inorganic matter. This balance is essential for maintaining effective treatment performance and ensuring the removal of contaminants from wastewater.

The fundamental constituents of an activated sludge waste management scheme are aeration tank, secondary vessel, and clarifier. The inbound effluent is incessantly fed into the aeration tank containing bacterial mass, in a desirable quantity, and retained in suspension.

Following the preliminary treatment, the wastewater enters the aeration tank, where oxygen is supplied through mechanized agitation. This aeration enhances microbial activity, enabling microorganisms to oxidize organic matter into water and carbon dioxide. This stage plays a crucial role in biological treatment by promoting the breakdown of organic pollutants. The activated sludge process is highly efficient, achieving a biological oxygen demand (BOD) removal efficiency of 75–90% (Tortora 2010). By reducing the BOD levels, the system significantly improves water quality before the treated effluent is discharged or subjected to further purification.

The activated sludge process relies on microbial activity to break down organic matter in wastewater. The carbonaceous organic matter present in wastewater serves as an energy source for microbial growth and cell production. During this process, microorganisms convert organic compounds into new cell tissue and oxidized end-products, including carbon dioxide (CO_2), sulfate (SO_4^{2-}), nitrate (NO_3^-), and phosphate (PO_4^{3-}) (Spellman 1997).

The main purpose of most activated sludge systems is to break down carbonaceous biochemical oxygen demand (BOD). However, advanced systems can also support ammonia oxidation through nitrification and nitrate reduction to nitrogen gas via denitrification. The efficiency of an activated sludge system depends on critical operational factors, including the solid retention time (SRT), which indicates how long the sludge remains in the system, and the food-to-microorganism (F:M) ratio, which measures the amount of organic matter available per unit of microbial biomass. Another critical parameter is the hydraulic detention time, which influences the extent of wastewater treatment (Grey 2002).

Constant aeration, agitation, and recirculation in the activated sludge system create an optimal environment for microbial activity while inhibiting the growth of larger organisms. The microbial community in activated sludge includes bacteria, fungi, protozoa, rotifers, and nematodes, though their composition varies across different systems. Among these organisms, bacteria play the most crucial role in degrading organic matter, which makes them the dominant group in wastewater treatment (Grey 2002).

The processes of synthesis of new cells and respiration occur simultaneously during the microbial metabolic processes. Energy is released from organic matter to low energy-containing compounds such as carbon dioxide, water, and nitrogen products. The activated sludge and primary settling reduce the population of intestinal pathogenic organisms and BOD by 85–90%.

Biological Treatment Process

Biological treatment processes are integrated in treatment processes that allow for the natural degradation of organic materials or their removal to take place. Most decentralized wastewater treatment systems have incorporated biological treatment processes in the removal of toxic waste from wastewater. Examples of decentralized wastewater treatment systems include constructed wetlands, membrane bioreactors, and anaerobic digestion systems (Capodaglio 2017).

Constructed Wetlands

Constructed wetlands are cost-effective systems for wastewater treatment, efficiently removing biochemical oxygen demand (BOD) and chemical oxygen demand (COD) from biodegradable organic compounds. These engineered wetlands mimic natural processes, using microbial activity, vegetation, and substrates to degrade pollutants. Anaerobic fermentation in low-redox environments generates biogas, mainly methane and carbon dioxide, contributing to energy recovery. Widely used in industrial, domestic, and agricultural settings, constructed wetlands offer a sustainable alternative to conventional treatment methods. They are particularly effective in the Mediterranean countries and Europe, reducing wastewater discharge impacts and supporting ecosystem services such as biodiversity conservation and water reuse.

Their efficiency depends on design, hydraulic retention time, and environmental factors. Further research is needed to enhance their performance and expand applicability. As a nature-based solution, constructed wetlands play a vital role in sustainable wastewater management and environmental conservation.

Membrane Bioreactors

The membrane bioreactor (MBR) technology integrates biological wastewater treatment with membrane filtration, offering an efficient and sustainable solution for contaminant removal. MBRs combine activated sludge treatment with membrane separation, where microorganisms degrade pollutants, and membranes physically remove suspended solids and pathogens. The use of microfiltration or ultrafiltration membranes results in high-quality effluent suitable for reuse or discharge (Capodaglio et al. 2016; Abegglen et al. 2008). MBRs outperform conventional wastewater treatment systems by enhancing pollutant removal, reducing footprint, and improving effluent quality. They effectively eliminate nutrients such as nitrogen and phosphorus, minimizing eutrophication risks in receiving waters. Additionally, they remove micropollutants, including pharmaceutical residues and endocrine-disrupting chemicals, through biodegradation and membrane filtration (De Cazes et al. 2014). The system's high biomass concentration allows for smaller reactor sizes, reducing land use requirements. MBR-treated water meets stringent discharge standards, making it suitable for non-potable reuse applications. However, challenges such as membrane fouling can impact efficiency and maintenance costs. Research focuses on optimizing membrane materials, operational strategies, and hybrid treatment approaches to improve performance and cost-effectiveness. MBRs are widely applied in municipal and industrial wastewater treatment, hospital effluent management, and water reuse initiatives. Their continued development can contribute to water conservation and environmental sustainability, particularly in regions facing water scarcity and pollution challenges.

(iv) *Anaerobic Digesters*

Anaerobic digestion systems are cost-effective, low-energy bioreactors widely used for wastewater treatment. These systems rely on microbial activity to break down organic matter in the absence of oxygen, forming a dense sludge blanket at the bottom of the reactor. The up-flow velocity promotes bacterial aggregation and enhances sludge retention, improving treatment efficiency. One of the most effective anaerobic digestion technologies is the Up-flow Anaerobic Sludge Blanket (UASB) reactor, designed to treat high-strength wastewater. The UASB system achieves up to 80% chemical oxygen demand (COD) removal from concentrated black water with an initial COD concentration of approximately 9800 mg/L under ambient conditions. Additionally, it produces biogas rich in methane (CH_4), generating up to 1.8 m^3 CH_4 per cubic meter of treated wastewater. This dual benefit of wastewater treatment and bioenergy recovery makes UASB reactors a sustainable and practical solution for managing wastewater (de Graaff et al. 2010).

4.3 Microbial Collections Involved in Wastewater Bioremediation

The Global Water Microbiome Consortium (GWMC) analyzed 16S rRNA gene sequences from 1200 activated sludge (AS) samples collected from 269 wastewater treatment plants (WWTPs) and identified approximately one billion bacterial phenotypes. The 16SrRNA revealed the several microbial consortium in the activated sludge to be Plactomycetota, Nitrospirota, Bdellovibrionota, Protobacteria, Bacteroidota, Firmicutes, Verrucomicrobiota, Bdellovibrionota, and Myxococcota (Begmatov et al. 2022).

4.3.1 Bacteria

Bacteria are a censorious necessity for the effluent management scheme formulated in the early twentieth century and are the major component of water management globally. The microbes used to treat the effluent are jointly called activated sludge (Davies 2005). The organic carbon in the wastewater stimulates the proliferation of bacteria growth, which results in natural cleaning before discharge. The removal of nitrogen in the effluents concerns processes such as denitrification, nitrification, and anaerobic ammonia oxidation (anammox). The bacteria involved in the nitrification process are ammonia oxidizer *Nitrosococcus mobilis* and nitrogen oxidizing bacteria *Nitrosospira* sp. (Daims et al. 2006). The bacteria in the denitrification process can be either heterotrophic or autotrophic.

Different bacterial taxa exhibit variations in nitrogen metabolism depending on the carbon source available. Anoxybacillus contaminans is known to perform both aerobic denitrification and nitrification when supplied with glucose, while Zoologea spp. achieved 78.86% removal of nitrate using the substrate known as sodium acetate (Rajta et al. 2020). Additionally, sulfur-oxidizing bacteria, including Sulfurimonas, Thioalkalivibrio, Thiobacillus, and Paracoccus, facilitate the process of autotrophic denitrification (Cui et al. 2019). It has also been documented that anammox is carried out by Planctomycetes (Daims et al. 2006).

4.3.2 Aerobic and Anaerobic Bacteria

Aerobic and anaerobic bacteria predominate in different wastewater types depending on their sources.

Aerobic bacteria are predominantly used in new treatment plants in an aerated environment. Free oxygen within the water system is used to degrade the pollutants in the wastewater and then metabolizes it into energy for growth and reproduction. The optimization of the bacterial function and metabolism is dependent on continuous supply of oxygen externally.

Anaerobic bacteria reduce the volume of sludge and increase the production of methane gas from the sludge. They are very important in sludge treatment and the methane gas serves as an alternative source of energy for wastewater treatment, for which energy demand is high.

Anaerobic processes are commonly employed for treating wastewater with a high concentration of organic matter. Compared to aerobic processes, they offer several advantages, including the ability to function effectively in highly polluted water and a lower nutrient requirement for microbial growth. This makes anaerobic treatment particularly suitable for wastewater with insufficient nutrients to support aerobic bacterial activity.

Anaerobic bacteria are more advantageous than aerobic bacteria as they acquire more oxygen from its food source than aerobic bacteria and help in the removal of phosphorous from the wastewater treatment plant, thus ensuring optimization of the treatment. The anaerobic bacteria involved in wastewater fermentation liquid generated during the activated sludge process include *Sulfurospirillum* sp. Bacteria such as *Clostridium* sp., *Thauera* sp., *Arcobacter* sp. and *Azoarcus* sp. have also been reported.

Anaerobic processes are highly sensitive to fluctuations in temperature and pH, as these factors impact both the operational conditions and the by-products recovered at the end of the process. Additionally, anaerobic treatment of wastewater often requires post-treatment, as the microorganisms involved are less effective in removing inorganic nutrients.

4.3.3 Autotrophic and Heterotrophic Bacteria

The characteristic and environmental conditions of the wastewater are susceptible to changes by bacteria as the wastewater constituents serve as feedstock for the metabolic processes of the bacteria, leading to the production of various metabolic products.

4.3.4 Autotrophic Bacteria

Autotrophs are important in the universal nitrogen cycle and wastewater treatment system, particularly in nitrification. Employing in situ molecular and physiological assays, researchers have identified natural bacterial autotrophic communities involved in this process, including ammonia-oxidizing and nitrite-oxidizing bacteria. In activated sludge treatment systems, autotrophs perform an important function, especially in the nitrification process.

However, research evidence on the operational strategies, growth mechanisms, and products of their degradation remains very few compared to available findings on heterotrophs (Li et al. 2008). In activated sludge systems, nitrite inhibition is a major issue in autotrophic degradation, particularly for oxidizing bacteria, which demands significant energy for its maintenance (Zhang et al. 2022).

4.3.5 Heterotrophic Bacteria

Heterotrophic bacteria make use of varieties of macronutrients (organic and inorganic) for their cell development (Orchard et al. 2010). These comprise dissolved inorganic phosphate PO_4^{3-} and the organic phosphorus compounds derived from biomass such as glucose-6-phosphate, adenosine triphosphate (ATP), 2- aminoethyl phosphonic acid. These compounds are essential carbon sources needed for bioremediation and the microbial growth processes (Sisma-Ventura and Rahav 2019).

The phylum Proteobacteria is the most predominant (21–62%) group and the phylum Betaproteobacteria is the class of the phylum with the most abundance, essentially concerned with nutrient and organic removal. The phylum Proteobacteria and the class Betaproteobacteria are predominant in wastewater treatment systems for organic and nutrient degradation. The class includes the genera *Azoarcus, Nitrosomonas, Pseudomonas, Thauera, Nitrospira,* and *Thiobacillus* (Khanichaidecha et al. 2019).

4.3.6 Fungi

Fungi is a vital microorganism in the breaking down of organic pollutants known as mycoremediation. It can survive in the conditions of low pH and nitrogen. It is an important component of the sludge with the capacity to degrade the sludge due to its filamentous structure. Examples include Bdelloid rotifers that can absorb dispersed bacteria and organic matter (Lapinski and Tunnacliffe 2003). Fungal bioremediation works through several key mechanisms, including the growth of mycelia, the broad action range of fungal enzymes, and the reliance of fungal species on pollutants as a food source or substrate for growth. Fungal enzymes that have been used as effective pollutant biodegraders include laccases, peroxidases, and cytochrome P450 monooxygenases (Kumari et al. 2021).

4.3.7 Microalgae

Microalgae consist of cyanobacteria (blue-green algae), and prokaryotic and eukaryotic autotrophic protists (true algae) (Larsdotter 2006). They play a crucial role in the purification of water by either directly absorbing contaminants or by transforming them into harmless substances (Dhanker et al. 2021). According to Wollmann et al. (2019), certain algal species support organic matter degradation by bacteria by producing oxygen through the process of photosynthesis. This process of natural oxygenation lessens both energy costs and the need for mechanical aeration. Microalgae require phosphorus, carbon, and nitrogen for biomass production. They assimilate nitrogen in the form of nitrite, nitrate, and ammonia, which contribute to their growth and nutrient removal efficiency. Microalgae also remove carbon dioxide as a photosynthetic substrate and heavy metals from wastewater (Ahmed et al. 2022).

Microalgae have been used in the conversion of solar energy to biomass due to their photosynthetic ability. Microalgae such as Chlamydomonas reinhardtii, Planothidium lanceolatum, Oscillatoria angustissima, Chlorella, and Spirulina spp. have been reported to take up nitrogen, phosphorous, and carbonaceous substances and possess the capability to remove heavy metals from water (Kumar et al. 2015; Ahmed et al. 2022). It is important to highlight that most microalgae species display mixotrophic way of nourishment based on light situations; they can also shift between auto- and heterotrophy. Microalgae and cyanobacteria are important organisms in wastewater treatment due to the production of oxygen as a by-product of photosynthesis.

Optimization of microalgae has been possible through recombinant DNA technology as their genes are susceptible to modification through genetic engineering. Thus, they have been used in wastewater treatment systems at the secondary or tertiary treatment stages. Microalgae remove pollutants from wastewater through a multistage mechanism. First, attachment to the microalgae's surface by the pollutant majorly heavy metals. Second, microalgae transform pollutants into insoluble compounds, facilitating their removal from wastewater through precipitation. This is followed by nutrient utilization to produce biomass, and nitrogen and phosphorous removal by assimilation (Tripathi et al. 2019). This multi-stage process generates valuable by-products, including biochar and algal oil. Microalgal genera include *Ankistrodesmus, Golenkinia, Euglena, Micractinium,* and *Chlorella pyrenoidosa* (Lam et al. 2024). The microalgae *Chlorella vulgaris* was effective in the removal of COD, $NH_{3,}$ -N, and orthophosphate from domestic wastewater and COD removal increased to 84% on the fifth day (Alazaiza et al. 2023).

In some instances, microalgae functions are limited by some factors. Low concentrations of nitrogen and phosphorus in wastewater reduce cell density of microalgae, thus increasing the cost and energy requirements for separation of biomass and impeding the efficacy of microbial bioremediation (Al-Jabri et al. 2021). Optimal microalgae bioremediation requires external supply of carbon dioxide and sunlight for their growth and biomass production. The seasonal variability in light intensity and temperature may impede the rate of bioremediation (Nourmohammadi et al. 2013; Al-Jabri et al. 2021).

4.3.8 Cyanobacteria

The Cyanobacteria *Tolypothrix ceylonica* has shown increased capacity to absorb heavy metals in wastewater and to discharge solids (El-Bestawy 2008*). Spirulina platensis* complemented the activities of *Chlorella* species by increasing dewaterability of biosolids compared to the use of only *Chollera* (Yuan et al. 2012). The use of *Oscillatoria*, *Nostoc*, *Synechococcus*, *Nodularia*, and *Cyanothece* in Tannery water enhanced its decolorization (Dubey et al. 2011), while O*scillatoria* and *Nostoc* species depleted PO_4^{-2}, NO_3^{-2} and SO_4^{-2}, and Cl^- from contaminated wastewater (Azarpira et al. 2014).

Some disadvantages in the applications of cyanobacteria for wastewater treatment comprise production of toxins by some cyanobacteria strains, which has public health consequences (Daly et al. 2007). Therefore, monitoring cyanobacterial growth indicators is essential to help prevent the spread of toxins. Another disadvantage is the high biomass production from the synthesis of extracellular polysaccharides, leading to cyanobacterial bloom. Cyanobacteria may stick to the surfaces of filtration units and hinder the supply of recycled water (Kwon et al. 2005).

4.3.9 Integrated Microalgal-Bacteria Wastewater Treatment System

An integrated microalgae-bacteria wastewater treatment system with several advantages has been proposed. Microalgae can absorb and use phosphorus, nitrogen, and carbon from wastewater effluents for their growth. They help reduce both inorganic and organic forms of nitrogen, phosphorus, and carbon, thereby improving wastewater quality (Wollmann et al. 2019). This provides a cost-effective and less energy-consuming system for the treatment of wastewater effluents. Additionally, oxygen is generated by the process of photosynthesis, which becomes useful for aerobic bacteria and CO_2 is absorbed.

The removal of nitrogen and phosphorus from wastewater involves different operational compartments, which inevitably leads to increased energy consumption. Thus, the algal-bacterial consortium offers the advantage of a single-stage treatment for inorganic N and P removal, which enables no switch between different operating environments (Gouveia et al. 2016).

However, the integrated algal-bacterial system has several disadvantages. Microalgal growth is more dependent on light than the activated sludge process. Additionally, photosynthesis by microalgae increases the medium's pH, which can negatively affect the bacterial consortium involved in the activated sludge process. Furthermore, the slow growth rate of microalgae results in a longer solid retention time (SRT) compared to the requirements of heterotrophic organisms. Furthermore, treating large volumes of wastewater requires a significant volume of microalgae, adding to the system's complexity (Katam and Bhattacharyya 2021; Mathew et al. 2022).

4.3.10 Protozoa

(Madoni 2003). Protozoans are active in activated sludge processes and they maintain a good balance in the biological ecosystem by removing excess bacteria, and promote floc production and bacterial activity through the secretion of growth-stimulating substances (Madoni 2011). Protozoans can convert carbon to carbon dioxide in activated sludge and reduce mineral nutrient after the carbon source is accelerated by bacteria.

Protozoans also contribute to the activation of nitrogen cycling (Petropoulos and Gilbride 2005; Pogue and Gilbride 2007). Their ability to enhance the rate of cell nitrification stems from their influence on bacterial cell growth. They release both inorganic and organic by-products into their environment, mainly nitrogen, phosphorus, and organic carbon. They also produce stimulatory compounds that contribute to the pool of dissolved organic carbon and support the health and growth of bacterial communities.(Jurgens and Matz 2002).

4.4 Examples of Pollutants Degraded in Microbial Wastewater Treatment

Predominantly in wastewater are found major contaminants including nutrients such as nitrogen and phosphorus, hydrocarbons, heavy metals, and microbes. A well-studied bacteria for the removal of heavy metal from groundwater and industrial wastewater contaminated with arsenic is *Pseudomonas* sp., for instance, As (III) compound (Orellana et al. 2018). Arsenate oxidase produced by some bacterial strains can oxidize toxic arsenic from its more harmful form, As(III), to the less toxic As(V) (Sanyal et al. 2016).

Hydrocarbons including petrol and polycyclic aromatic hydrocarbons have been remediated by the genus *Pseudomonas* (*P. putida*, *P. mendocina*), *Flavobacterium*, and *Acinetobacter*. Diesel oil has been degraded by *Bacillus cereus*. Cobalt, copper, chromium, zinc, and lead have been broken down by *Geobacter* species, *Lyssinibacillus sphaericus* CBAM5, *Bacillus safensis* (JX126862).

Aerobic bacteria and hydrophilic bacteria have demonstrated azo dye decolorization in aerobic conditions in the presence of oxygen insensitive or aerobic azo reductases whose activities are impaired by oxygen (Sarkar et al. 2017). Triarylmethane dyes have been decolorized by *Aeromonas hydrophila* in textile effluent with decolorization efficiency of 72–76%. Factors responsible for this decolorization efficiency include the pH of the solution, temperature, dye coloration, and oxygenation (Imran et al. 2015).

4.5 Mechanisms Involved in Microbial Treatment of Wastewater

The breakdown of organic compounds by microorganisms results in carbon mineralization into CO_2 and water or methane (Abdel-Shafy and Mansour 2018).

Biological nutrient removal (BNR) is driven by bacterial growth and metabolism (Pollard and Greenfield 1997). The organic carbon in wastewater induces the production of bacterial cell biomass, which is known as activated sludge (Davies 2005). Nitrogen removal involves denitrification, nitrification, and anaerobic ammonia oxidation (anammox) processes. During nitrification, aerobic chemo-lithoautotrophic ammonia-oxidizing bacteria (AOB) and nitrite-oxidizing bacteria (NOB) facilitate a two-step process. In wastewater treatment plants, Nitrosococcus mobilis and Nitrosospira sequentially convert ammonia into nitrite and then into nitrate (Daims et al. 2006).

Denitrification in the heterotrophic phase causes nitrate reduction through carbonaceous organic matter, which are electron donors. Bacterial species metabolize nitrogen differently using different carbon sources. Autotrophic denitrification is performed by sulfur-oxidizing bacteria (SOB), which use compounds such as sulfide, thiosulfate, and elemental sulfur to reduce nitrate and nitrite. Proteobacteria from the genera Paracoccus, Sulfurimonas, Thiobacillus, and Thioalkalivibrio have been identified as key contributors to these processes (Cui et al. 2019).

For anammox, dinitrogen gas is produced from nitrite and ammonia, which requires oxygen supply for partial oxidation of ammonia to nitrite. Inhibition of NOB that uses nitrite as a substrate is also necessary. This enables the depletion of nitrogen without adding any extra organic carbon, thus enhancing high-ammonia and low-carbon content in effluents. The genus Planctomycetes are responsible for the process of ammonax and are chemolithoautotrophs in nature (Daims et al. 2006).

Nitrate reduction follows an assimilatory, dissimilatory, and denitrification pathway. Nitrate is converted into biomass through the assimilatory pathways, into ammonium through the dissimilatory pathway, and into dinitrogen through denitrification (Rajta et al. 2020).

Phosphorus removal from water occurs through enhanced biological phosphorus removal (EBPR), a process that uses phosphorus-accumulating organisms (PAOs) and glycogen-accumulating organisms (GAOs). These microorganisms store polyphosphate within their cells, facilitating phosphorus removal. The extracellular enzymatic hydrolysis phase allows bacterial cells to break down organic matter in wastewater into smaller particles, ranging in size from 103 atomic mass units (amu) to 100 millimeters. Smaller particles below this size range are used by cells while larger sizes are removed by sedimentation.

De-polymerization reactions help degrade waste materials and are achieved by hydrolytic enzymes and lyases (Morgenroth et al. 2002). The waste is absorbed into the cell and broken down, and is useful for energy synthesis, which fuels motility, ion gradient maintenance, transport of materials, and protein/nucleic acid turnover (Hao et al. 2010).

4.5.1 Aerobic Treatment

Aerobic processes require the presence of oxygen in a solid or liquid treatment process. Aerobic phases of the growth include oxidation and synthesis, followed by endogenous respiration. These phases depend on mixing or aeration, which can be energy intensive.

The following equation shows how organic matter is consumed during oxidation to release CO_2 and the synthesis of new cells.

$$\text{Organic matter} + O_2 + \text{nutrients}^{®} \rightarrow CO_2 + H_2O + C_5H_7NO_2 + \text{other products} \tag{4.1}$$

Where $C_5H_7NO_2$ are the newly synthesized cells.

Degradable organic matter or the cellular content in wastewater is broken down and depleted during endogenous respiration by microorganisms in order to maintain their metabolic reaction. This is shown below.

$$C_5H_7NO_2 + 5O_2 \rightarrow 5CO_2 + 2H_2O + NH_3 + \text{energy} \tag{4.2}$$

4.5.2 Anaerobic Treatment

Oxygen depletion leads to anaerobic degradation in anaerobic and facultative waste stabilization ponds, septic tanks, and settling tanks. Sludge treatment undergoes anaerobic fermentation and less sludge is generated during this process from the less energetically favorable nature of anaerobic metabolisms.

Anaerobic digestion is an intricate procedure that occurs in multiple stages, including hydrolysis, fermentation, acidogenesis, acetogenesis, and methanogenesis. During hydrolysis, complex organic compounds and particulate matter in waste are broken down into simpler, more bioavailable forms. Simultaneously lipids, proteins, and polysaccharides are transformed into fatty acids, amino acids and monosaccharides.

Through acidogenesis, additional degradation of sugars, amino acids, and fatty acids into methanogenic substrates occurs. Methanogens enable the process of a group of archaea to split acetate into methane and carbon dioxide, while another group generates methane by using hydrogen and carbon dioxide. The optimal temperature for methanogenesis is at (30–38 °C) and (49–57 °C) for mesophilic and thermophilic temperatures, respectively.

Methanogenic degradation are presented below (Madigan et al. 2004):

$$4H_2^+ CO_2 \rightarrow CH_4 + 2H_2O$$

$$CH_3OH + H_2 \rightarrow CH_4 + H_2O$$

$$CH_3COO^- + H_2O \rightarrow CH_4 + HCO_3$$

The association between the acidogens and the methanogens is syntrophic. The acidogens produce hydrogen, which is used by the methanogens. This in turn provides an ideal partial pressure for the acidogens. Any reduction in the process results in an increase in volatile fatty acids, leading to a lowered pH and methanogenic process. Thus, the digester is referred to as going "sour." Therefore, the growth phase of methanogens is a limiting step in the process. Thus, pH monitoring and operation must be put in place to ensure an optimal process.

4.5.3 Nitrogen Cycling

Nitrogen cycling is vital for the reduction in ammonia nitrogen. Indiscrete release of nitrogen into the environment constitutes a public health hazard. Inorganic forms of nitrogen serve microbial growth and nitrogen predominantly available as ammonia is released during the hydrolysis process.

4.5.4 Nitrification

Ammonia-oxidizing bacteria oxidize ammonia into nitrite, which is rapidly followed by oxidation of nitrite to nitrate by nitrite-oxidizing bacteria.

$$2NH_2 + 3O_2 + \text{Nitrosomonas} \rightarrow 2NO_2^- + 2H_2O + 2H^+$$

$$2NO_2^- + O_2 + \text{Nitrobacter} \rightarrow 2NO_3$$

Optimal conditions for the nitrification processes include a pH range of 7.5 to 8.0, dissolved oxygen concentration above 1 mg/L, and alkalinity as calcium carbonate ($CaCO_3$) per gram of ammonia nitrogen (as nitrogen) converted into nitrate. pH ranges below 6.0 inhibit the nitrification process (Metcalf and Eddy 2003).

The equation below summarizes the nitrification process.

$$2NH_2 + 3O_2 + \text{Nitrosomonas} \rightarrow 2NO_2 - + 2H_2O + 2H^+$$

$$2NO_2 - + O_2 + \text{Nitrobacter} \rightarrow 2NO_3$$

4.5.5 Denitrification

Nitrate reduced to nitrogen gas escapes into the atmosphere during the denitrification process. In anoxic environments, where oxygen is limited, nitrate serves as the electron acceptor. Heterotrophic and autotrophic aerobic bacteria are active at the optimal pH range of 7.0–8.0 to provide the required BOD for denitrification to take place (Metcalf and Eddy 2003).

The process of denitrification follows the series of intermediate steps shown below.

$$NO_3^- \rightarrow NO_2^- \rightarrow NO \rightarrow N_2O \rightarrow N_2$$

Where:

NO_3^- = nitrate
NO_2^- = nitrite
NO = nitrogen oxide
N_2O = nitrous oxide
N_2 = nitrogen gas

4.5.6 Phosphorus Cycle

Factors including sedimentation, complexation, mineralization, precipitation, pH, redox potential, and plant uptake influence the removal of phosphorus during wastewater treatment.

Molecules contain acid or base forms of orthophosphoric acid (H_3PO_4) or phosphate (PO_4^{3-}), or organically bound phosphorus (e.g., nucleic acids, phospholipids, and phosphorylated proteins). Organic materials are mineralized and released during degradation processes and phosphate is absorbed into the sludge. During microbiological treatment, microorganisms utilize up to 10–30% of phosphorus, which can be augmented by biological dephosphatation, or through chemical precipitation with $FeCl_3$ or $Al_2(SO_4)^3$ or $FeSO_4$ applied in wastewater treatment.

4.6 Mechanism of Contaminant Removal by Microalgae

Microalgae converts CO_2 into reduced carbon compounds like sugars in the presence of water and light in a process called oxygenic photosynthesis (Benedetti et al. 2018). This process is driven by four multi-subunit membrane protein complexes:

Photosystem I, Photosystem II, Cytochrome b6f, and F-ATPase. These complexes carry out light-dependent reactions that include the oxidation of water to produce oxygen, the reduction of $NADP^+$ to NADPH (nicotinamide adenine dinucleotide phosphate), and the synthesis of ATP. The two photosystems oxidize water and reduce NADPH discretely, and the cytochrome transfers electrons between the photosystems while producing the PMF (proton motive force) for ATP synthesis from F-ATPase. The stroma of the chloroplast actively takes up CO_2 from the environment during the Calvin cycle, while the ATP and NADPH from the light reaction generate carbohydrates (Nelson and Shem 2004) in the absence of light. *Synechocystis aquatilis* (1500 mg/L/day), *Botryococcus braunii* (1100 mg/L/day), *Chlorella vulgaris* (865 mg/L/day), and *Synechococcus* spp. are good CO_2 fixers of different microalgae species (Singh and Singh 2014).

Ammonium (NH_4) is a highly assimilated nitrogen source in green algae as other sources are partially or completely inhibited in its presence. Ammonium assimilation undergoes the glutamine synthase (GS)–glutamate synthase cycle. However, concentrations of ammonium above 100 mg/L are converted into ammonia, which is toxic to microalgae and inhibits microalgal photosynthesis (Su 2021). The organic nitrogen appears in the form of amino acids or urea, which varies among bacterial species due to its dependence on transport abilities and degradation enzymes. An example is found in *Stichococcusbacillaris*, which possesses systems for the active transport of multiple amino acids, while members of the order Volvocales contain active transport systems for arginine but lack transport systems for other amino acids (Hellebust and Ahmad 1989).

Microalgae uses three mechanisms for heavy metal removal, which is pH dependent. These mechanisms involve extracellular precipitation, which requires living cells, as well as cell surface sorption or complexation by both living and dead cells. Additionally, intracellular accumulation plays a crucial role in the metabolic processes of microalgae. Certain microalgae species exhibit high heavy-metal removal capabilities. For example, Chlamydomonas reinhardtii efficiently removes cadmium, while live cells of Planothidium lanceolatum can absorb substantial amounts of Cd^{2+}, reaching 275.51 mg/g. Spirogyra spp. and Oscillatoria angustissima are promising candidates for cobalt removal, whereas Chlorella and Spirulina spp. effectively remove hexavalent chromium (Cr^{6+}), with Spirulina achieving a chromium uptake rate of 333 mg/g. Spirulina species have the capacity to degrade copper at 389 mg/g with nickel and zinc, at concentrations of 1378 mg/g. Degradation of heavy metals ranges under different pH values (Kumar et al. 2015).

Phosphorus is important for microalgae for the production of nucleic acids, DNA (deoxyribonucleic acid), and RNA (ribonucleic acid), ATP for energy, and phospholipids of the membrane. These molecules are taken up in various forms including PO_4, HPO_4, H_2PO_4, and Phosphite (PO_3). High light intensity induces the generation of acid-soluble polyphosphate (ASP) converts it to proteins and DNA. After internalization, the extracellular phosphorous compounds are changed to cellular phosphate phospholipids and RNA formation (Su 2021).

During tertiary treatment of wastewater, the microalgae *C. vulgaris* is important in the degradation of N and P compounds and heavy metals (de-Bashan et al. 2002).

Approximately, 88 and 89% of the total nitrogen and phosphorus contents are degraded by *Coelastrum microporum*, respectively.

Varying light conditions affect the mode of nutrition of many algae, enabling them to interchange between autotrophic and heterotrophic modes of nutrition. This is referred to as the mixotrophic mode of nutrition and it enables the microalgae to use organic carbon, thus, facilitating depletion of organic material in wastewater. However, only a few of them including *C. reinhardtii* can grow as a heterotroph. Abundance of sunlight enables glucose to be broken down via the Embden Meyerhof Pathway (EMP) or the pentose phosphate pathway (PPP) in the dark heterotrophically. The pyruvate forms acetyl CoA and undergoes the TCA (tricarboxylic acid) cycle under aerobic conditions (Su 2021).

4.7 Determining Factors in Wastewater Treatment Efficacy

The wastewater industry worldwide is shifting toward resource recovery and water recycling centers. This transition is driven by increasing operational costs and stringent regulations on effluent discharge (Wu et al. 2019). Researchers and engineers are developing innovative technologies to achieve high effluent quality while minimizing energy consumption. Wastewater treatment is key to sustainable environmental health and sustainable use of water resources. Many complex factors contribute to the efficacy of wastewater treatment systems, and a comprehensive insight into these factors is vital for optimizing treatment processes and meeting increasingly stringent environmental regulations.

4.7.1 Influent Characteristics

The composition and properties of the incoming wastewater significantly impact the treatment efficacy. Key parameters include biochemical oxygen demand (BOD), chemical oxygen demand (COD), total suspended solids (TSS), pH, temperature, and the presence of toxic substances. Variations in these parameters can impact the operations of biological treatment processes and the efficiency of chemical treatments.

4.7.2 Treatment Technology Selection

The selection of treatment technology is vital in determining the overall efficacy. The activated sludge process, membrane bioreactors (MBR), sequencing batch reactors (SBR), trickling filters, anaerobic digestion—each technology has its strengths and limitations, and the selection should be based on influent characteristics, effluent quality requirements, and operational constraints.

4.7.3 Operational Parameters

Proper control of operational parameters is essential for maintaining treatment efficacy. Th factors to be considered include hydraulic retention time (HRT), solids retention time (SRT), dissolved oxygen (DO) levels, mixed liquor suspended solids (MLSS) concentration, and the food to microorganism (F/M) ratio. Optimizing these parameters can enhance the performance of biological treatment processes and improve the overall treatment efficacy.

4.7.4 Environmental Conditions

Environmental factors can significantly influence treatment efficacy, particularly in biological treatment systems. Important considerations include temperature fluctuations, seasonal variations in influent composition, and extreme weather events. These factors can affect microbial activity, reaction rates, and overall treatment performance.

4.7.5 Process Control and Monitoring

Effective process control and monitoring are crucial for maintaining treatment efficacy. This includes real-time monitoring of key parameters, implementation of advanced control systems, and regular performance evaluation and optimization. Proper monitoring and control enable timely adjustments to maintain optimal treatment conditions.

4.7.6 Nutrient Removal Efficiency

Many wastewater treatment plants are required to remove nutrients, particularly nitrogen and phosphorus, to prevent eutrophication in the receiving water bodies. Factors affecting nutrient removal include the C:N:P ratio in the influent, the presence of appropriate microbial populations, and adequate anoxic and anaerobic zones in the treatment process. Balancing these factors is crucial for achieving high nutrient removal efficiency.

4.7.7 Emerging Contaminants

Emerging contaminants, such as pharmaceuticals, personal care products, and microplastics, are increasing the challenges to wastewater treatment efficiency. Factors to consider include detection and quantification methods, development of advanced treatment technologies, and regulatory frameworks for emerging contaminants. Addressing these challenges is becoming increasingly significant for comprehensive treatment efficacy.

Many factors contribute to the efficacy of wastewater treatment, ranging from influent characteristics to operational parameters and emerging contaminants. Strategically considering all these factors will ensure treatment optimization, which will meet the requirements for environmental standards. Continuous research and innovation in this field are key to addressing evolving threats and improving overall treatment efficacy.

4.8 Recent Advancements in Microbial Wastewater Treatment

Microbial wastewater treatment, a cornerstone of sustainable water management, has witnessed significant strides in recent years. These advancements have been driven by a combination of technological innovations, heightened environmental concerns, and the imperative to optimize resource utilization.

4.8.1 Bioaugmentation and Biostimulation

Bioaugmentation includes the seeding of specific strains of microorganisms or microbial consortia into a purification system to improve the breakdown of desired pollutants (Chettri et al. 2024). This approach is predominantly deployed in cases where the indigenous microbial community is inadequate to degrade certain compounds, such as recalcitrant organic pollutants or emerging contaminants like pharmaceuticals. By supplementing the microbial community with specialized degraders, bioaugmentation can improve treatment efficiency and accelerate the bioremediation process (Herrero et al. 2015; Chettri et al. 2024). Biostimulation focuses on enhancing environmental conditions, such as pH, temperature, and nutrient availability, to promote existing microbial growth and activity (Adams et al. 2015).

Bioaugmentation entails adding particular strains of microorganisms or microbial consortia into a treatment system to ensure the breakdown of the selected contaminants (Chettri et al. 2024). This method works especially well for substances like emerging pollutants like pharmaceuticals or stubborn organic pollutants, which the local microbial community is unable to degrade. Bioaugmentation can boost treatment effectiveness and quicken the bioremediation process by adding specialized degraders to the microbiota (Herrero et al. 2015; Chettri et al. 2024). Conversely, biostimulation aims to enhance the development and activity of the current microbial population by optimizing environmental factors such as pH, temperature, and nutrition availability (Adams et al. 2015).

The potential of bioaugmentation has been further enhanced by advancements in recombinant DNA technology and molecular biology, which enable production of genetically modified microorganisms (GMOs) with specialized abilities to break down particular contaminants (Bustamante-Torres et al. 2024). These modified microorganisms may have improved metabolic pathways, which would enable the breakdown of harmful substances more quickly and thoroughly. But a wider use of this technology has been constrained by worries about the discharge of genetically modified organisms into the environment and possible ecological effects (Saxena et al. 2020).

4.8.2 Microbial Fuel Cells (MFCs)

According to Esfandyari et al. (2024), microbial fuel cells are a state-of-the-art advancement in microbial wastewater treatment that provide the double advantages of pollutant removal and energy generation. Electroactive microbes in wastewater decompose organic materials to release electrons, which are then collected by an electrode in an MFC to produce electricity. This procedure creates a sustainable energy source in addition to cleaning the wastewater. Even though MFC technology is still in its experimental stages, it has enormous potential for wastewater treatment in the future, especially for off-grid and decentralized applications.

Enormous energy is being consumed during conventional wastewater treatment processes, which creates a major issue that MFCs may be able to solve. MFCs are budding technologies with the capacity to lower the overall energy footprint of wastewater treatment plants by transforming the chemical energy found in wastewater into electrical energy (Ranjan et al. 2022; Esfandyari et al. 2024). Numerous researches are now being conducted to increase the MFCs' cost-effectiveness and scalability, which will increase its appeal for widespread use.

This cutting-edge aerobic wastewater treatment technique combines membrane filtration with biological treatment. Organic pollutants are broken down by microorganisms in MBR systems, and the treated effluent is then sent over a membrane that serves as a physical barrier to trap bacteria, suspended particles, and other contaminants. When compared to conventional activated sludge processes, MBRs offer a higher-quality effluent produced during the filtration phase, making it a very promising technology (Bera et al. 2022). Because the treated water can meet strict discharge or reuse regulations, MBRs are especially well-suited for applications involving the reuse of water. Similarly, compared to the massive amounts of sludge produced by conventional wastewater treatment, the MBR systems function with higher biomass concentrations, which results in lower sludge formation. Because of these characteristics, MBRs are the preferred method for treating wastewater from cities and businesses, particularly when a high standard of effluent quality is needed or when reusing and recycling water is of utmost importance.

4.8.3 Sequencing Batch Reactor (SBR)

Here is another type of aerobic wastewater treatment technology that runs in a batch mode as opposed to the continuous mode. In an SBR system, wastewater is treated in a single tank through a series of time-sequenced stages. Every stage serves a distinct purpose that enables the biological treatment to be managed. This system's flexibility and control, which allow for the timing of each step to be adjusted to improve treatment for certain conditions and make it adaptable to various wastewater types, providing a comparative advantage over conventional systems (Askari et al. 2024). Because of the careful regulation of conditions throughout the treatment process, it also has a higher rate of removal of nutrients and organic materials. SBRs are generally advised for use in certain industrial settings and in small-scale municipal wastewater treatment plants. This technology is an effective and economical choice.

4.8.4 Novel Microbial Communities

Engineered microbial consortia are artificial communities of microorganisms that have been specifically created to carry out particular tasks in the treatment of wastewater, like breaking down particular contaminants or increasing the overall treatment efficacy (Yaashikaa et al. 2022). Synthetic biology and metabolic engineering principles are used in the construction of these microbial consortia. Since particular contaminants can be targeted for degradation, this technique has great promise. Therefore, resistant pollutants such as heavy metals, pesticides, drug residues, or complex chemical compounds can be further degraded by this approach (Yaashika et al. 2022). Its ability to operate in a variety of environmental settings, including variations in temperature, pH, and pollution load, makes it a dependable treatment method.

4.8.5 Immobilized Microorganisms

Significant progress has been made in the field of microbial wastewater treatment, especially with the use of immobilized microbe technology. According to Zhou et al. (2008), this approach has demonstrated superiority over conventional suspended systems, like activated sludge, by providing benefits including increased biomass concentration, improved metabolic activity, and increased resistance to harmful chemicals. Microorganisms that are immobilized offer various advantages over those that are suspended. Higher biomass loading and retention times are made possible by them, and as a result, biodegradation rates and operational stability are enhanced. Additionally, easier solid-liquid separation is made possible by this method, which is essential for effective wastewater treatment (Bouabidi et al. 2019).

Various methods are used to immobilize microorganisms, including adsorption, covalent bonding, cross-linking, and encapsulation (Najim et al. 2024). Adsorption is the simplest method, relying on physical interactions between microorganisms and carrier surfaces, though it can lead to desorption during operation. Covalent bonding enhances stability but may reduce bioactivity after coupling. Cross-linking involves covalent bonds between biomolecules but requires careful control (Zhou et al. 2008). Encapsulation in polymers such as polyacrylamide and polyvinyl alcohol is common, though diffusion limitations can occur.

Carriers are chosen based on characteristics like non-toxicity, mechanical strength, and cost-effectiveness (Bouabidi et al. 2019). They are broadly classified into inorganic materials, like porous glass and ceramics, known for their durability and stability, and organic polymers, which include both natural polymers such as alginate and synthetic options like polyurethane (Zhou et al. 2008). The latter are noted for their mechanical performance and resistance to biodegradation.

4.9 Conclusion

Advancements in aerobic and anaerobic processes, along with innovations in microbial community engineering, have significantly improved wastewater treatment systems. Membrane bioreactors (MBRs) and sequencing batch reactors (SBRs) offer enhanced treatment efficiency and effluent quality in aerobic systems. Hybrid anaerobic/aerobic systems and hydrothermal carbonization (HTC) contribute to energy savings and resource recovery in anaerobic processes. Meanwhile, novel microbial approaches such as engineered consortia and metagenomics allow for more specific and tailored treatment, addressing complex pollutants and optimizing system performance. Together, these innovations represent a shift toward more sustainable, efficient, and resilient wastewater treatment technologies.

References

Abdel-Raouf N, Al-Homaidan AA, & Ibraheem I (2012) Microalgae and wastewater treatment. Saudi J Biol Sci, 19(3), 257-275

Abdel-Shafy HI, Mansour MSM (2018) Microbial degradation of hydrocarbons in the environment: an overview. In: Kumar V, Kumar M, Prasad R (eds) Microbial action on hydrocarbons. Springer, Singapore. https://doi.org/10.1007/978-981-13-1840-5_15

Abdiyev K, Azat S, Kuldeyev E, Ybyraiymkul D, Kabdrakhmanova S, Berndtsson R, Khalkhabai B, Kabdrakhmanova A and Sultakhan S, 2023 Review of slow sand filtration for raw water treatment with potential application in less-developed countries. Water, 15(11), p.2007

Abegglen CH, Ospelt M, Siegrist H (2008) Biological nutrient removal in a small-scale MBR treating household wastewater. Water Res, vol 42. Academic Press, pp 338–371

Adams GO, Fufeyin PT, Okoro SE, Ehinomen I (2015) Bioremediation, biostimulation and bioaugmention: a review. Int J Environ Bioremediat Biodegrad 3(1):28–39

Adesiyan IM, Bisi-Johnson M, Aladesanmi OT, Okoh AI, Ogunfowokan AO (2018) Concentrations and human health risk of heavy metals in rivers in Southwest Nigeria. J Health Pollut 8(19):180907

Ahmed SF, Mofijur M, Parisa TA, Islam N, Kusumo F, Inayat A et al (2022) Progress and challenges of contaminate removal from wastewater using microalgae biomass. Chemosphere 286(Pt 1):131656. https://doi.org/10.1016/j.chemosphere.2021.131656

Alazaiza MYD, He S, Su D, Abu Amr SS, Toh PY, Bashir MJK (2023) Sewage water treatment using *Chlorella vulgaris* microalgae for simultaneous nutrient separation and biomass production. Separations 10:229. https://doi.org/10.3390/separations10040229

Al-Jabri H, Das P, Khan S, Thaher M, Abdul Quadir M (2021) Treatment of wastewaters by microalgae and the potential applications of the produced biomass—a review. Water 13:27. https://doi.org/10.3390/w1301002

Anekwe IMS, Adedeji J, Akpasi SO, Kiambi SL (2022) Available technologies for wastewater treatment. Wastewater Treatment 33

Arun A, & Eyini M (2011) Comparative studies on lignin and polycyclic aromatic hydrocarbons degradation by basidiomycetes fungi. Bioresour Technol, 102(17), 8063-8070

Askari SS, Giri BS, Basheer F, Izhar T, Ahmad SA, Mumtaz N (2024) Enhancing sequencing batch reactors for efficient wastewater treatment across diverse applications: a comprehensive review. Environ Res 11:119656

Azarpira H, Behdarvand P, Dhumal K, Pondhe G (2014) Comparative studies on phycoremediation of sewage water by using blue green algae. Int J Biosci 4:58–64

Bajpai M, Katoch SS, Chaturvedi NK (2019) Comparative study on decentralized treatment technologies for sewage and graywater reuse—a review. Water Sci Technol 80:11

Begmatov S, Dorofeev AG, Kadnikov V, Beletsky A, Pimenov N, Ravin V, Mardanov AV (2022) The structure of microbial communities of activated sludge of large-scale wastewater treatment plants in the city of Moscow. Sci Rep 12:3458. https://doi.org/10.1038/541598-022-07132-4

Benedetti M, Vecchi V, Barera S, Osto L (2018) Biomass from microalgae: the potential of domestication towards sustainable biofactories. Microb Cell Factories 17:173. https://doi.org/10.1186/s12934-018-1019-3

Bera SP, Godhaniya M, Kothari C (2022) Emerging and advanced membrane technology for wastewater treatment: a review. J Basic Microbiol 62(3–4):245–259

Bouabidi ZB, El-Naas MH, Zhang Z (2019) Immobilization of microbial cells for the biotreatment of wastewater: a review. Environ Chem Lett 17:241–257

Bustamante-Torres M, Torres O, Abad-Sojos S, Pardo S, Bucio E (2024) Application of genetically modified microorganisms for bioremediation of polluted environments. In: Genetically engineered organisms in bioremediation. CRC Press, pp 18–51

Capodaglio AG (2017) Integrated, decentralized wastewater management for resource recovery in rural and peri-urban areas. Resources 6:22. https://doi.org/10.3390/resources6020022

Capodaglio AG, Hlavínek P, Raboni M (2016) Advances in wastewater nitrogen removal by biological processes: state of the art review. Rev Ambient Água 11(2):251–267

Capodaglio AG, Callegari A, Cecconet D, & Molognoni D (2017) Sustainability of decentralized wastewater treatment technologies. Water Pract Technol, 12(2), 463-477

Chettri D, Verma AK, Verma AK (2024) Bioaugmentation: an approach to biological treatment of pollutants. Biodegradation 35(2):117–135

Chowdhary P, Bharagava RN, Mishra S, Khan N (2020) Role of industries in water scarcity and its adverse effects on environment and human health. Environmental Concerns and Sustainable Development: Volume 1: Air, Water and Energy Resources, pp 235–256

Cui YX, Biswal BK, Guo G, Deng YF, Huang H, Chen GH et al (2019) Biological nitrogen removal from wastewater using sulphur-driven autotrophic denitrification. Appl Microbiol Biotechnol 103:6023–6039. https://doi.org/10.1007/s00253-019-09935-4

Daims H, Taylor MW, Wagner M (2006) Wastewater treatment: a model system for microbial ecology. Trends Biotechnol 24:483–489. https://doi.org/10.1016/j.tibtech.2006.09.002

Daly RI, Ho L, Brookes JD (2007) Effect of chlorination on Microcystis aeruginosa cell integrity and subsequent microcystin release and degradation. Environ Sci Technol 41:4447–4453

Davies PS (2005) The biological basis of wastewater treatment. Strathkelvin Instruments Ltd, Glasgow

De Cazes M, Abejón R, Belleville MP, Sanchez-Marcano J (2014) Membrane bioprocesses for pharmaceutical micropollutant removal from waters. Membranes 4:692–729

de Graaff MS, Temmink H, Zeeman G, Buisman CJN (2010) Anaerobic treatment of concentrated black water in a UASB reactor at a short HRT. Water 2010(2):101–119. https://doi.org/10.3390/w201010

de-Bashan LE, Moreno M, Hernandez JP, Bashan Y (2002) Removal of ammonium and Phosphorus ions from synthetic wastewater by the microalgae *Chlorella vulgaris*coimmobilized in alginate beads with microalgae growth promoting bacterium Azospirillumbrasilense. Water Res 36:2941–2948. https://doi.org/10.1016/S0043-1354(01)00522-X

Dhanker R, Hussain T, Tyagi P, Singh KJ, Kamble SS (2021) The emerging trend of bio-engineering approaches for microbial nanomaterial synthesis and its applications. Front Microbiol 12:638003. https://doi.org/10.3389/fmicb.2021.638003

Dubey SK, Dubey J, Mehra S, Tiwari P, Bishwas AJ (2011) Potential use of cyanobacterial species in bioremediation of industrial effluents. Afr J Biotechnol 10:1125–1132

El-Bestawy E (2008) Treatment of mixed domestic–industrial wastewater using cyanobacteria. J Ind Microbiol Biotechnol 35:1503–1516

Esfandyari M, Jafari D, Azami H (2024) Microbial fuel cells for energy production in wastewater treatment plants—a review. Biofuels 15(6):743–753

Foroutan R, Mohammadi R, Farjadfard S, Esmaeili H, Saberi M, Sahebi S, ... & Ramavandi B (2019) Characteristics and performance of Cd, Ni, and Pb bio-adsorption using Callinectes sapidus biomass: real wastewater treatment. Environ Sci Pollut Res, 26, 6336-6347

Ghoshal G (2021) Microbial fuel cell: recent advancements. In: Bioremediation, nutrients, and other valuable product recovery. Elsevier, pp 83–113

Gouveia L, Graça S, Sousa C, Ambrosano L, Ribeiro B, Botrel EP et al (2016) Microalgae biomass production using wastewater: treatment and costs scale-up considerations. Algal Res 16:167–176. https://doi.org/10.1016/j.algal.2016.03.010

Gray FN (2002) Water technology: an introduction for environmental scientists and engineers. Butterworth-Heinemann, Oxford, pp 35–80

Guimarães JT, Souza AL, Brígida AIS, Furtado AA, Chicrala PC, Santos VR, Alves RR, Luiz DB and Mesquita EF 201. Quantification and characterization of effluents from the seafood processing industry aiming at water reuse: A pilot study. J Water Process Eng, 26, pp.138-145

Hao XD, Wang QL, Zhu JY, Loosdrecht MCMV (2010) Microbiological endogenous processes in biological wastewater treatment systems. Crit Rev Environ Sci Technol 40:239–265. https://doi.org/10.1080/10643380802278901

Haripriyan U, Gopinath KP, Arun J, Govarthanan M (2022) Bioremediation of organic pollutants: a mini review on current and critical strategies for wastewater treatment. Arch Microbiol 204(5):286

Harleman DR, & Murcott S (1999) The role of physical-chemical wastewater treatment in the mega-cities of the developing world. Water Sci Technol, 40(4-5), 75-80

Hellebust JA, & Ahmad I (1989) Regulation of nitrogen assimilation in green microalgae. Biol Ocean, 6(3-4), 241-255

Herrero M, & Stuckey DC (2015) Bioaugmentation and its application in wastewater treatment: a review. Chemosphere, 140, 119-128

Imran M, Crowley DE, Khalid A, Hussain S, Mumtaz MW, Arshad M (2015) Microbial biotechnology for decolorization of textile wastewaters. Rev Environ Sci Biotechnol 14(1):73–92. https://doi.org/10.1007/s11157-014-9344-4

Jurgens K, Matz C (2002) Predation as a shaping force for the phenotypic and genotypic composition of planktonic bacteria. Antoine Van Leeuwenhoek 81:413–434

Karri RR, Ravindran G, Dehghani MH (2021) Wastewater—sources, toxicity, and their consequences to human health. In: Soft computing techniques in solid waste and wastewater management. Elsevier, pp 3–33

Katam K, Bhattacharyya D (2021) Improving the performance of activated sludge process with integrated algal biofilm for domestic wastewater treatment: System behavior during the start-up phase. Bioresour Technol Rep 13:100618. https://doi.org/10.1016/j.biteb.2020.100618

Khanichaidecha W, Nakaruk A, Ratananikom K, Eamrat R, Kazama F (2019) Heterotrophic nitrification and aerobic denitrification using pure-culture bacteria for wastewater treatment. J Water Reuse Desalination 9(1):10–17

Kumar KS, Dahms HU, Won EJ, Lee JS, Shin KH (2015) Microalgae—a promising tool for heavy metal remediation. Ecotoxicol Environ Saf 113:329–352. https://doi.org/10.1016/j.ecoenv.2014.12.019

Kumari R, Singh A, Yadav AN (2021) Fungal enzymes: degradation and detoxification of organic and inorganic pollutants. In: Yadav AN (ed) Recent trends in mycological research. Fungal biology. Springer, Cham. https://doi.org/10.1007/978-3-030-68260-6_5

Kwon B, Park N, Cho J (2005) Effect of algae on fouling and efficiency of UF membranes. Desalination 179:203–214

Lam TQK, Sreekeshava KS, Bhargavi C, Ganesh CR, Ambale NS, Do TMD (2024) Exploring the potential of green microalgae-based phycoremediation treated wastewater for sustainable concrete production. Adv Civil Eng 17:8564202. https://doi.org/10.1155/2024/8564202

Lapinski J, Tunnacliffe A (2003) Reduction of suspended biomass in municipal wastewater using bdelloid rotifers. Water Res 37:9

Larsdotter K (2006) Wastewater treatment with microalgae—a literature review. Vatten 62:31–38

Li Y, Liu Y, Shen L, Chen F (2008) DO diffusion profile in aerobic granule and its microbiological implications. Enzym Microb Technol 43:349–354

Madigan MT, Martinko JM, Parker J, Fernández CR, Pérez MS (2004) Brock's microbiology. Pearson, Education, Inc, New Jersey

Madoni P (2003) Protozoa as indicators of wastewater treatment efficiency. In: Mara D, Horan N (eds) The handbook of water and wastewater microbiology. Academic Press, London, pp 361–371

Madoni P (2011) Protozoa in wastewater treatment processes: a minireview. Italian J Zool 78(1):3–11. https://doi.org/10.1080/11250000903373797

Mathew MM, Khatana K, Vats V, Dhanker R, Kumar R, Dahms HU, & Hwang JS (2022) Biological approaches integrating algae and bacteria for the degradation of wastewater contaminants—a review. Front Microbiol, 12, 801051

Metcalf L, Eddy HP (2003) Wastewater engineering treatment and reuse. McGraw-Hill, New York

Mishra RK (2023) Fresh water availability and its global challenge. Brit J Multidiscip Adv Stud 4(3):1–78

Morgenroth E, Kommedal R, & Harremoës P (2002) Processes and modeling of hydrolysis of particulate organic matter in aerobic wastewater treatment–a review. Water Sci Technol, 45(6), 25-40

Najim AA, Radeef AY, al-Doori I, Jabbar ZH (2024) Immobilization: the promising technique to protect and increase the efficiency of microorganisms to remove contaminants. J Chem Technol Biotechnol 99:1707

Nelson N, & Ben-Shem A (2004) The complex architecture of oxygenic photosynthesis. Nat Rev Mol Cell Biol, 5(12), 971-982

Nishat Y, Danish M, Siddiqui N, & Hussain T (2022) Removal of emerging pollutants from the environment through microbes, In Relationship Between Microbes and the Environment for Sustainable Ecosystem Services, Volume 2 (pp. 181-203). Elsevier

Nourmohammadi D, Esmaeeli MB, Akbarian H, & Ghasemian M (2013) Nitrogen Removal in a Full-Scale Domestic Wastewater Treatment Plant with Activated Sludge and Trickling Filter. Environ Public Health, 2013(1), 504705

Ødegaard H, & Karlsson I (1994) Chemical wastewater treatment—value for money. In Chemical Water and Wastewater Treatment III: Proceedings of the 6th Gothenburg Symposium 1994 June 20–22, 1994 Gothenburg, Sweden (pp. 191-209). Springer Berlin Heidelberg

Orchard ED, Benitez-Nelson CR, Pellechia PJ, Lomas MW, & Dyhrman ST (2010) Polyphosphate in Trichodesmium from the low-phosphorus Sargasso Sea. Limnology and Oceanography, 55(5), 2161-2169

Orellana R, Macaya C, Bravo G, Dorochesi F, Cumsille A, Valencia R, Rojas C and Seeger M, 2018 Living at the frontiers of life: extremophiles in Chile and their potential for bioremediation. Front Microbiol, 9, p.2309

Owoseni M, Okoh AI (2022) New insights into chlorine disinfectant usage inwater quality: a need for review of existing guidelines. In: Bioenergy and environmental biotechnology for sustainable development, 1st edn. eBook ISBN 9781003180289, Taylor and Francis, CRC Press, pp 159–179. https://doi.org/10.1201/9781003180289-8

Petropoulos P, Gilbride KA (2005) Nitrification in activated sludge batch reactors is linked to protozoan grazing of the bacterial population. Can J Microb 51:791–799

Pogue AJ, Gilbride KA (2007) Impact of protozoan grazing on nitrification and the ammonia- and nitrite-oxidizing bacterial communities in activated sludge. Can J Microb 53:559–571

Pollard PC, Greenfield PF (1997) Measuring in situ specific bacterial growth rates and population dynamics in wastewater. Water Res 31:1074–1082. https://doi.org/10.1016/S0043-1354(96)00326-0

Raghav N, Nigam R, Mathur S, Singh D, Ranjan R (2022) Microbial degradation of wastewater. In: Recent trends in wastewater treatment. Springer International Publishing, Cham, pp 145–170

Rajta A, Bhatia R, Setia H, Pathania P (2020) Role of heterotrophic aerobic denitrifying bacteria in nitrate removal from wastewater. J Appl Microbiol 128:1261–1278. https://doi.org/10.1111/jam.14476

Rana N, Gupta P (2022) Microbes: an eco-friendly tool in wastewater treatment. In: Synergistic approaches for bioremediation of environmental pollutants: recent advances and challenges. Academic Press, pp 161–183

Ranjan P, Maithani D, Suyal DC, Singh AK, Giri K, Sharma VK, Soni R (2022) Microbial fuel cells for wastewater treatment. Bioremediation of Environmental Pollutants: Emerging Trends and Strategies, pp 53–74

Samsami S, Mohamadizaniani M, Sarrafzadeh MH, Rene ER, & Firoozbahr M (2020) Recent advances in the treatment of dye-containing wastewater from textile industries: Overview and perspectives. Process safety and environmental protection, 143, 138-163

Sanyal SK, Mou TJ, Chakrabarty RP, Hoque S, Hossain MA, Sultana M (2016) Diversity of arsenite oxidase gene and arsenotrophic bacteria in arsenic affected Bangladesh soils. AMB Expr 6:21. https://doi.org/10.1186/s13568-016-0193-0

Saravanan A, Kumar PS, Jeevanantham S, Karishma S, Tajsabreen B, Yaashikaa PR, Reshma B (2021) Effective water/wastewater treatment methodologies for toxic pollutants removal: processes and applications towards sustainable development. Chemosphere 280:130595

Sarkar S, Banerjee A, Halder U, Biswas R, Bandopadhyay R (2017) Degradation of synthetic azo dyes of textile industry: a sustainable approach using microbial enzymes. Water Conserv Sci Eng 2:121–131. https://doi.org/10.1007/s41101-017-0031-5

Saxena G, Kishor R, Saratale GD, Bharagava RN (2020) Genetically modified organisms (GMOs) and their potential in environmental management: constraints, prospects, and challenges. Bioremediation of Industrial Waste for Environmental Safety: Volume II: Biological Agents and Methods for Industrial Waste Management, pp 1–19

Shah AI, Dar MUD, Bhat RA, Singh JP, Singh K, Bhat SA (2020) Prospectives and challenges of wastewater treatment technologies to combat contaminants of emerging concerns. Ecol Eng 152:105882

Singh SP, Singh P (2014) Effect of CO_2 concentration on algal growth: a review. Renew Sust Energ Rev 38:172–179. https://doi.org/10.1016/j.rser.2014.05.043

Silva-Bedoya LM, Sánchez-Pinzón MS, Cadavid-Restrepo GE, & Moreno-Herrera CX (2016) Bacterial community analysis of an industrial wastewater treatment plant in Colombia with screening for lipid-degrading microorganisms. Microbiol Res, 192, 313-325

Sisma-Ventura G, Rahav E (2019) DOP stimulates heterotrophic bacterial production in the oligotrophic southeastern Mediterranean coastal waters. Front Microbiol 10:1913

Spellman FR (1997) Microbiology for water/wastewater operators. Technomic Publishing Co Inc., Lancaster

Su Y (2021) Revisiting carbon, nitrogen and, phosphorus metabolisms in microalgae for wastewater treatment. Sci Total Environ 768:144590. https://doi.org/10.1016/j.scitotenv.2020.144590

Tchobanoglous G, Burton F, Stensel D (2003) Wastewater engineering: treatment and reuse, 4th edn. McGraw-Hill Incorporation, New York

Templeton MR, Butler D (2011) An Introduction to Wastewater Treatment. Ventus Publishing ISBN: 978-87-7681-843-2

Tortora GJ, Berdell RF, Christine LC (2010) Microbiology: an introduction, 10th edn. ISBN 10:0-321-55007-2; ISBN 13: 978-0-321-55007-1

Tripathi R, Gupta A, Thakur IS (2019) An integrated approach for phycoremediation of wastewater and sustainable biodiesel production by green microalgae, *Scenedesmus* sp. ISTGA1. Renew Energy 135:617–625

van Haandel A, van der Lubbe J (2007) Handbook biological wastewater treatment. Quist Publishing, Leidschendam

Van Vliet MT, Flörke M, & Wada Y (2017) Quality matters for water scarcity. Nature Geoscience, 10(11), 800-802

Wang X, Zhang Y, Zhang T, Zhou J, & Chen M (2016) Waste activated sludge fermentation liquid as carbon source for biological treatment of sulfide and nitrate in microaerobic conditions. Chem Eng J, 283, 167-174

Wong KA, Lam SM, & Sin JC (2019) Wet chemically synthesized ZnO structures for photodegradation of pre-treated palm oil mill effluent and antibacterial activity. Ceramics International, 45(2), 1868-1880

Wei Y, Van Houten RT, Borger AR, Eikelboom DH, & Fan Y (2003) Minimization of excess sludge production for biological wastewater treatment. Water res, 37(18), 4453-4467

Wollmann F, Ackermann SDJU, Bley T, Walther T, Steingroewer J, Krujatz F (2019) Microalgae wastewater treatment: biological and technological approaches. Eng Life Sci 19:860–871. https://doi.org/10.1002/elsc.201900071

Wu L, Ning D, Zhang B et al (2019) Global diversity and biogeography of bacterial communities in wastewater treatment plants. Nat Microbiol 4(7):1183–1195. https://doi.org/10.1038/s41564-019-0426-5

Yaashika PR, Devi MK, Kumar PS (2022) Engineering microbes for enhancing the degradation of environmental pollutants: a detailed review on synthetic biology. Environ Res 214:113868

Yuan X, Wang M, Park C, Sahu AK, & Ergas SJ (2012) Microalgae growth using high-strength wastewater followed by anaerobic co-digestion. Water Environ Res, 84(5), 396-404

Zhang X, Davidson EA, Mauzerall DL, Searchinger TD, Dumas P, & Shen Y (2015) Managing nitrogen for sustainable development. Nature, 528(7580), 51-59

Zhang M, Tan Y, Fan Y, Gao J, Liu Y, Lv X, Ge L, Wu J (2022) Nitrite accumulation, denitrification kinetic and microbial evolution in the partial denitrification process: The combined effects of carbon source and nitrate concentration. Bioresour Technol 361:127604

Zhou, Wen Zhong, and Jian Xing Li (2004) "Sewage heat source pump system's application examples and prospect analysis in China."

Zhou L, Li G, An T, Fu J, Sheng G (2008) Recent patents on immobilized microorganism technology and its engineering application in wastewater treatment. Recent Patent Eng 2(1):28–35

Phytoremediation in Restoring Infected or Damaged Water Bodies

5

Aarti Dwivedi and Vinay Dwivedi

Abstract

Eco-friendly and cost-effective biotechnological approaches such as phytoremediation are gaining prominence in restoring polluted water bodies containing excessive pollutants in the form of heavy metals, pesticides, and excess nutrients. Phytoremediation incorporates the natural ability of the plants to absorb, accumulate, degrade, or stabilize the pollutants from aquatic environments. Aquatic macrophytes such as Eichhornia crassipes (water hyacinth), Typha latifolia (cattails), and Lemna minor (duckweed) play a significant role with their high biomass, fast growth, and capacities for pollutant uptake. New breakthroughs in genetic engineering and microbial symbiosis have improved the phytoremediation potential of plants through increased resistance to contaminants and increased pollutant removal efficiency. Despite potential vastness, problems that exist include those of disposal of plants at the end of the process, efficiency at certain times of the year, and location-specific constraints. This chapter discusses, in relation to rehabilitating contaminated water bodies, the mechanisms involved in phytoremediation that include phytoextraction, rhizofiltration, and phytodegradation and then points to its integration with sustainable water management approaches for long-term environmental restoration and ecosystem health.

A. Dwivedi
Department of Chemistry, Amity School of Engineering & Technology, Amity University, Gwalior, Madhya Pradesh, India
e-mail: vdwivedi@gwa.amity.edu

V. Dwivedi (✉)
Amity Institute of Biotechnology, Amity University, Gwalior, Madhya Pradesh, India

N. Dwivedi et al. (eds.), *Wastewater Treatment Through Nature-Based Solutions*,
https://doi.org/10.1007/978-981-96-6699-7_5

Keywords

Phytoremediation · Phytoextraction · Rhizofiltration · Ecosystem health

5.1 Introduction

Global Water Pollution Overview: Water pollution is one of the most critical environmental challenges we face in the twenty-first century. The rapid pace of industrialization, urban growth, and intensified farming practices have led to alarming levels of contamination in our freshwater resources. The United Nations (2023) reports that around 80% of the world's wastewater is released into the environment without treatment, which significantly harms rivers, lakes, and oceans. This section delves into the complex nature of water pollution, examining its sources, effects, and the regions most affected.

5.1.1 Industrial Contributions to Water Pollution

Industries play a major role in water pollution, releasing vast amounts of harmful chemicals and waste. Key sectors such as textiles, chemicals, pharmaceuticals, mining, and energy production contribute pollutants like heavy metals (including lead, mercury, and cadmium), synthetic dyes, oil residues, and persistent organic pollutants (POPs) (Xu et al. 2022).

For example:

- The Ganga River in India suffers from severe pollution due to untreated industrial waste, including chromium and arsenic from tanneries.
- In China, rivers such as the Yangtze and Pearl are significantly affected by wastewater from manufacturing centers.

The absence of strict regulations in many developing countries worsens this problem, enabling industries to dump toxic waste into nearby water sources.

5.1.2 Agricultural Contributions to Water Pollution

Agricultural runoff is a significant contributor to water pollution, introducing nitrates, phosphates, pesticides, and herbicides into aquatic ecosystems (Lin et al. 2024). The overuse of fertilizers and pesticides has resulted in nutrient loading, which leads to eutrophication in lakes and reservoirs. Eutrophication causes algal blooms that deplete oxygen levels in the water, endangering aquatic life (Springer Open 2023).

- The Mississippi River in the United States transports agricultural runoff to the Gulf of Mexico, creating an annual "dead zone" due to low oxygen levels.
- Similar nutrient pollution impacts Europe's Baltic Sea and Asia's Mekong River, affecting fisheries and biodiversity. Moreover, livestock farming adds pathogens,

sediments, and pharmaceuticals to water systems, posing health risks to communities downstream.

5.1.3 Urban Contributions to Water Pollution

Urban areas are significant contributors to water pollution due to untreated sewage, plastic waste, and urban runoff. Rapid population growth and insufficient wastewater treatment facilities in cities, especially in developing countries, have strained water resources.

- In Sub-Saharan Africa, untreated sewage from informal settlements pollutes rivers, leading to outbreaks of waterborne diseases like cholera and typhoid.
- Plastics and microplastics from urban areas contaminate rivers and oceans, with research showing that 80% of ocean plastics come from land-based sources (Lebreton et al. 2017).

5.1.4 Global Hotspots of Water Pollution

Certain areas have become significant hotspots for water pollution, driven by high population density, industrial activities, and intensive agriculture:

- South Asia: Rivers such as the Ganga, Yamuna, and Brahmaputra are heavily polluted from both industrial and domestic sources.
- China: The rapid pace of industrialization has drastically impacted the water quality of the Yangtze and Yellow rivers.
- Africa: A combination of water scarcity and pollution poses challenges for major rivers like the Nile and Limpopo.
- Latin America: The Amazon river is threatened by deforestation, mining activities, and agricultural runoff. The ongoing pollution of these water bodies disrupts ecosystems, diminishes biodiversity, and jeopardizes the livelihoods that rely on clean water.

5.1.5 Impacts on Ecosystems, Human Health, and the Economy

Water pollution has significant effects on ecosystems, public health, and the economy:

- Ecosystems: Pollutants interfere with food chains, destroy habitats, and lead to a decline in aquatic biodiversity. Coral reefs, for instance, are especially susceptible to nutrient pollution and sediment buildup.
- Human Health: Contaminated water can spread diseases like cholera, dysentery, and hepatitis. Heavy metals such as lead and mercury can accumulate in fish, entering the human food chain and leading to neurological and developmental issues.

- Economic Impact: Water pollution raises healthcare expenses, lowers agricultural productivity, and harms fisheries. In 2019, the World Bank estimated that poor water quality could cost some countries as much as 2.5% of their GDP each year (Desbureaux et al. 2019).

5.1.6 Global Hotspots of Water Pollution Limitations of Traditional Remediation Methods

Traditional water remediation techniques, which include chemical, physical, and mechanical approaches, have been widely used to address water pollution. However, these methods come with significant limitations that affect their long-term effectiveness and sustainability.

5.1.6.1 Chemical Treatment Methods

Chemical treatments, such as coagulation, precipitation, and oxidation, are commonly employed to eliminate heavy metals, organic pollutants, and pathogens (Song et al. 2021). While they can be effective, these methods have several drawbacks:

- High Costs: The chemicals used in these treatment processes can be quite expensive, especially for large-scale applications.
- Secondary Pollution: The residues and by-products from chemical treatments may lead to additional environmental issues. For instance, chlorination, a common disinfection method, can produce harmful by-products like trihalomethanes (THMs).
- Energy-Intensive: Techniques like advanced oxidation require a significant amount of energy, which increases their carbon footprint.

5.1.7 Physical Treatment Methods

Physical methods such as sedimentation, filtration, and adsorption are frequently used to separate pollutants from water. However, these methods also have notable limitations:

- Limited Scope: They tend to be effective only for specific pollutants, such as suspended solids or certain metals.
- Maintenance Issues: Filters and adsorbents need regular replacement or regeneration, which adds to operational costs.

5.1.8 Mechanical Treatment Methods

Mechanical approaches like aeration, dredging, and skimming are used to treat water bodies and manage pollutants.

- High Infrastructure Costs: Establishing and maintaining treatment plants and equipment can be prohibitively expensive for resource-limited areas.
- Environmental Disturbance: Techniques like dredging can disrupt aquatic ecosystems by resuspending sediments and releasing trapped contaminants. These limitations highlight the urgent need for alternative, sustainable methods of water remediation.

Phytoremediation has emerged as a promising green technology for tackling water pollution. This eco-friendly method employs plants and their associated microbial communities to degrade, absorb, or stabilize pollutants, providing a sustainable alternative to traditional techniques. Phytoremediation functions through natural processes such as:

- Phytoextraction: The uptake of pollutants into plant tissues
- Phytodegradation: The breakdown of contaminants using plant enzymes
- Phytostabilization: The immobilization of pollutants in soils or sediments
- Rhizofiltration: The absorption or adsorption of pollutants by plant roots
- Phytovolatilization: The conversion of pollutants into volatile forms that are released into the atmosphere

These processes are supported by specific plant species, often referred to as hyperaccumulators, which show a high tolerance for contaminants. The idea of phytoremediation can be traced back to the late twentieth century when researchers noted the natural ability of certain plants to flourish in contaminated areas. Over time, advancements in plant biology, genetics, and ecology have improved phytoremediation techniques, making them suitable for large-scale use. Phytoremediation directly contributes to SDG-6, which aims to "ensure availability and sustainable management of water and sanitation for all." By improving water quality and restoring ecosystems, phytoremediation helps achieve several targets under SDG-6, including:

- Reducing water pollution
- Increasing the reuse of treated water
- Protecting and restoring aquatic ecosystems

Moreover, phytoremediation is in line with other SDGs, such as SDG-13 (Climate Action) and SDG-15 (Life on Land), by advocating for sustainable and nature-based solutions to environmental issues.

5.1.9 Benefits of Phytoremediation

- Cost-Effectiveness: Involves low infrastructure costs and minimal upkeep
- Eco-Friendly: Promotes biodiversity and decreases dependence on chemical solutions
- Multi-Functionality: Improves water quality, restores ecosystems, and captures carbon simultaneously

The increasing focus on phytoremediation signifies a shift toward sustainable water management practices that combine environmental restoration with socioeconomic advantages.

5.2 Mechanisms of Phytoremediation

Phytoremediation is an eco-friendly, plant-based method for addressing water and soil pollution. It works by absorbing, transforming, stabilizing, or volatilizing pollutants through various processes. Recent progress in biotechnology, genetics, and microbial interactions has improved the effectiveness of these processes, allowing for wider applications in different environmental settings.

5.2.1 Phytoextraction

Phytoextraction is the process where plants absorb heavy metals and pollutants, accumulating them in their biomass, which can then be harvested. This method is extensively researched for cleaning up contaminated soils and water sources.

5.2.1.1 Genetic Engineering Advances in Hyperaccumulators

Recent studies have concentrated on genetically modifying plants to boost their phytoextraction capabilities. Hyperaccumulators such as *Thlaspi caerulescens* and *Arabidopsis halleri* have been engineered to express metal transporter genes, enhancing their ability to gather heavy metals.

- Overexpression of Metal Transporters: Research indicates that introducing genes like IRT1 (iron transporter) and HMA4 (heavy metal ATPase) improves the uptake and storage of metals such as cadmium, zinc, and nickel.
- CRISPR-Cas9 Applications: This genome-editing technology has been used to alter regulatory pathways in plants, increasing their tolerance and extraction abilities.

  A 2023 study by Zhang et al. found that CRISPR-modified Brassica juncea showed a 30% increase in cadmium accumulation compared to wild-type plants.
- Synthetic Biology: The engineering of plants with synthetic pathways to target specific pollutants is gaining traction. For instance, a new pathway that incorporates bacterial genes into poplar trees has enhanced the bioaccumulation of arsenic (Chen et al. 2022).

5.2.2 Limitations of Phytoextraction

1. Biomass Disposal Challenges: The disposal of harvested biomass containing concentrated pollutants poses challenges. Common methods like incineration or secure landfilling can lead to secondary pollution or require substantial energy.

2. Bioavailability Barriers: The effectiveness of phytoextraction is influenced by the bioavailability of pollutants. Many heavy metals are strongly bound to soil particles, making them difficult for plant roots to access. Research is ongoing into strategies like chelator application and rhizosphere engineering to overcome this issue.
3. Slow Process: Phytoextraction is a time-consuming method and may require extensive periods to achieve significant results.

5.2.3 Phytodegradation

Phytodegradation refers to the process where organic pollutants are broken down enzymatically within plant tissues or in the surrounding rhizosphere. This method is particularly effective in cleaning up pesticides, hydrocarbons, and industrial solvents.

5.2.3.1 Specific Enzymatic Pathways

Key enzymes, including laccases, peroxidases, and nitroreductases, are crucial in phytodegradation. Recent research has aimed at optimizing these enzymatic pathways through genetic modifications and microbial interventions.

- Example 1: *Populus deltoides* has been genetically altered to overproduce laccase enzymes, which boosts its capacity to degrade polycyclic aromatic hydrocarbons (PAHs).
- Example 2: Plants that have been engineered with bacterial nitroreductase genes have shown improved degradation of explosives such as TNT and RDX.

5.2.3.2 Recent Advances in Genetically Modified Plants

Genetically modified plants have demonstrated significant potential for enhanced phytodegradation. For instance:

- A study by Kumar et al. (2022) modified tobacco plants to express a fungal cytochrome P450 enzyme, allowing for the effective breakdown of persistent herbicides like atrazine.
- Hybrid poplars that express microbial glutathione S-transferase genes have been successfully used to degrade trichloroethylene (TCE), a prevalent groundwater contaminant.

5.2.3.3 Rhizospheric Microbial Augmentation

The rhizosphere, the area of soil affected by plant roots, is a hub of microbial activity. Recent advancements in microbial augmentation have shown considerable promise:

- Bioaugmentation: Introducing pollutant-degrading bacteria or fungi into the rhizosphere can speed up phytodegradation. For example, Pseudomonas putida enhances hydrocarbon breakdown when grown alongside grasses in oil-contaminated areas.

- Symbiotic Engineering: Developing plant-microbe partnerships, such as mycorrhizal fungi with pollutant-degrading abilities, has shown potential in improving phytodegradation efficiency.

5.2.4 Phytostabilization

Phytostabilization is a technique that focuses on immobilizing pollutants in soil or sediments, effectively preventing their movement through leaching or erosion. This approach is particularly beneficial for addressing contamination issues at mining sites and wetlands.

5.2.4.1 Applications in Mining Sites and Wetlands

Phytostabilization has shown significant effectiveness in areas affected by heavy-metal pollution:

- Mining Sites: Species such as *Vetiveria zizanioides* and *Salix viminalis* are frequently used to stabilize tailings that contain arsenic, lead, and cadmium. A study conducted in Spain in 2023 found that using willow trees for phytostabilization reduced heavy-metal leaching from mine tailings by 40%.
- Wetlands: Constructed wetlands that incorporate Typha and Phragmites species have effectively stabilized pollutants in sediments while also improving ecosystem services.

5.2.4.2 Comparison with Other In Situ Stabilization Methods

Phytostabilization presents distinct advantages over chemical stabilization methods:

- Cost-Effectiveness: In contrast to chemical amendments, phytostabilization requires minimal ongoing input once the plants are established.
- Eco-Friendliness: The use of plants enhances biodiversity and aids in carbon sequestration, unlike chemical methods, which may pose long-term toxicity risks.

Nonetheless, phytostabilization tends to be a slower process and may not provide the rapid pollutant containment needed in urgent situations.

5.2.5 Rhizofiltration

Rhizofiltration is the process where plant roots absorb or adsorb pollutants in hydroponic systems or constructed wetlands. This method is particularly effective in removing heavy metals, radionuclides, and emerging contaminants such as pharmaceuticals.

5.2.5.1 Innovations in Constructed Wetland Designs

Recent developments in wetland engineering have significantly improved the efficiency of rhizofiltration:

- Hybrid Wetlands: The combination of horizontal and vertical flow systems enhances pollutant removal rates by optimizing oxygen levels and hydraulic retention times.
- Engineered Substrates: Adding materials like biochar or zeolites to wetland substrates boosts the adsorption capacity for metals and pharmaceuticals.
- Multi-Plant Systems: Using a variety of plant species, including *Eichhornia crassipes* (water hyacinth) and *Lemna minor* (duckweed), helps ensure a broader spectrum of pollutant removal. Real-World Applications.
- Heavy Metals: In India, rhizofiltration has been effectively used to extract lead and cadmium from industrial wastewater with the help of Brassica juncea.
- Radionuclides: Sunflowers (*Helianthus annuus*) have been employed to absorb radioactive cesium and strontium from contaminated water after the Chernobyl and Fukushima nuclear incidents.
- Pharmaceutical Pollutants: Constructed wetlands across Europe have successfully lowered the levels of antibiotics and endocrine disruptors found in hospital effluents.

5.2.6 Phytovolatilization

Phytovolatilization refers to the process where plants absorb contaminants and convert them into volatile forms that are then released into the atmosphere. This method is especially effective in cleaning up volatile organic compounds (VOCs) and certain heavy metals.

5.2.6.1 Mechanisms and Recent Advances

The main mechanisms involved in phytovolatilization include:

- The absorption of contaminants through the roots of plants.
- The enzymatic transformation of these pollutants into volatile forms within the plant tissues.
- The release of these transformed pollutants through the stomata in the leaves. Recent research has made significant strides in the volatilization of heavy metals:
  - Mercury: Transgenic plants that express the bacterial merA and merB genes can convert ionic mercury into elemental mercury, which is less toxic and more volatile.
  - Arsenic: Genetic modifications in rice plants to produce arsenite methyltransferase enzymes have shown effective arsenic volatilization in contaminated soils.

5.2.6.2 Benefits and Risks

- Benefits: Phytovolatilization helps prevent the accumulation of pollutants in soil or water, which reduces the need for disposing of biomass.
- Risks: There is a concern about secondary pollution, as the volatilized pollutants could affect air quality or re-condense back into the environment.

To address these risks, careful monitoring and site selection are essential.

5.2.6.3 Promising Research

A study conducted in 2022 by Li et al. found that transgenic poplar trees were able to volatilize 80% of the arsenic they absorbed, leading to a significant reduction in soil contamination. However, the study also highlighted the importance of establishing buffer zones to minimize potential impacts on air quality.

5.3 Applications of Phytoremediation in Water Body Restoration

Phytoremediation is gaining traction as a method for restoring polluted water bodies, providing a sustainable alternative to traditional remediation techniques. It effectively targets a variety of contaminants, including heavy metals, organic pollutants, excess nutrients, and pathogens. This section delves into these applications, highlighting recent case studies and advancements in the field.

5.3.1 Removal of Heavy Metals

Heavy metals like cadmium (Cd), lead (Pb), mercury (Hg), arsenic (As), and chromium (Cr) are among the most dangerous pollutants found in water due to their toxicity, persistence, and potential for bioaccumulation.

5.3.1.1 Impact of Heavy Metals in Water

Heavy metals present significant threats to both aquatic ecosystems and human health:

- They disrupt metabolic processes in aquatic organisms, leading to stunted growth, reproductive issues, and even death.
- Long-term exposure in humans can result in serious health problems, including neurological disorders, kidney damage, and cancer.

For example, arsenic-contaminated groundwater in Bangladesh has led to widespread health emergencies.

Case Studies 1. China's Mining Regions: The Zijinshan Copper Mine area has experienced severe heavy-metal pollution. A 2022 study highlighted the successful use of *Phragmites australis* (common reed) and *Typha latifolia* (cattails) to extract cadmium and lead from wastewater, achieving removal efficiencies of up to 85%.

2. India's Industrial Zones: In Gujarat, industrial effluents heavily laden with chromium were treated using *Brassica juncea* (Indian mustard). Over three cycles, the levels of chromium in soil and water decreased by 50%, as noted by Patel et al. (2023).

5.3.1.2 Innovations in Heavy Metal Removal

- Nanotechnology Integration: Incorporating metal-oxide nanoparticles into the rhizosphere has shown promise in enhancing the uptake of heavy metals.
- Bioengineered Plants: Transgenic poplar trees that express enhanced phytochelatin synthase enzymes have been reported to accumulate 30% more cadmium and zinc.

5.3.2 Treatment of Organic Pollutants

Organic pollutants, such as hydrocarbons, pesticides, and pharmaceuticals, have a significant effect on water quality and aquatic ecosystems. Phytoremediation is a promising method for their removal and degradation.

5.3.2.1 Advancements in Degrading Hydrocarbons

Oil spills and industrial discharges introduce hydrocarbons into water bodies. Recent developments include:

- Hydrocarbon-Degrading Plants: *Scirpus robustus* (alkali bulrush) and *Eichhornia crassipes* (water hyacinth) have shown remarkable effectiveness in breaking down polycyclic aromatic hydrocarbons (PAHs) in oil-contaminated waters.
- Case Study: A wetland system in Nigeria using *Typha domingensis* achieved an 80% reduction in total petroleum hydrocarbons (TPHs) within a year (Olawale et al. 2023).

5.3.2.2 Breakdown of Pesticides and Pharmaceuticals

- Pesticides: *Vetiveria zizanioides* (vetiver grass) has demonstrated potential in degrading persistent pesticides like atrazine and chlorpyrifos.
- Pharmaceuticals: Constructed wetlands with Phragmites australis have proven effective in eliminating diclofenac and other pharmaceutical residues, reaching up to 90% removal efficiency in a European study (Birk et al. 2022).

5.3.2.3 Bioengineered Plants for Enhanced Degradation

Genetically modified plants with improved enzymatic pathways are becoming important contributors:

- *Arabidopsis thaliana*, modified to express bacterial laccase enzymes, showed increased degradation of synthetic dyes and pesticides.
- A 2023 study introduced fungal genes into rice plants, allowing for the effective breakdown of organophosphate pesticides in irrigation water.

5.3.3 Control of Eutrophication and Excess Nutrients

Excess nutrients like nitrogen and phosphorus can lead to eutrophication, resulting in algal blooms, oxygen depletion, and degradation of aquatic ecosystems.

Phytoremediation serves as an effective method for removing these nutrients and aiding in ecosystem recovery.

5.3.3.1 Global Instances of Eutrophication and Recovery

- Lake Taihu, China: Phytoremediation initiatives using *Eichhornia crassipes* and *Hydrilla verticillata* have significantly lowered phosphorus levels, enhancing water clarity and oxygen content.
- Chesapeake Bay, USA: Floating treatment wetlands planted with Juncus effusus (soft rush) have decreased nitrogen runoff from agricultural sources by more than 50%, according to a 2022 report.

5.3.3.2 Nutrient Cycling and Ecological Implications

Plants contribute in two ways: by eliminating excess nutrients and promoting nutrient cycling:

- Nitrogen Assimilation: Aquatic plants such as Lemna minor (duckweed) take up nitrate and ammonium, transforming them into biomass.
- Phosphorus Recovery: Species like Typha can sequester phosphorus, which can then be extracted from harvested biomass for use as fertilizer.

5.3.3.3 Constructed Wetlands and Floating Islands

- Constructed wetlands are increasingly used for nutrient management. A study conducted in Malaysia found that wetlands featuring *Thalia geniculata* reduced phosphorus levels in agricultural runoff by 70%.
- Floating treatment wetlands are becoming more common in urban water bodies. For example, a project in Bangalore, India, in 2023 employed floating islands with native plants to address eutrophication in contaminated lakes.

5.4 Case Studies: Phytoremediation in Action

Phytoremediation has been successfully applied across diverse environmental settings, ranging from municipal wastewater treatment to managing the impacts of mining. This section explores real-world applications, highlighting their operational challenges, innovations, and scalability potential.

5.4.1 Constructed Wetlands

Constructed wetlands (CWs) are specially designed ecosystems that replicate the functions of natural wetlands to effectively remove pollutants, using the combined efforts of plants, microorganisms, and substrates. Municipal Wastewater Treatment CWs are widely used for municipal wastewater treatment because they are cost-effective and environmentally friendly.

- Europe: In Denmark, horizontal subsurface flow wetlands have been incorporated into rural wastewater treatment systems. A 2023 study indicated that CWs planted with Phragmites australis (common reed) successfully removed 80–90% of nitrogen and phosphorus from wastewater. Additionally, the system decreased pathogen loads by over 95%, meeting strict EU water quality standards. Poland has implemented vertical flow constructed wetlands for small communities, using Typha latifolia. These wetlands handle 100 cubic meters of wastewater daily, achieving an 88% reduction in biochemical oxygen demand (BOD).
- Asia: In India, a pilot CW in Hyderabad employed *Canna indica* and *Schoenoplectus californicus* to treat municipal wastewater. A 2022 study revealed that the system accomplished a 75% reduction in chemical oxygen demand (COD) and an 85% decrease in nitrate levels. China's wetland parks feature hybrid CWs for urban wastewater treatment. The CW at Beijing Wetland Park managed to reduce heavy metals by 70% and nutrient loads by 80% in a large-scale municipal pilot.

5.4.1.1 Operational Challenges and Scalability

- Challenges: Over time, clogging of wetland substrates can hinder efficiency. It is challenging to maintain optimal hydrology and healthy vegetation across various climates.
- Scalability: Researchers are exploring modular designs to enhance scalability, especially in densely populated urban areas. Hybrid systems that combine vertical and horizontal flow CWs have shown potential for large-scale applications.

5.4.2 Floating Treatment Wetlands (FTWs)

FTWs represent a cutting-edge phytoremediation method that uses floating platforms with aquatic plants. They provide a versatile solution for water bodies where conventional constructed wetlands may not be applicable. Advances in FTW Design Recent developments include:

- Biomimetic Designs. By imitating natural root structures, synthetic root mats boost microbial colonization and enhance pollutant absorption.
- Hybrid Systems: The integration of FTWs with aeration and microbial inoculants to accelerate degradation rates.
- Case Studies Urban Waterways: In Bangkok, Thailand, FTWs featuring *Vetiveria zizanioides* (vetiver grass) were implemented in canals to combat eutrophication. A study from 2022 revealed that these FTWs achieved a removal rate of up to 60% for total nitrogen and phosphorus, helping alleviate algal blooms. In Singapore, floating wetlands with Pistia stratiotes (water lettuce) were established in stormwater reservoirs. These systems effectively reduced heavy metals like cadmium and lead by 80%, demonstrating their effectiveness in managing urban stormwater.

- Agricultural Runoff: In the U.S. Midwest, FTWs were deployed in ponds to capture agricultural runoff. A report from 2023 highlighted a system in Iowa that successfully removed 70% of nitrate and 60% of pesticide residues from runoff using *Lemna minor* (duckweed). Challenges and Innovations.
- Challenges: Securing FTWs in high-flow water bodies. Managing plant biomass and variations in seasonal performance.
- Innovations: Combining FTWs with renewable energy sources (e.g., solar-powered aeration). Employing biodegradable platforms is to reduce waste.

5.4.3 Mine-Impacted Water Bodies

Mining operations produce waste materials such as mine tailings and acid mine drainage (AMD), which present serious environmental and health hazards. Phytoremediation techniques in these situations aim to stabilize contaminants and restore ecological integrity.

Phytoremediation of Mine Tailings Plants play a crucial role in stabilizing and extracting heavy metals from mine tailings, which helps minimize pollutant leaching into surrounding water bodies.

- South Africa: In Witwatersrand, *Helianthus annuus* (sunflower) and *Sorghum bicolor* (sorghum) were used on gold mine tailings. A study conducted in 2023 indicated that these plants decreased the mobility of arsenic and lead by 50% while also producing biomass that holds economic value.
- Peru: At the Cerro Verde copper mine, native grasses (*Festuca dolichophylla*) were employed to stabilize tailings, leading to a 60% reduction in windborne dust and heavy metal runoff.

5.4.3.1 Treatment of Acid Mine Drainage (AMD)

AMD is known for its low pH and high levels of toxic metals and sulfates. Phytoremediation provides a natural method to neutralize acidity and immobilize metals.

- Southeast Asia: o In Indonesia's coal mining regions, constructed wetlands planted with *Phragmites karka* and *Scirpus grossus* achieved notable reductions in AMD acidity and sulfate levels, as reported in 2023.

In the Philippines, a pilot study using *Typha latifolia* in AMD-contaminated water demonstrated an 80% reduction in iron and manganese concentrations within six months.

5.4.3.2 Innovations in Mine-Impacted Phytoremediation

- Microbial Augmentation: Enhancing the microbial communities in the rhizosphere has led to improved metal uptake and sulfate reduction.
- Biochar Integration: The addition of biochar to substrates has been shown to stabilize pH levels and promote plant growth in areas affected by AMD.

5.4.3.3 Challenges and Future Directions

- Challenges: The extreme conditions found in mine-impacted areas pose significant challenges to plant survival and growth. Long-term maintenance and monitoring require substantial resources.
- Future Directions: Genetically modifying plants to enhance their tolerance to acidic conditions and improve heavy metal uptake. Integrating phytoremediation with renewable energy systems such as solar powered irrigation may enhance its activity.

5.5 Challenges and Limitations of Phytoremediation

While phytoremediation shows great potential as a sustainable and cost-effective method for cleaning up contaminated environments, it also encounters several challenges and limitations that restrict its wider use. These challenges include environmental factors, the complexity of dealing with mixed pollutants, issues related to biomass disposal, and concerns about scalability.

5.5.1 Environmental Factors

The effectiveness of phytoremediation is significantly affected by environmental conditions such as soil quality, water availability, and climate. Climate change has introduced new variables that can greatly influence the success of phytoremediation efforts.

5.5.1.1 Effects of Climate Change on Phytoremediation

1. Rising Temperatures:
   - Increased temperatures can boost microbial activity in the rhizosphere, speeding up degradation processes. However, excessive heat may put stress on plants, leading to reduced growth and lower contaminant uptake.
   - A 2022 study conducted in Mediterranean regions revealed that heatwaves diminished the ability of Phragmites australis to remove nitrates from wetlands by 30%.
2. Increased rainfall can lead to waterlogging, which diminishes oxygen levels in the root zone and restricts the absorption of pollutants. Conversely, drought conditions can impede plant growth and the movement of pollutants driven by transpiration. A 2023 study highlighted that drought-tolerant species like *Vetiveria zizanioides* demonstrated resilience in arid areas but required supplemental irrigation to effectively eliminate pollutants.

5.5.1.2 Extreme Weather Events

Storms and floods can disrupt phytoremediation systems, resulting in the release of stored contaminants. For instance, during the 2022 floods in Pakistan, floating treatment wetlands were compromised, leading to the remobilization of heavy metals into adjacent water bodies.

Adaptation Strategies:

- Using genetically engineered plants that exhibit greater resilience to heat and drought.
- Integrating phytoremediation with renewable energy-powered irrigation systems to mitigate water stress.

5.5.2 Mixed Pollutants

Water bodies often contain a mix of contaminants, including heavy metals, organic pollutants, and pathogens, which interact in ways that complicate phytoremediation efforts.

5.5.2.1 Interactions Between Contaminants

- Synergistic and Antagonistic Effects: Some pollutants, such as heavy metals and hydrocarbons, can interact synergistically, enhancing degradation. However, antagonistic effects are more common, where one pollutant obstructs the uptake or breakdown of another. A 2023 study in industrial wetlands in China found that elevated cadmium levels reduced the enzymatic activity of rhizospheric microbes responsible for hydrocarbon degradation by 40%.
- Plant Health and Toxicity: The presence of mixed pollutants can exceed toxicity thresholds for plants, leading to stunted growth or even death. For example, a 2022 pilot study indicated that high levels of both arsenic and organic solvents inhibited the growth of Helianthus annuus.

5.5.2.2 Mitigation Approaches

- Implementing bioaugmentation with pollutant-degrading microbes to specifically target certain contaminants.

5.5.3 Biomass Disposal

The buildup of contaminants in plant tissues presents a major challenge, as improper disposal can result in secondary pollution.

5.5.3.1 Safe Disposal of Pollutant-Rich Biomass

1. Incineration:
   - This method is often used for biomass that is heavily contaminated. While incineration reduces the volume of waste, it can release pollutants into the air unless advanced filtration systems are in place.

   - A study conducted in the UK in 2022 revealed that modern incineration facilities with flue gas treatment systems successfully neutralized 95% of heavy metal emissions.
2. Landfilling:
   - Landfilling is a cost-effective disposal method, but it carries the risk of leachate contaminating groundwater.
   - Research from 2023 emphasized the effectiveness of biopolymer liners in landfills to prevent leachates from phytoremediation biomass.
3. Biochar Production:
   - Converting pollutant-rich biomass into biochar through pyrolysis offers a promising solution. When applied to contaminated land, biochar can immobilize heavy metals and enhance soil health.
   - A 2023 study demonstrated that biomass from Eichhornia crassipes (water hyacinth) treated with cadmium was transformed into biochar, which immobilized 85% of cadmium when used on acidic soils.
4. Biogas Generation:
   - The anaerobic digestion of biomass can generate renewable energy while also reducing its volume. However, the presence of contaminants like heavy metals can hinder microbial activity, making pretreatment necessary.

5.5.4 Scaling Up

While phytoremediation works well in small-scale settings, applying it to larger or more heavily contaminated areas presents significant challenges.

5.5.4.1 Challenges in Large-Scale Applications

1. Site Heterogeneity:
   - Large sites often show variations in soil and water conditions, necessitating customized strategies.
   - For example, a project in the Amazon basin in 2022 demonstrated the difficulties of standardizing phytoremediation methods across regions with differing contaminant profiles.
2. Timeframe:
   - Phytoremediation is a slower process compared to chemical or mechanical methods, making it less ideal for urgent cleanup situations.
   - A 2023 study indicated that using Brassica juncea to remove heavy metals from a mining site in Peru took over five years to achieve meaningful reductions.
3. Cost and Resources:
   - The initial costs for setting up large-scale systems, including irrigation and monitoring infrastructure, can be quite high.

5.5.4.2 Innovations for Scalability

Hybrid Systems: Merging phytoremediation with other technologies, such as bioreactors or nanomaterials, has shown promise. For instance, nanomaterial-enhanced wetlands in India cut remediation time by 40%.

Drones and Internet of things (IoT): Drones are being used to monitor extensive phytoremediation sites, while IoT devices keep track of environmental conditions in real time, enhancing plant health and pollutant removal.

Policy and Funding: Encouraging phytoremediation through government subsidies and international funding has proven effective. A 2022 report noted that EU subsidies facilitated large-scale wetland construction in Poland.

5.6 Conclusion

The successful use of phytoremediation methods, such as constructed wetlands, floating treatment wetlands, and hyperaccumulators, demonstrates how effective green technologies can be in tackling water pollution. These examples showcase phytoremediation's ability to restore ecosystems, enhance water quality, and support the goals of SDG-6. Ongoing research and investment in these technologies will improve their application and effectiveness in addressing environmental issues.

5.6.1 Challenges and Considerations

Although phytoremediation presents a promising solution for restoring water bodies, it faces several challenges. Various factors can affect the success of this green technology, including climate conditions, water chemistry, the types of contaminants involved, and the ecological setting where the remediation takes place (Ali et al. 2013; Ijaz et al. 2020).

5.6.2 Influence of Environmental Factors

Climate significantly influences the success of phytoremediation. Changes in temperature, rainfall, and seasonal patterns can affect plant growth, pollutant absorption, and microbial activity in the rhizosphere. For instance, extreme temperatures might hinder plant growth or lessen the effectiveness of contaminant uptake (Zhao et al. 2021). Water chemistry factors, such as pH, salinity, and the presence of competing ions, also play a role in determining the bioavailability of pollutants and the plants' capacity to absorb them (Rani et al. 2021).

The types of contaminants involved can further complicate the phytoremediation process. Heavy metals, for example, differ in their toxicity and mobility, with some being more difficult to remediate than others (Zhang et al. 2023). Moreover, organic pollutants may need specific plant enzymes for degradation, and not all plants have these capabilities (Li et al. 2022a, b, c). The presence of mixed contaminants can also lead to synergistic or antagonistic effects, making the remediation process even more complex (Ali et al. 2013).

5.6.3 Long-Term Maintenance and Biomass Disposal

One major challenge is the long-term upkeep of phytoremediation systems. It is vital to regularly monitor the health of the plants, their growth, and the levels of contaminants to ensure the system continues to function effectively over time (Ijaz et al. 2020). Additionally, the disposal of harvested biomass that contains concentrated pollutants must be handled with care. If not disposed of properly, it can lead to secondary pollution, negating the advantages of phytoremediation. Thus, developing safe disposal methods or finding alternative uses for the biomass, such as in bioenergy production or composting, is essential (Khalid et al. 2022).

5.6.3.1 Multidisciplinary Approaches

Tackling these challenges necessitates a multidisciplinary approach that merges scientific expertise, engineering practices, and community involvement. Researchers and practitioners need to work together closely to refine the design and execution of phytoremediation systems, ensuring they are sustainable in the long run and beneficial to the environment (Zhao et al. 2021). Community participation is also vital for the success of phytoremediation projects.

Involving local stakeholders can foster awareness and support for these initiatives, which can lead to improved management practices and a sense of shared responsibility for the remediation efforts (Hillenbrand et al. (2023). Moreover, combining traditional ecological knowledge with scientific research can improve the design of phytoremediation systems, making them more effective and culturally appropriate (Rani et al. 2021).

References

Ali H, Khan E, Sajad MA (2013) Phytoremediation of heavy metals—concepts and applications. Chemosphere 91(7):869–881. https://doi.org/10.1016/j.chemosphere.2013.01.075

Birk J, Schmidt A, Müller K (2022) Efficacy of constructed wetlands with *Phragmites australis* for the removal of pharmaceutical contaminants. Environ Sci J 29(2):456–468. https://doi.org/10.1016/j.esj.2022.01.012

Chen X, Li Y, Zhang J (2022) Synthetic pathways in poplar trees for enhanced arsenic bioaccumulation. J Biotechnol Environ Res 34(2):156–167. https://doi.org/10.1016/j.jber.2022.01.012

Desbureaux S, Damania R, Rodella A-S, Russ J, Zaveri E (2019) The impact of water quality on GDP growth: evidence from around the world. World Bank. https://doi.org/10.1596/33071

Hillenbrand, T., Tettenborn, F., Bloser, M., Rechenberg, J., & others. (2023). Engaging Stakeholders to Solve Complex Environmental Problems Using the Example of Micropollutants. Water, 15(19), 3441. https://doi.org/10.3390/w15193441Ijaz A, Iqbal J, Ahmad Z, Rehman SU (2020) Phytoremediation: recent advances in plant-based strategies for the removal of heavy metals and emerging organic pollutants. Environ Sci Pollut Res 27:24859–24881. https://doi.org/10.1007/s11356-020-08905-5

Khalid MU, Iqbal A, Aslam S (2022) Floating treatment wetlands: a novel approach for sustainable urban water management. Water 14(18):2906. https://doi.org/10.3390/w14182906

Kumar A, Patel R, Sharma S (2022) Genetic engineering of tobacco plants expressing fungal cytochrome P450 for atrazine degradation. J Agric Biotechnol 15(3):456–467. https://doi.org/10.1016/j.jab.2022.03.00

Lebreton LCM, Van der Zwet J, Damsteeg JW, Slat B, Andrady A, Reisser J (2017) River plastic emissions to the world's oceans. Nat Commun 8:15611. https://doi.org/10.1038/ncomms15611

Li J, Dong H, Zhang Y, Lin Y (2022a) Water pollution and associated human health risks: a review. Environ Res 212:113252. https://doi.org/10.1016/j.envres.2022.113252

Li X, Zhang Y, Wang L (2022b) Transgenic poplar trees engineered for arsenic volatilization and their impact on soil contamination. J Environ Biotechnol 28(4):1345–1358. https://doi.org/10.1016/j.jeb.2022.04.015

Li Z, Zhang Y, Zhang W, Zhou H, Liu M (2022c) Phytoremediation of heavy metal pollution: a promising approach for revegetation of metal-polluted land. Front Plant Sci 11:359. https://doi.org/10.3389/fpls.2020.00359

Lin S-S, Shen S-L, Zhou A, Lyu H-M (2024) Assessment and management of industrial impacts on water quality. Front Environ Sci 12:103456. https://doi.org/10.3389/fenvs.2024.103456

Olawale A, Adegbite O, Smith J (2023) Utilization of Typha domingensis in wetland systems for the reduction of petroleum hydrocarbons. J Environ Manag 35(4):1123–1136. https://doi.org/10.1016/j.jenvman.2023.04.045

Patel R, Sharma A, Gupta S (2023) Chromium reduction in soil and water through bioremediation: a three-cycle study. J Environ Sci Technol 50(4):1234–1245. https://doi.org/10.1016/j.jest.2023.01.045

Rani A, Kumar A, Kumar S (2021) Phytoremediation of water pollutants: a green approach for sustainable environment. Environ Technol Innov 24:101823. https://doi.org/10.1016/j.eti.2021.101823

Song H, Zhang Z, Li J (2021) Challenges and advances in chemical treatment technologies for hazardous pollutants in wastewater. Front Environ Chem 11:129874. https://doi.org/10.3389/fenvc.2021.129874

SpringerOpen (2023) Agricultural runoff and eutrophication: Nitrate and phosphate loading in aquatic ecosystems. Appl Biol Chem 66(4):1245. https://doi.org/10.1186/s42774-023-00567-9

United Nations (2023) Global status on wastewater treatment and its environmental impacts. United Nations Environmental Reports 2023(1):12–15. https://doi.org/10.1016/j.unenv.2023.03.012

Xu Y, Huang H, Wang Y (2022) Effects of water pollution on human health and disease heterogeneity: a review. Front Public Health 10:987123. https://doi.org/10.3389/fpubh.2022.987123

Zhang J, Zhao H, Li Q, Liu S (2023) Recent advances in phytoextraction of heavy metals: a review. Environ Sci Pollut Res 30:8321–8345. https://doi.org/10.1007/s11356-023-25877-0

Zhao J, Wang Y, Wei Y (2021) Advances in phytoremediation technology: a review on ecological, environmental, and social challenges. Environ Technol Rev 10(1):34–47. https://doi.org/10.1080/21622515.2021.1883644

Urban Circularity Challenges Via Nature-Based Solution

6

Shruti Awasthi, Bellam Darshitha, and Naveen Dwivedi

Abstract

Urban regions are centers of environmental stress, suffering from resource depletion, waste generation, water management inefficiencies and climate change. Advancing urban circularity is a system centered on resource efficiency, waste minimization, and ecosystem integration—is crucial for achieving sustainable urban development. Nature-based solutions (NBS) offer these solutions in an innovative, multifunctional manner. Nature-based solutions (NBS), such as green roofs, wetlands, urban forests, and permeable pavements, leverage natural ecological processes to restore ecological balance and address critical urban challenges including water management, nutrient recovery, and energy efficiency. These approaches also enhance resilience and support the sustainable regeneration of resources. Implementing NBS in urban development planning boosts the socioeconomic sustainability of cities by alleviating environmental challenges, improving adaptation, and increasing resilience. This chapter underscores the potential of NBS in transforming urban systems into adaptive, inclusive, and sustainable environments aligned with circular economy principles.

Keywords

Nature-based solutions (NBS) · Urban circularity · Green infrastructure · Circular economy · Resource management · Sustainable cities

S. Awasthi · B. Darshitha
Department of Life Sciences, Garden City University, Bangalore, Karnataka, India

N. Dwivedi (✉)
Department of Biotechnology Engineering and Food Technology, University Institute of Engineering, Chandigarh University, Mohali, Punjab, India

N. Dwivedi et al. (eds.), *Wastewater Treatment Through Nature-Based Solutions*,
https://doi.org/10.1007/978-981-96-6699-7_6

6.1 Introduction

Cities are now at the center of global environmental issues due to the twenty-first century's growing urbanization. Despite making up only 2% of the planet's land area, urban areas produce an equal amount of garbage, consume more than 60% of its energy, and emit 70% of its greenhouse gas emissions (Langergraber et al. 2021; Ristić Trajković et al. 2024). Resource shortages, waste accumulation, climate change, and biodiversity loss are some of the issues brought on by this unprecedented increase. These problems are made worse by the linear "take-make-dispose" economic model, which emphasizes the need for different strategies for urban development. Circular economy (CE) is defined as an economic system that substitutes reducing, alternatively reusing, recycling, and recovering resources in production/distribution and consumption processes for the "end-of-life" idea. With the goal of achieving sustainable development, it works at the micro-level (products, businesses, and consumers), meso-level (eco-industrial parks), and macro-level (city, region, country, and beyond) to create social equity, economic prosperity, and environmental quality for the benefit of present and future generations (Calisto Friant et al. 2020; Kirchherr et al. 2017; Vanhuyse 2024).

The CE serves as a model for cities that may separate resource use from urban development, combining economic well-being goals with environmental pressure removal and tackling socioeconomic issues in cities (Gravagnuolo et al. 2019; Marchesi and Tweed 2021; Murray et al. 2017; Sánchez Levoso et al. 2020; Vanhuyse 2024). The circular economy is a straightforward yet effective approach that seeks to reduce waste production and virgin material input by completing ecological and economic resource flow cycles. The shift to a CE takes into account both conventional policies and a shift in waste disposal practices in response to government waste management pledges. This transformation is focused on a broad departure from existing methods and takes into account product design that permits efficient material reuse and recycling. To enforce the suggested solutions, however, legal and implementation procedures are needed (Dawson 2019; Negrete-Cardoso et al. 2022).

CE states that landfill disposal should be the final option and burning for energy the second-to-last. This keeps the product's value chain and life cycle as high-quality and valuable as feasible for as long as possible while also making it as energy-efficient as possible. It makes financial and commercial sense to use the value created as long as possible once a raw material has been harvested, purified, and produced at the typical costs; that is, to retain the product's function, service, and use-value in the economy for as long as feasible (Korhonen et al. 2018). A resilient, sustainable, and healthy urban environment will result from a circular flow system that uses nature-based solution (NBS) to manage nutrients and resources within the urban biosphere (Atanasova et al. 2021; Langergraber et al. 2020). A system approach that takes into consideration the larger economic, social, and environmental systems in which commercial supply chains function is necessary for the

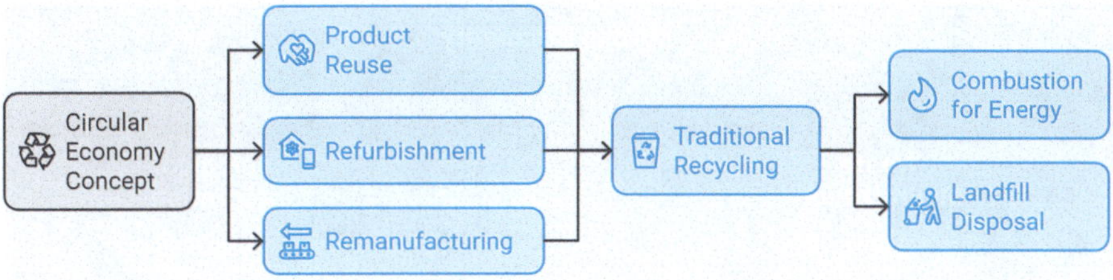

Fig. 6.1 A circular flow system that implements NBS

implementation of the circular economy. By taking into account how management choices may affect different stakeholder groups, such as the impoverished and other marginalized societal groups that are particularly susceptible to the negative health effects of waste proliferation, circular economy design should aim to be inclusive (Fiksel et al. 2021) (Fig. 6.1).

6.2 Defining Urban Circularity

It is believed that the circular economy can assist cities in transitioning from linear to circular resource management (Atanasova et al. 2021). Inspired by the ideas of the circular economy, urban circularity places a strong emphasis on protecting ecosystems and using resources sustainably. It suggests switching from linear models to resource-recycling and regenerating closed-loop systems.

This strategy fits in well with nature-based solutions (NBS), which use natural processes to address urban problems. Examples of co-benefits that go beyond urban circularity include urban forests for carbon sequestration-based solutions, wetlands for water filtering, and green roofs for urban cooling. They increase human well-being, biodiversity, and resistance to climate-related shocks like heat waves and floods. Urban planning that incorporates natural systems can help cities move toward a more resilient and sustainable future (Langergraber et al. 2021; Ristić Trajković et al. 2024) (Fig. 6.2).

The idea of a circular city is particularly significant in the NBS framework. Starting from a systems perspective, it seeks to transform the linear resource flow into a circular one in order to produce a closed loop for every natural or manufactured product. The freshly broadened notions of circular economics and urban circular development now include resource looping (Geissdoerfer et al. 2017; Paiho et al. 2020; Tsaligopoulos et al. 2022; Williams 2019, 2020, 2021). A relatively recent invention in the sustainable development paradigm, the circular city emerged as a result of the circularity principle's application to environmental sustainability. The system approach to city management served as the foundation for its development (Ristić Trajković et al. 2024) (Figs. 6.3 and 6.4).

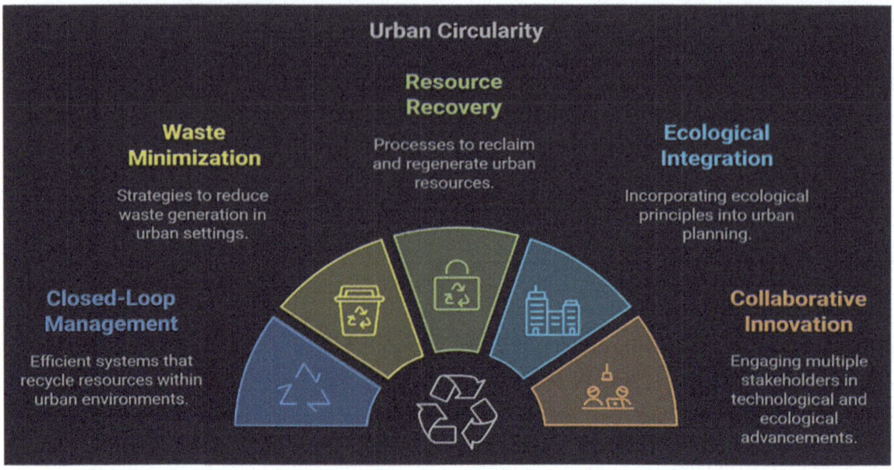

Fig. 6.2 Urban circularity

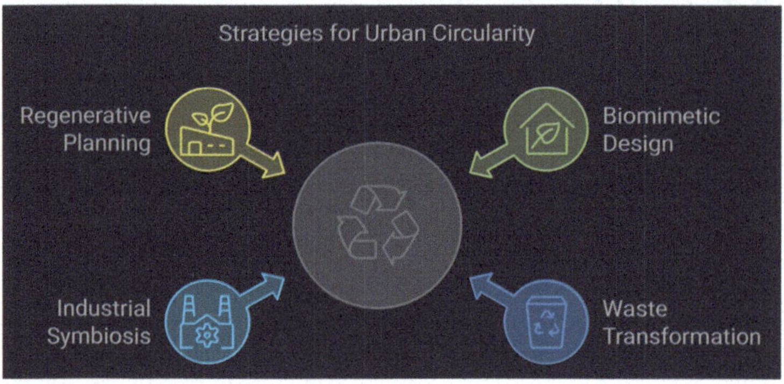

Fig. 6.3 Strategies for urban circularity

6.2.1 Challenges in Achieving Urban Circularity

The following are the main urban difficulties that have been identified, or Urban Circularity Challenges (UCC) in this transition: (1) reestablishing and preserving the water cycle; (2) treating, recovering, and reusing water and waste; (3) recovering nutrients; (4) recovering and reusing materials; (5) producing food and biomass; (6) recovering energy efficiency; and (7) recovering building systems. These issues might be resolved by using NBS in cities because it offers numerous advantages and is multipurpose (Atanasova et al. 2021) (Fig. 6.5).

In addition to lacking natural settings, urban regions frequently experience air pollution, heat island effects, and a reduction in biodiversity. These problems can be lessened by incorporating green and blue infrastructure (GBI) and NBS into urban planning, which will make cities resilient and sustainable places. Numerous

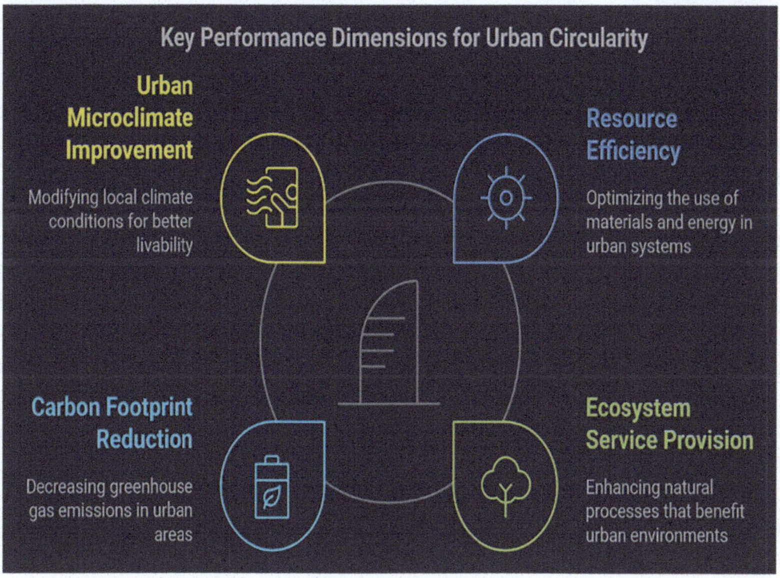

Fig. 6.4 Performance dimensions for urban circularity

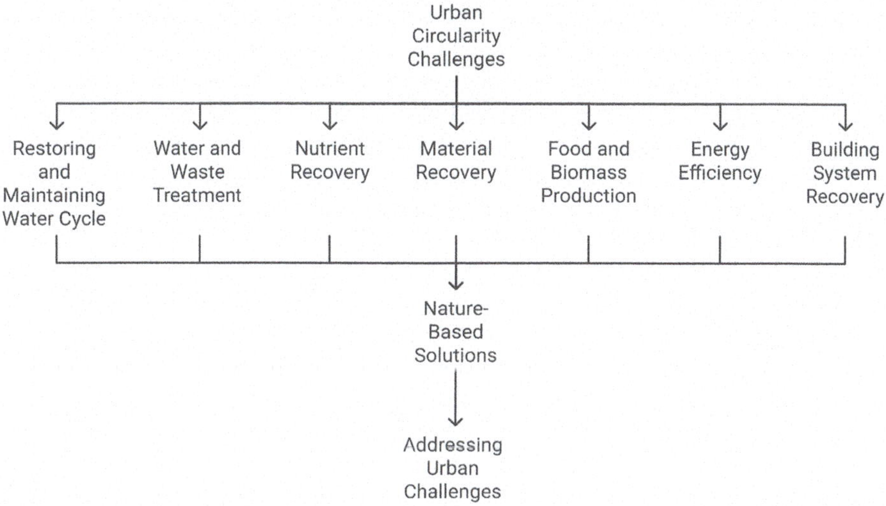

Fig. 6.5 Challenges in urban circularity

advantages result from these solutions, such as better biodiversity, temperature control, air and water quality, and aesthetic value (Haase 2021; Liu et al. 2021; Marselle et al. 2021; Pinto et al. 2023).

From the CE aspect, in order to restore and preserve the natural water cycle processes, the water management services must incorporate and implement rainwater management services, which increase retention and infiltration in the city by simulating natural processes. After being carried to treatment facilities, wastewater—that is, water that has been enhanced with nutrients and organic matter from food consumption—is released back into the environment, leaving behind by-products that are rich in nutrients and organic matter, such as sludge. Managing wastewater to properly reuse water and by-products rather than importing freshwater and fertilizers is a challenge. The water services should change from treatment for disposal to management for reuse from a CE perspective. The separation of various waste streams (urine, feces, and gray water) is one of the most effective methods of recovering nutrients; yet, it necessitates significant adjustments to the infrastructure and water management. While materials from demolition sites are frequently dumped in landfills, modern building materials mostly depend on nonrenewable resources. Dedicated reuse of demolition waste as filter media, the utilization of recycled building materials like wood, metal, or plastic, and the creation of bio-composite additives for the enormous quantity of concrete and isolating materials generated are all examples of nature-based solutions to this problem. By supporting a circular urban metabolism and adhering to CE principles, urban farming can be utilized as a component of the systemic solution for sustainable resource management. However, cities continue to face difficulties in implementing urban food production as a systemic solution. By employing the CE idea, urban food production is carried out using local resources, which entails recycling water and nutrients from waste streams. This practice faces numerous obstacles, ranging from technical and legal ones to social acceptance and health safety issues. Among the anticipated objectives that can be accomplished with NBS in a circular concept are energy-efficient structures, mitigation of the urban heat island effect and, as a result, lowering the need for cooling in buildings, and heat and energy recovery from various waste streams. Materials, energy, waste, biodiversity, health and well-being, human culture, and society are all positively impacted simultaneously by circular constructions (Atanasova et al. 2021).

There are numerous circularity tactics available to cut down on waste creation and the use of materials and natural resources. Depending on how circular they are, they can be ranked in order of priority (Potting et al. 2017) (Figs. 6.6 and 6.7).

The circular economy's 9Rs framework offers a thorough method for reducing waste, making the best use of available resources, and promoting sustainability. Every element stands for a distinct approach to creating a closed-loop system.

Refuse Refuse services or goods that are not sustainable or necessary to avoid wasting resources, for example, declining single-use plastic items.

Rethink Rethink how items are utilized and think of more sustainable or efficient alternatives, for example, using shared transportation services instead of owning a car.

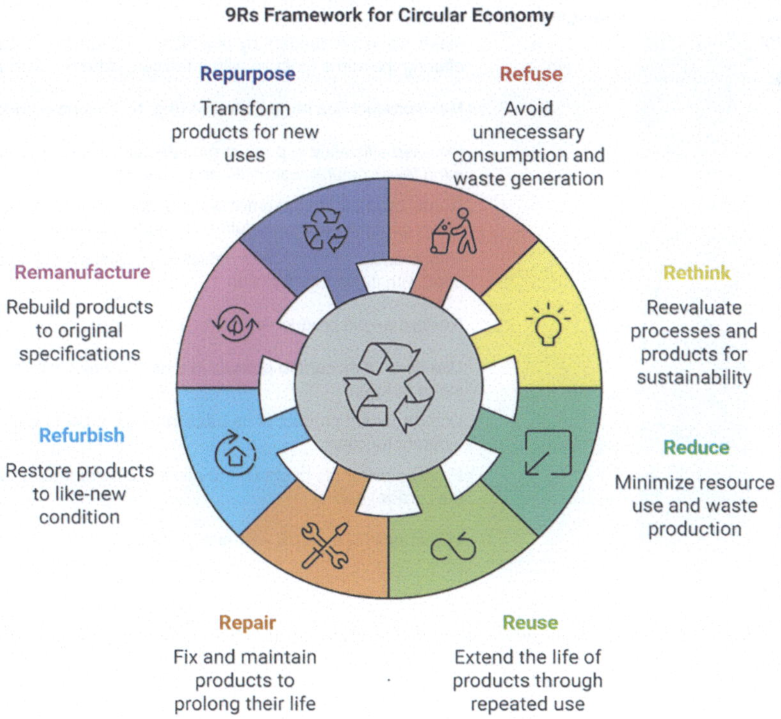

Fig. 6.6 9R framework

Reduce Reduce the amount of energy, materials, and waste produced and consumed, for example, manufacturing lightweight packaging.

Reuse Reusing a product for the same or different applications without making major changes can help it last longer, for example, refilling glass bottles or repurposing old jars.

Repair To prolong the useful life of worn-out or damaged products, fix them, for example, repairing electronics rather than discarding them.

Refurbish Renewing a product puts it back in good shape, usually for future sale, for example, upgrading old furniture or reconditioning smartphones.

Remanufacture Rebuilding a product to its original specs using new, restored or repurposed parts, for example, refitting machinery parts in industrial equipment.

Repurpose Modify a product to serve a purpose other than the one for which it was designed, for example, converting used tires into playground equipment.

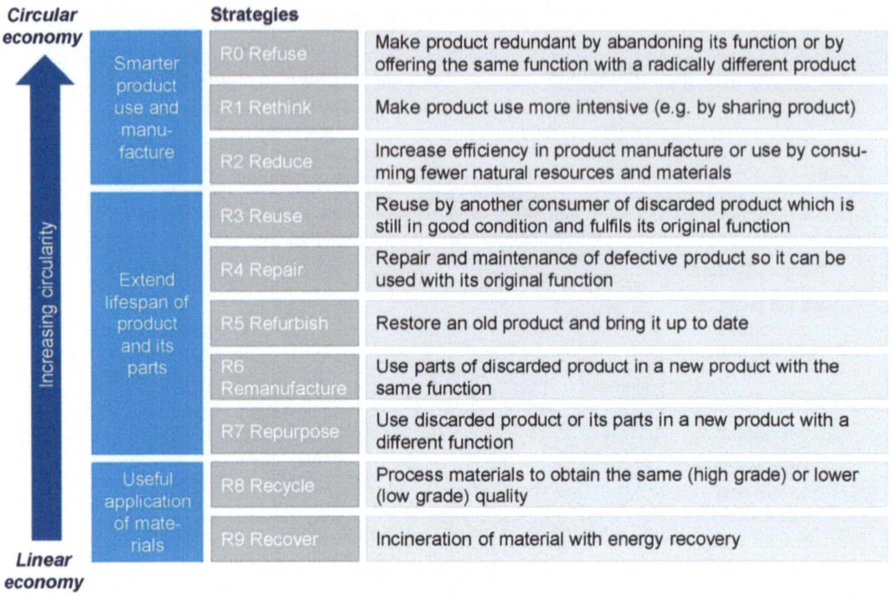

Fig. 6.7 9Rs Framework strategies. (Source: Kirchherr et al. (2017), Plotting et al. (2017), et RLI (2015))

Recycle Reducing waste and preserving resources by processing items to produce new goods, for example, melting down aluminum cans to produce new ones.

Recover Extract energy or valuable resources from nonrecyclable garbage, for example, waste-to-energy conversion through incineration.

The key difference between all R-lists is the quantity of circularity methods they propose. Usually, they offer a variety of tactics arranged from low circularity (high R-number) to high circularity (low R-number). Because fewer products are required to perform the same purpose, R0 and R1 methods reduce the consumption of natural resources and commodities used in a product chain. Consequently, even while R0 and R1 may not always entail enhancing the reuse of products and components or the reapplication of recycled materials, they are typically also regarded as circularity methods (Potting et al. 2017).

Higher-order Rs like Refuse, Rethink, and Reduce should be prioritized according to this framework since they have the most effects on waste reduction and resource conservation. Although they are essential, lower-order Rs like Recycle and Recover should preferably be used as a last option.

One of the most crucial sustainable development goals (SDGs), universal access to clean water and sanitation, will not be accomplished by 2030. Despite official declarations, the United Nations failed to meet one of its Millennium Development Goals for 2015, which was to cut in half the percentage of the world's population without sustained access to safe drinking water and basic sanitation.

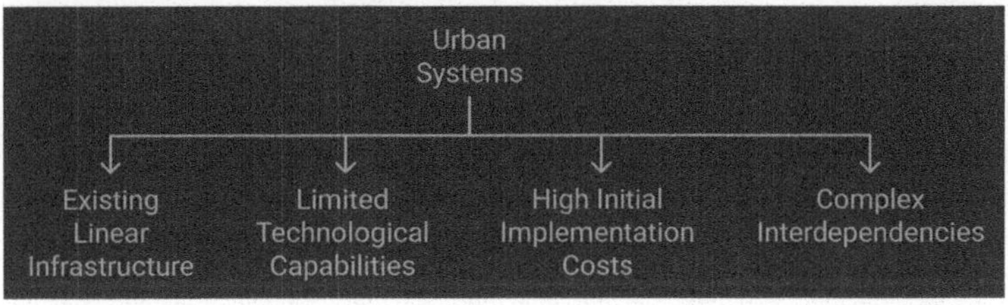

Fig. 6.8 Urban systems

For decades, billions of people in Latin America, Asia, and Africa have had limited access to these essential services. Despite having smaller numbers, vulnerable and impoverished groups particularly Indigenous people face similar issues in the United States, Canada, Australia, and the EU (Tortajada 2024) (Fig. 6.8).

6.2.2 Implementing Nature-Based Solutions to Address Urban Circularity Challenges

NBS are well known for its capacity to effectively address emerging urban issues like sustainability, resilience, and climate change adaptation. NBS is a relatively new environmental idea that has recently gained traction as constructive solutions to societal issues and has the potential to concurrently achieve social, economic, and environmental goals (Ramusino et al. 2017).

In order to address a variety of issues pertaining to ecosystem resilience, climate change, human health, and general well-being, NBS refers to interventions that use nature or simulate natural processes to address urban challenges. Integrating natural elements into urban planning and design enhances the quality of the urban environment. These solutions accomplish this by combining tried-and-true methods for preserving ecosystem services, protecting biodiversity, and putting green infrastructure (GI) into place (Pinto et al. 2023).

Gray, green, and digital infrastructure; wise investments in wastewater plant construction and maintenance, and the know-how to run them sustainably and efficiently including energy sources and costs are all necessary for effective wastewater management. In order to identify drinking water abstraction locations and any water quality issues, which in turn informs policy and decision-making, catchment management and an understanding of sewer sheds areas of land where sewers flow to a single endpoint are essential.

Reusing water is an effective way to help achieve the human right to equitable access to water of appropriate quantity and quality. This is only feasible, though, if long-term planning incorporates inter-sectoral policies, rules, and instruments for inclusive governance, financial support, public opinion, and regulatory compliance (Tortajada 2024).

One international approach to addressing the problem of the global water issue is circular management. The mainstreaming of circularity in wastewater management can be accelerated by mutual learning through information exchange among G20 nations and beyond (Ensuring Water Secure Cities 2024).

Existing ideas and methods that use and improve nature, like ecosystem-based adaptation (EbA), green infrastructure (GI), ecosystem services (ESS), and nature-based solutions (NBS), have gained traction because they address issues (like disaster risk reduction, water management, climate mitigation and adaptation, degradation and loss of natural capital etc.) in a more sustainable manner than traditional hard engineering. Although the four concepts all adhere to the multifunctionality principle, NBS can be seen of as an umbrella for the others, with biodiversity at its heart and a strong solution-oriented focus (Nika et al. 2020; Pauleit et al. 2017).

Freshwater resources in India are under a lot of stress. Only 28% of the 72,000 million liters of spent water (domestic sewage) produced daily by Indian cities in 2021 were treated (CPCB 2021). The residual untreated wastewater is released into lakes and rivers, contaminating the water and affecting populations, particularly those downstream. In India, the most important authorities in charge of managing urban residential water use are Urban Local Bodies (ULBs) or urban local bodies.

Cities bear the primary duty for the development and upkeep of used water infrastructure and service delivery. Reusing treated used water, or Treated Used Water (TUW), is a crucial part of managing urban water resources. In addition to producing financial gains for the implementing body, it can lessen the strain on already limited freshwater supplies. A number of Indian towns, including Surat, Bangalore, Ahmedabad, Indore, and Visakhapatnam, are effectively carrying out reuse initiatives. However, the country has not yet mainstreamed Treated Used Water (TUW) reuse. Recently, the Indian government has launched a number of urban-focused projects aimed at reducing water pollution, enhancing the treatment of spent water, and encouraging its reuse. Some of the important ones include the Namami Gange Programme and the Atal Mission for Rejuvenation and Urban Transformation (AMRUT) (Enabling Circular Economy in Used Water Management in India 2024) (Fig. 6.9).

6.2.3 Green Infrastructure

Partially realized through nature-based solutions, green and blue infrastructure (GBI), also known as green infrastructure (GI), is essential to the effective execution of regulations. A carefully designed network of natural and semi-natural areas with extra environmental features to offer a variety of ecosystem services is known as "green infrastructure," according to the European Commission (Communication from the Commission to the European Parliament, The Council, The European Economic and Social Committee and The Committee of the Regions Green Infrastructure (GI)—Enhancing Europe's Natural Capital 2013; Pinto et al. 2023) (Fig. 6.10).

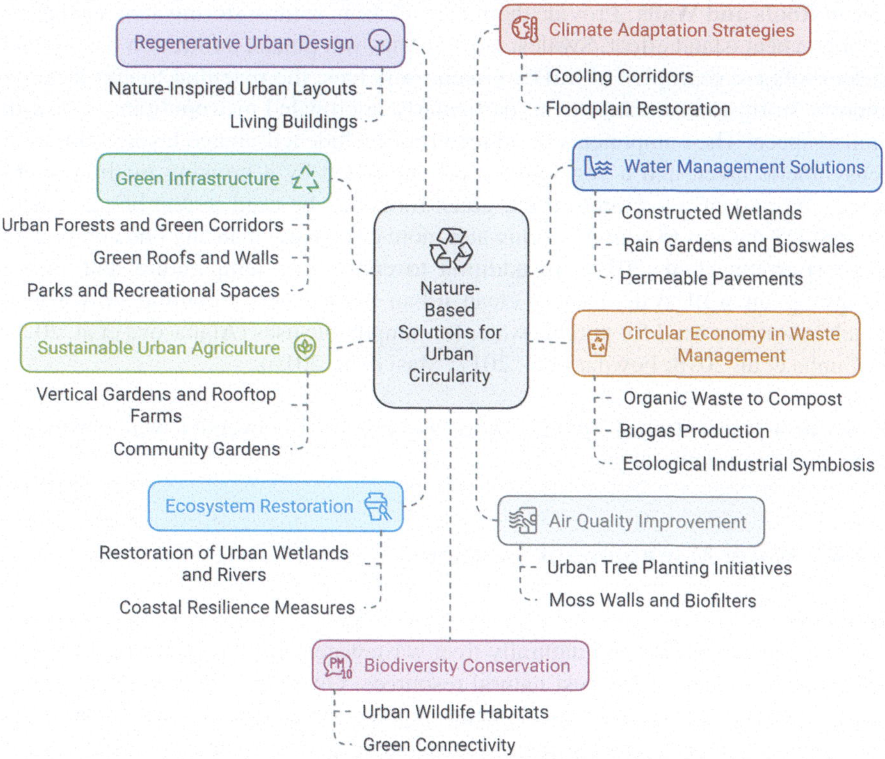

Fig. 6.9 Nature-based solutions to address urban circularity challenges

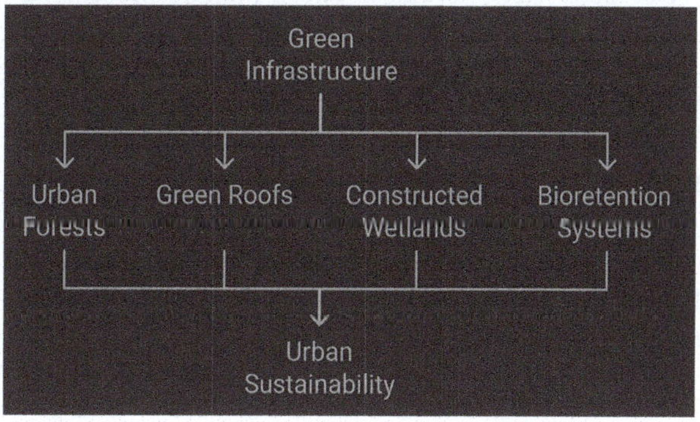

Fig. 6.10 Green infrastructure

Urban Forests and Green Corridors Improve air quality, increase biodiversity, and sequester carbon. In order to plan and manage trees and green spaces in and near human populations and give advantages like NBS, urban forestry offers a common vocabulary, sets of science-based techniques, and experience (Escobedo et al. 2019; Raymond et al. 2017).

Green Roofs and Walls Provide thermal insulation, control stormwater, and lessen the urban heat island effect. Swales, rain gardens, wet ponds, infiltration basins, and green roofs are examples of SuDS. Green roofs have the potential to significantly improve stormwater management, particularly in crowded metropolitan areas with limited space. The components of a green roof include a drainage layer, a filtration geosynthetic layer, and a vegetative layer supported by a growing medium (substrate). As part of a treatment train, green roofs can be used in combination with other SuDS devices to control stormwater more effectively than any one component alone (Berretta et al. 2014). In addition to controlling temperature and saving energy, green wall systems may clean urban wastewater, including gray water, which lowers the need for potable water for nonpotable uses (Atanasova et al. 2021; da Cunha et al. 2018; Fowdar et al. 2017; Masi et al. 2016).

Parks and Recreational Spaces Serve as hubs for biodiversity, water management, and community engagement.

6.2.4 Water Management Solutions

Wetlands

Create wildlife habitats and naturally treat wastewater. In urban settings, wetlands are regarded as one of the best natural resources. Given their many benefits, wetlands are among the most crucial elements of green–blue infrastructure. Urban areas are more socially and environmentally sustainable when wetlands are used to their full potential. Specifically, urban wetlands are an essential component of urban ecosystems. Urban wetlands are essential suppliers to human societies and offer a variety of ecosystem services. These include preserving coastal areas, improving the quality of the water, cutting down on air pollution, sequestering carbon, and leisure and recreation (Alikhani et al. 2021; Boyer and Polasky 2004; Das and Basu 2020; Mitsch et al. 2013; Quin et al. 2015; Temmerman et al. 2013; Verhoeven et al. 2006).

Rain Gardens and Bioswales Recharge groundwater, filter contaminants, and control stormwater discharge.

Permeable Pavements To lessen surface runoff and flooding, let rainwater seep into the earth.

6.2.5 Sustainable Urban Agriculture

Vertical Gardens and Rooftop Farms Make use of urban areas to grow food while improving insulation and air quality.

Community Gardens Reduce food waste while fostering social cohesiveness and food security.

6.2.6 Ecosystem Restoration

Restoration of Urban Wetlands and Rivers Rejuvenate aquatic habitats and incorporate them into urban environments to support biodiversity and reduce flooding.

Coastal Resilience Measures In coastal areas, use mangroves and other natural barriers to lessen the effects of urbanization.

6.2.7 Circular Economy in Waste Management

Organic Waste to Compost Convert organic waste from cities into compost for use in farming.

Biogas Production Utilize garbage that is biodegradable to generate renewable energy.

Ecological Industrial Symbiosis To reduce waste, encourage industries to share resources.

6.2.8 Air Quality Improvement

Urban Tree Planting Initiatives Increase tree cover strategically to mitigate the urban heat island effect and absorb pollutants.

Moss Walls and Biofilters Enhance indoor and outdoor air quality by filtering air pollutants using moss and other plants.

6.2.9 Regenerative Urban Design

Nature-Inspired Urban Layouts For effective resource management, include biomimetic designs that mimic natural processes.

Living Buildings Build buildings that incorporate water recycling, sustainable energy sources, and green roofs.

6.2.10 Climate Adaptation Strategies

Cooling Corridors Incorporate greenery to help cool cities, particularly during hot weather.

Floodplain Restoration Restoring natural floodplains will help control urban flooding.

6.2.11 Biodiversity Conservation

Urban Wildlife Habitats Incorporate habitats such as frog ponds, bee hotels, and birdhouses into urban landscapes.

Green Connectivity Link green areas to promote genetic diversity and wildlife migration.

6.2.12 Integration of NBS with Technology

AI for Ecosystem Monitoring Make data-driven decisions about the performance of green infrastructure to maximize NBS.

Smart Green Systems Using Internet of things (IoT)-based waste management and irrigation systems to guarantee resource efficiency.

6.2.13 Evaluating the Contribution of Nature-Based Solutions to Urban Circularity

NBS's urban green infrastructure (UGI) has the potential to benefit urban areas in a number of ways, including mitigating the effects of climate change and environmental degradation. UGI can significantly aid in the implementation of any policies whose objectives may be accomplished by NBS since it uses natural solutions to provide economic, ecological, and social benefits. The United Nations General Initiative (UGI) is a network of natural and human-managed ecosystems that maintain and enhance the advantages of ecosystems, increase biodiversity, and benefit human populations (Štrbac et al. 2023).

6.3 Future Implications of Nature-Based Solutions in Urban Circular Systems

In line with the circularity principles, nature-based solutions (NBS) have become a revolutionary way to solve urban problems. They are essential to the future of urban development because of their role in promoting resilience, sustainability, and regeneration.

6.3.1 Reshaping Urban Metabolism

The movement of materials and energy through cities is referred to as urban metabolism. Because NBS promote regenerative processes and close resource loops, they have the ability to drastically change this system (Fig. 6.11).

Resource Recovery NBS can improve the transformation of waste into resources. In addition to recycling wastewater, urban wetlands and artificial ecosystems can provide biomass and biogas for fertilizer or energy.

Decentralized Systems By serving as decentralized rainwater gathering hubs, urban trees and green roofs lessen reliance on centralized water infrastructure and mitigate urban flooding (Cities and Circular Economy for Food 2019). Future developments in urban metabolism may enable cities to operate as self-sufficient systems with smooth integration of biological and technology resource flows.

6.3.2 Climate Change Mitigation and Adaptation

By giving cities, the resources for both adaptation and mitigation, NBS works as a natural buffer against the effects of climate change.

Carbon Sequestration Wetland restoration, green walls, and urban forests can all help achieve global climate goals by storing large amounts of carbon (Cohen-Shacham et al. 2016) (Fig. 6.12).

Resilience to Extreme Weather By lowering the risk of flooding during extreme weather events and eliminating urban heat islands, green infrastructure improves resilience.

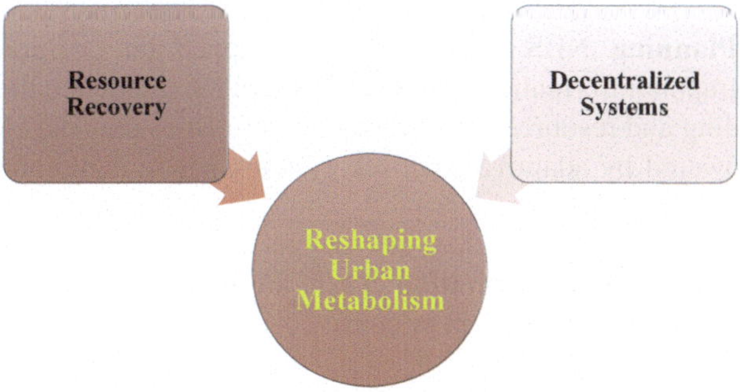

Fig. 6.11 Reshaping urban metabolism

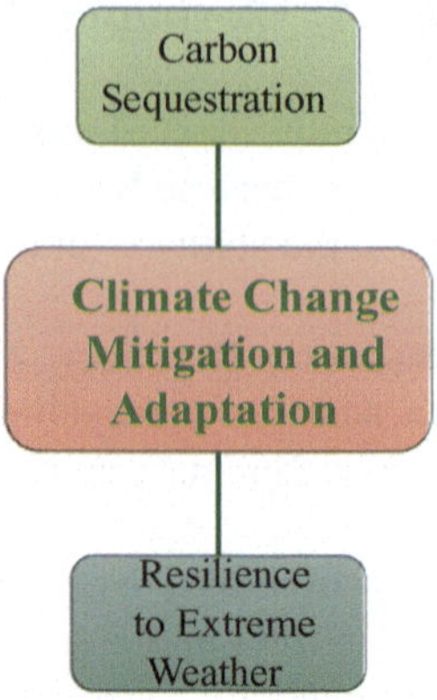

Fig. 6.12 Climate change mitigation and adaptation

In order for cities to prosper in the face of shifting environmental conditions, ecological engineering and urban design will become more and more important in the future of climate adaption.

6.3.3 Technological Integration with NBS

The next generation of NBS is being shaped by the intersection of environment and technology. Better planning, monitoring, and optimization of these solutions are made possible by cutting-edge technologies like artificial intelligence (AI), remote sensing, and the Internet of Things (IoT) (Fig. 6.13).

AI-Enabled Ecosystem Modeling AI can forecast the results of NBS actions, model ecological dynamics, and maximize the efficiency of designs.

Smart Green Infrastructure Real-time management is made possible by sensors installed in urban forests or green roofs that can track environmental variables including temperature, soil moisture, and air quality.

Data-Driven Planning NBS can be precisely mapped and positioned thanks to advanced geographic information systems (GIS), which guarantee their efficacy in resilience building and resource optimization. Future cities may develop into living laboratories powered by adaptive feedback systems where nature and technology coexist peacefully.

Economic and Social Transformation There are important social and economic impacts of NBS integration in urban systems:

Fig. 6.13 Technological integration with NBS

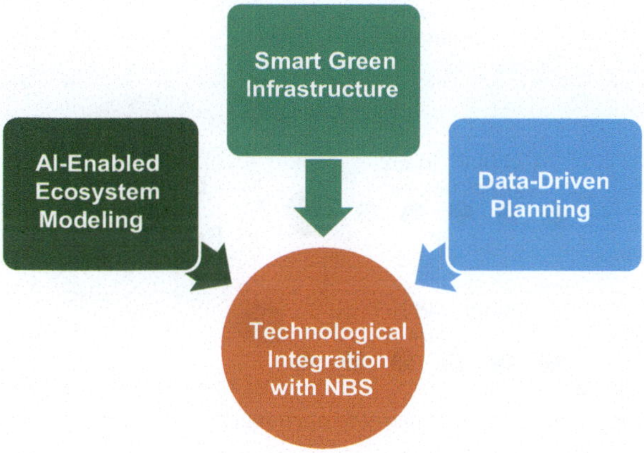

Fig. 6.14 Economic and social transformation

Job Creation and Green Economy: Employment in fields like ecological restoration, urban farming, and green infrastructure upkeep can result from the widespread use of NBS.

Community Engagement: NBS promotes participatory government, in which communities actively manage urban resources, by encouraging local stewardship (Cities and Circular Economy for Food 2019).

Cost-Effectiveness: NBS frequently involves less capital investment than typical gray infrastructure while providing multipurpose advantages like enhanced biodiversity, recreation, and health (Cohen-Shacham et al. 2016). NBS will probably play a key role in inclusive urban development in the future, bridging the gap between socioeconomic fairness and ecological sustainability (Fig. 6.14).

6.3.4 Urban Circularity Through Regenerative Design

The future of sustainable cities is represented by regenerative urban design, which is led by NBS:

Biomimicry in Urban Planning: Cities can be planned to mimic biological processes like energy flow and nutrient cycle by taking inspiration from natural ecosystems.

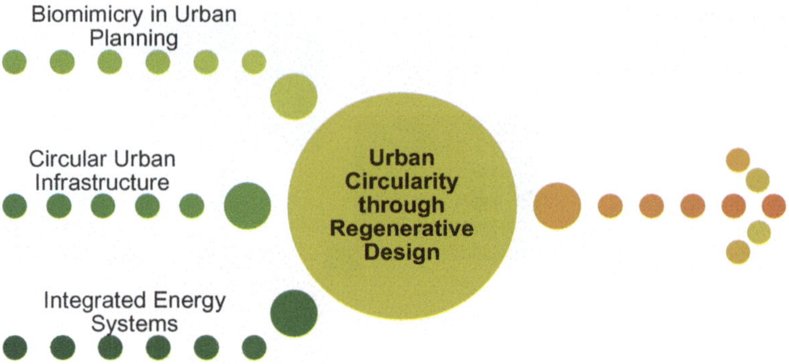

Fig. 6.15 Urban circularity through regenerative design

Circular Urban Infrastructure: In order to support a circular economy, infrastructure such as eco-industrial parks can use NBS to transform waste streams into useful resources.

Integrated Energy Systems: By incorporating energy resilience into urban circularity, NBS can support the production of renewable energy, such as bioenergy from algal systems (Fig. 6.15).

6.3.5 Emerging Research and Innovations

NBS's future will be fueled by innovative and cutting-edge research. Regenerative methods that maintain and improve the health of urban ecosystems will be given top priority in future urban planning.

Genetically Modified Organisms (GMOs) GMO research for ecosystem services and urban agriculture may increase NBS's efficacy.

Synthetic Biology In urban settings, combining synthetic biology and NBS can improve processes like carbon capture and bioremediation (Cohen-Shacham et al. 2016).

Circular Materials Circularity will be included into urban economies through the development of sustainable building and manufacturing materials from urban biomass. To guarantee that these technologies are in line with sustainability principles, governance frameworks and ethical concerns will be necessary (Fig. 6.16).

6.3.6 Transdisciplinary Collaboration

Collaboration between disciplines, sectors, and stakeholders is essential for the effective implementation of NBS:

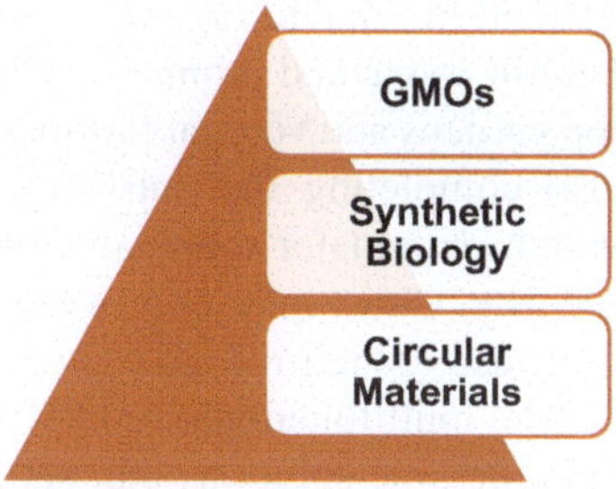

Fig. 6.16 Emerging research and innovations

Policy and Governance: Interagency cooperation and robust governance structures are necessary for incorporating NBS into urban policy frameworks (Cities and Circular Economy for Food 2019).

Public-Private Partnerships: Governments, corporations, and communities working together can pool resources and knowledge to scale NBS solutions (Cohen-Shacham et al. 2016).

Knowledge Sharing: The sharing of creative solutions and best practices for urban sustainability can be facilitated by global networks and knowledge-sharing platforms. The combined knowledge and efforts of various stakeholders will be the lifeblood of future urban systems.

6.4 Conclusions

Urban circularity is a crucial strategy for building resilient, sustainable cities in the face of growing environmental issues, and fast urbanization. Maintaining a closed-loop system in cities where resources are reused, waste is minimized, and environmental consequences are decreased is the goal of urban circularity. This framework's incorporation of nature-based solutions (NBS) offers a practical way to address several urban issues at once while promoting social, economic, and environmental sustainability.

Green roofs, bioswales, urban wetlands, and artificial wetlands are examples of nature-based solutions that complement natural processes to improve and repair urban ecosystems. NBS is essential to the management of water resources because it replicates the natural hydrological cycle. Permeable pavements and urban wetlands are two examples of solutions that can improve water quality and mitigate flood risks by capturing precipitation, reducing surface runoff, and improving groundwater recharge. Beyond regulating stormwater, NBS also helps treat and repurpose wastewater as part of their role in water management. For instance, constructed wetlands can be used to recycle treated water for nonpotable or agricultural purposes in addition to purifying water by eliminating impurities. These interventions significantly alleviate pressure on centralized treatment systems, supporting a more decentralized and efficient water infrastructure.

In a similar vein, NBS promotes nutrient recovery and reuse, which is crucial for creating an urban circular food system. Initiatives for composting and urban agriculture enable organic waste to be converted into useful nutrients, lowering landfill usage and promoting regional food production. As instances of NBS, rooftop gardens and vertical farming offer chances to cultivate food in urban areas while also completing the nutrient cycle. This reduces food transportation expenses, which are a major source of carbon emissions, in addition to addressing urban food security. Cities may improve their sustainability by cutting waste and making the most use of their resources by incorporating NBS into their urban environments.

The multifunctionality of NBS is one of its strongest points. NBS offers affordable, flexible solutions that support social and environmental advantages by tackling multiple urban issues at once. Green infrastructure, for example, serves as a tool for climate adaptation while simultaneously enhancing urban aesthetics, fostering biodiversity, and offering inhabitants recreational opportunities. By lowering the urban heat island effect, improving air quality, and encouraging carbon sequestration, these strategies can greatly increase urban resilience to climate change, especially when applied widely.

NBS improves urban residents' quality of life in addition to the environment. It has been demonstrated that integrating nature into urban areas enhances social cohesion, lowers stress levels, and improves mental health. In this way, NBS benefits citizens' physical and mental health in addition to enhancing the ecological health of cities. The relationship between human health and environmental sustainability highlights the comprehensive advantages of NBS and supports its potential as a crucial urban development approach in the future.

NBS can be effectively incorporated into urban planning and design to encourage urban circularity, as shown in places like Rotterdam, Singapore, and Copenhagen (Calisto Friant et al. 2023). These cities have demonstrated that using natural solutions to achieve long-term urban sustainability is not only an idealistic strategy but also a workable and efficient one. Cities can transition to more regenerative systems—where resources are continuously recycled back into the economy, waste is reduced, and environmental effects are lessened—by adopting NBS.

To sum up, nature-based solutions are a revolutionary strategy for promoting urban circularity. Cities may improve their resistance to climate change and address important urban issues like water management, waste treatment, material recovery, and food production by collaborating with nature rather than fighting it. Urban areas may become more self-sufficient, flexible, and sustainable with the integration of NBS, helping to create a more circular and regenerative future for everybody.

References

Alikhani S, Nummi P, Ojala A (2021) Urban wetlands: a review on ecological and cultural values. Water 13(22):22. https://doi.org/10.3390/w13223301

Atanasova N, Castellar JAC, Pineda-Martos R, Nika CE, Katsou E, Istenič D, Pucher B, Andreucci MB, Langergraber G (2021) Nature-based solutions and circularity in cities. Circ Econ Sustain 1(1):319–332. https://doi.org/10.1007/s43615-021-00024-1

Berretta C, Poë S, Stovin V (2014) Reprint of 'Moisture content behaviour in extensive green roofs during dry periods: the influence of vegetation and substrate characteristics'. J Hydrol 516:37–49. https://doi.org/10.1016/j.jhydrol.2014.04.001

Boyer T, Polasky S (2004) Valuing urban wetlands: a review of non-market valuation studies. Wetlands 24(4):744–755. https://doi.org/10.1672/0277-5212(2004)024[0744:VUWARO]2.0.CO;2

Calisto Friant M, Vermeulen WJV, Salomone R (2020) A typology of circular economy discourses: navigating the diverse visions of a contested paradigm. Resour Conserv Recycl 161:104917. https://doi.org/10.1016/j.resconrec.2020.104917

Calisto Friant M, Reid K, Boesler P, Vermeulen WJV, Salomone R (2023) Sustainable circular cities? Analysing urban circular economy policies in Amsterdam, Glasgow, and Copenhagen. Local Environ 28(10):1331–1369. https://doi.org/10.1080/13549839.2023.2206643

Cities and Circular Economy for Food (2019, January 1). https://www.ellenmacarthurfoundation.org/cities-and-circular-economy-for-food

Cohen-Shacham E, Walters G, Janzen C, Maginnis S (eds) (2016) Nature-based solutions to address global societal challenges. IUCN International Union for Conservation of Nature. https://doi.org/10.2305/IUCN.CH.2016.13.en

Communication from the commission to the European Parliament, The Council, The European Economic and Social Committee, and The Committee of the regions Green Infrastructure (GI)—Enhancing Europe's Natural Capital (2013). https://eur-lex.europa.eu/legal-content/EN/TXT/?uri=CELEX:52013DC0249

da Cunha JAC, Arias CA, Carvalho P, Rysulova M, Canals JM, Pérez G, Gonzalez MB, Morató JF (2018) "WETWALL"—an innovative design concept for the treatment of wastewater at an urban scale. Desalin Water Treat 109:205–220. https://doi.org/10.5004/dwt.2018.22143

Das A, Basu T (2020) Assessment of *peri*-urban wetland ecological degradation through importance-performance analysis (IPA): a study on Chatra wetland, India. Ecol Indic 114:106274. https://doi.org/10.1016/j.ecolind.2020.106274

Dawson L (2019) 'Our waste, our resources; a strategy for England'– switching to a circular economy through the use of extended producer responsibility. Environmental Law Review 21(3):210–218. https://doi.org/10.1177/1461452919851943

Enabling Circular Economy in Used Water Management in India. (2024, February 28). CEEW. https://www.ceew.in/publications/municipal-used-water-management-index-for-enabling-circular-economy-in-urban-wastewater-treatment

Ensuring Water Secure Cities (2024, October 17) CEEW. https://www.ceew.in/publications/ensuring-water-secure-cities

Escobedo FJ, Giannico V, Jim CY, Sanesi G, Lafortezza R (2019) Urban forests, ecosystem services, green infrastructure and nature-based solutions: Nexus or evolving metaphors? Urban For Urban Green 37:3–12. https://doi.org/10.1016/j.ufug.2018.02.011

Fiksel J, Sanjay P, Raman K (2021) Steps toward a resilient circular economy in India. Clean Techn Environ Policy 23(1):203–218. https://doi.org/10.1007/s10098-020-01982-0

Fowdar HS, Hatt BE, Breen P, Cook PLM, Deletic A (2017) Designing living walls for greywater treatment. Water Res 110:218–232. https://doi.org/10.1016/j.watres.2016.12.018

Geissdoerfer M, Savaget P, Bocken NMP, Hultink EJ (2017) The circular economy—a new sustainability paradigm? J Clean Prod 143:757–768. https://doi.org/10.1016/j.jclepro.2016.12.048

Gravagnuolo A, Angrisano M, Fusco Girard L (2019) Circular economy strategies in eight historic port cities: criteria and indicators towards a circular city assessment framework. Sustain For 11(13):Article 13. https://doi.org/10.3390/su11133512

Haase D (2021) Integrating ecosystem services, green infrastructure and nature-based solutions—new perspectives in sustainable urban land management. In: Weith T, Barkmann T, Gaasch N, Rogga S, Strauß C, Zscheischler J (eds) Sustainable land management in a European Context: a co-design approach. Springer International Publishing, pp 305–318. https://doi.org/10.1007/978-3-030-50841-8_16

Kirchherr J, Reike D, Hekkert M (2017) Conceptualizing the circular economy: an analysis of 114 definitions. Resour Conserv Recycl 127:221–232. https://doi.org/10.1016/j.resconrec.2017.09.005

Korhonen J, Nuur C, Feldmann A, Birkie SE (2018) Circular economy as an essentially contested concept. J Clean Prod 175:544–552. https://doi.org/10.1016/j.jclepro.2017.12.111

Langergraber G, Pucher B, Simperler L, Kisser J, Katsou E, Buehler D, Mateo MCG, Atanasova N (2020) Implementing nature-based solutions for creating a resourceful circular city. Blue-Green Systems 2(1):173–185. https://doi.org/10.2166/bgs.2020.933

Langergraber G, Castellar JAC, Pucher B, Baganz GFM, Milosevic D, Andreucci MB, Kearney K, Pineda-Martos R, Atanasova N (2021) A framework for addressing circularity challenges in cities with nature-based solutions. Water 13(17):17. https://doi.org/10.3390/w13172355

Liu Z, Cheng W, Jim CY, Morakinyo TE, Shi Y, Ng E (2021) Heat mitigation benefits of urban green and blue infrastructures: a systematic review of modeling techniques, validation and scenario simulation in ENVI-met V4. Build Environ 200:107939. https://doi.org/10.1016/j.buildenv.2021.107939

Marchesi M, Tweed C (2021) Social innovation for a circular economy in social housing. Sustain Cities Soc 71:102925. https://doi.org/10.1016/j.scs.2021.102925

Marselle MR, Lindley SJ, Cook PA, Bonn A (2021) Biodiversity and health in the urban environment. Curr Environ Health Reports 8(2):146–156. https://doi.org/10.1007/s40572-021-00313-9

Masi F, Bresciani R, Rizzo A, Edathoot A, Patwardhan N, Panse D, Langergraber G (2016) Green walls for greywater treatment and recycling in dense urban areas: a case-study in Pune. J Water Sanit Hyg Dev 6(2):342–347. https://doi.org/10.2166/washdev.2016.019

Mitsch WJ, Bernal B, Nahlik AM, Mander Ü, Zhang L, Anderson CJ, Jørgensen SE, Brix H (2013) Wetlands, carbon, and climate change. Landsc Ecol 28(4):583–597. https://doi.org/10.1007/s10980-012-9758-8

Murray A, Skene K, Haynes K (2017) The circular economy: an interdisciplinary exploration of the concept and application in a global context. J Bus Ethics 140(3):369–380. https://doi.org/10.1007/s10551-015-2693-2

Negrete-Cardoso M, Rosano-Ortega G, Álvarez-Aros EL, Tavera-Cortés ME, Vega-Lebrún CA, Sánchez-Ruíz FJ (2022) Circular economy strategy and waste management: a bibliometric analysis in its contribution to sustainable development, toward a post-COVID-19 era. Environ Sci Pollut Res Int 29(41):61729–61746. https://doi.org/10.1007/s11356-022-18703-3

Nika CE, Gusmaroli L, Ghafourian M, Atanasova N, Buttiglieri G, Katsou E (2020) Nature-based solutions as enablers of circularity in water systems: a review on assessment methodologies, tools and indicators. Water Res 183:115988. https://doi.org/10.1016/j.watres.2020.115988

Paiho S, Mäki E, Wessberg N, Paavola M, Tuominen P, Antikainen M, Heikkilä J, Rozado CA, Jung N (2020) Towards circular cities—conceptualizing core aspects. Sustain Cities Soc 59:102143. https://doi.org/10.1016/j.scs.2020.102143

Pauleit S, Zölch T, Hansen R, Randrup TB, Konijnendijk van den Bosch C (2017) Nature-based solutions and climate change—four shades of green. In: Kabisch N, Korn H, Stadler J, Bonn A (eds) Nature-based solutions to climate change adaptation in urban areas: linkages between science, policy and practice. Springer International Publishing, pp 29–49. https://doi.org/10.1007/978-3-319-56091-5_3

Pinto LV, Inácio M, Pereira P (2023) Green and blue infrastructure (GBI) and urban nature-based solutions (NbS) contribution to human and ecological well-being and health. Oxford Open Infrastructure and Health 1:ouad004. https://doi.org/10.1093/ooih/ouad004

Potting J, Hekkert M, Worrell E, Hanemaaijer A (2017) Circular economy: measuring innovation in the product chain

Quin A, Jaramillo F, Destouni G (2015) Dissecting the ecosystem service of large-scale pollutant retention: the role of wetlands and other landscape features. AMBIO44 44 Suppl 1:127–137. https://doi.org/10.1007/s13280-014-0594-8

Ramusino LC, Cortese M, Lennard Z (2017) Re-naturing the city: Nature4Cities project to elevate the concept of nature-based solutions. Proceedings 1(7):7. https://doi.org/10.3390/proceedings1070696

Raymond CM, Frantzeskaki N, Kabisch N, Berry P, Breil M, Nita MR, Geneletti D, Calfapietra C (2017) A framework for assessing and implementing the co-benefits of nature-based solutions in urban areas. Environ Sci Pol 77:15–24. https://doi.org/10.1016/j.envsci.2017.07.008

Ristić Trajković J, Krstić V, Milovanović A, Calheiros CSC, Ćujić M, Karanac M, Kazak JK, Di Lonardo S, Pineda-Martos R, Garcia Mateo MC, Milošević D, Milousi M, Niţă MR, Palermo SA, Piro P, Pirouz B, Siscan Z, Turco M, Vaccari M et al (2024) Moving towards a holistic approach to circular cities: obstacles and perspectives for implementation of nature-based solutions in Europe. Sustain For 16(16):16. https://doi.org/10.3390/su16167085

Sánchez Levoso A, Gasol CM, Martínez-Blanco J, Durany XG, Lehmann M, Gaya RF (2020) Methodological framework for the implementation of circular economy in urban systems. J Clean Prod 248:119227. https://doi.org/10.1016/j.jclepro.2019.119227

Štrbac S, Kašanin-Grubin M, Pezo L, Stojić N, Lončar B, Ćurčić L, Pucarević M (2023) Green infrastructure designed through nature-based solutions for sustainable urban development. Int J Environ Res Public Health 20(2):2. https://doi.org/10.3390/ijerph20021102

Temmerman S, Meire P, Bouma TJ, Herman PMJ, Ysebaert T, De Vriend HJ (2013) Ecosystem-based coastal defence in the face of global change. Nature 504(7478):79–83. https://doi.org/10.1038/nature12859

Tortajada C (2024) Reused water as a source of clean water and energy. Nature Water 2(2):102–103. https://doi.org/10.1038/s44221-024-00198-6

Tsaligopoulos A, Kyvelou SS, Chiotinis M, Karapostoli A, Klontza EE, Lekkas DF, Matsinos YG (2022) The sound of a circular city: towards a circularity-driven quietness. Int J Environ Res Public Health 19(19):19. https://doi.org/10.3390/ijerph191912290

Vanhuyse F (2024) The urban circularity assessment framework (UCAF): a framework for planning, monitoring, evaluation, and learning from CE transitions in cities. Circ Econ Sustain 4(2):1069–1092. https://doi.org/10.1007/s43615-023-00314-w

Verhoeven JTA, Arheimer B, Yin C, Hefting MM (2006) Regional and global concerns over wetlands and water quality. Trends Ecol Evol 21(2):96–103. https://doi.org/10.1016/j.tree.2005.11.015

Williams J (2019) Circular cities. Urban Stud 56(13):2746–2762. https://doi.org/10.1177/0042098018806133

Williams J (2020) The role of spatial planning in transitioning to circular urban development. Urban Geogr 41:915. https://doi.org/10.1080/02723638.2020.1796042. https://www.tandfonline.com/doi/abs/

Williams J (2021) Circular cities: what are the benefits of circular development? Sustainability 13(10):10. https://doi.org/10.3390/su13105725

Phyco-application for the Treatment of Industrial Wastewater

7

Gholamareza Abdi, Mukul Machhindra Barwant, Maheswari Behera, and Lakshmi Singh

Abstract

Algae have been used for fuels and wastewater treatment, among other things. The potential uses of phycoremediation have led researchers to focus more on developing algal-based technologies in order to better use their capacity for bioremediation and the production of by-products. Numerous algal species have been investigated and used in wastewater treatment. Phycoremediation is the method of employing algae to remove heavy metals and nutrients from sewage wastewater and industrial effluent. The fact that it is simple to grow, modify, and develop in an in vitro environment makes it useful for remediation research. Algae have been used for many purposes, such as the production of fuel, medications, and wastewater treatment. Phycoremediation is the process of utilizing algae to remove pollutants from wastewater, such as fertilizers, heavy metals, and other contaminants. This is done because algal species are often easy to grow, adapt, and control in an in vitro environment, making them seem like excellent organisms to use in remediation research. Phycoremediation is the process of using algae to biotransform or remove contaminants from wastewater.

G. Abdi (✉)
Department of Biotechnology, Persian Gulf Research Institute, Persian Gulf University, Bushehr, Iran
e-mail: abdi@pgu.ac.ir

M. M. Barwant
Department Botany, Sanjivani Rural Education Society, Sanjivani Arts Commerce and Science College Kopargaon, Ahmednagar, Maharashtra, India

Department of Botany, YashavantraoChavan Institute of Science, Satara, Maharashtra, India

M. Behera · L. Singh
Department of Botany, College of Basic Science and Humanities, Odisha University of Agriculture and Technology, Bhubaneswar, Odisha, India

N. Dwivedi et al. (eds.), *Wastewater Treatment Through Nature-Based Solutions*,
https://doi.org/10.1007/978-981-96-6699-7_7

The use of this technology offers an alternative to the current practice of treating industrial effluents utilizing physical and chemical processes, among other conventional approaches.

Keywords

Phycoremediation · Macroalgae · Microalgae · Wastewater Treatment

7.1 Introduction

Algae, derived from the Greek word "phyco" meaning "alga," are diverse aquatic organisms, primarily classified as eukaryotic, ranging from single cells to highly differentiated plants. Because they make up the base of the food chain and account for over 50% of all photosynthetic activity, algae are vital to the planet. Algae release carbon dioxide, which they can use for self-growth, and produce oxygen, which is used by organic materials to break down. The rate of photosynthesis in algae decreases in the absence of sunshine, and they also play a less significant role in extracting nutrients from organic sources. The term "phycotechnology" refers to the use of algae, such as cyanobacteria, microalgae, and macroalgae, in industrial processes and products (Bansal et al. 2018). In today's world, we are facing numerous economic and environmental challenges, with concerns such as the depletion of conventional energy sources, global warming, and water pollution being particularly pressing. These issues impact society in various ways. Although there are alternative solutions to fossil fuels, they come with certain limitations. Consequently, there is a significant global emphasis on producing renewable biofuels. Biodiesel stands out as a viable solution, as it does not contribute to the net accumulation of carbon dioxide (CO_2) in the atmosphere (Singh et al. 2016). The rapid pace of industrialization, while pivotal for economic growth, has resulted in the generation of large volumes of wastewater, often laden with harmful pollutants, including heavy metals, dyes, organic compounds, and nutrients such as nitrogen and phosphorus. Traditional methods for treating industrial wastewater, such as chemical treatments and filtration systems, are often energy-intensive, costly, and environmentally unsustainable. In response to these limitations, there is a growing interest in nature-based solutions for wastewater treatment, particularly the use of microalgae, commonly referred to as "phyco-application," as an effective and sustainable alternative. Microalgae offer a promising solution due to their ability to grow in nutrient-rich environments and their capacity to remove various contaminants from wastewater. These photosynthetic organisms can simultaneously treat wastewater and provide a source for valuable bioproducts such as biofuels, bioplastics, and fertilizers. The application of microalgae in wastewater treatment processes, therefore, presents an intersection of environmental protection and resource recovery, making it an attractive technology for sustainable development.

The globe is currently dealing with a number of economic and environmental issues, the most pressing of which are water contamination, global warming, and conventional energy depletion. The entire society is being impacted by these issues in many ways. Although there are numerous alternatives to fossil fuels, their drawbacks have led to a global emphasis on the generation of renewable biofuels. Because it does not add to the net buildup of carbon dioxide in today's world, we are facing numerous economic and environmental challenges, with concerns such as the depletion of conventional energy sources, global warming, and water pollution being particularly pressing. These issues impact society in various ways. Although there are alternative solutions to fossil fuels, they come with certain limitations. Consequently, there is a significant global emphasis on producing renewable biofuels. Biodiesel stands out as a viable solution, as it does not contribute to the net accumulation of CO_2 in the atmosphere, biodiesel can be a good solution to these issues (Singh et al. 2016). The use of microalgae in wastewater treatment has its roots in the natural ability of these organisms to thrive in nutrient-rich aquatic environments. Microalgae, such as *Chlorella* and *Spirulina*, can effectively assimilate nitrogen, phosphorus, and other pollutants as nutrients, thus cleaning the water through a process known as bioremediation. This process is powered by sunlight, enabling microalgae to utilize photosynthesis to produce oxygen, which in turn supports the breakdown of organic matter by aerobic bacteria in the wastewater. The concept of phyco-application gained momentum in the late twentieth century as an eco-friendly approach to managing nutrient pollution in water bodies affected by agricultural runoff and industrial waste. In recent decades, advances in biotechnology and environmental engineering have expanded the potential of microalgae-based treatment systems. Research has demonstrated that microalgae can remove heavy metals, pharmaceuticals, and other persistent organic pollutants from wastewater, making them suitable for a wide range of industrial applications. Furthermore, the biomass produced during the treatment process can be harvested and converted into high-value products such as biodiesel, animal feed, and biochemicals, creating a circular economy model.

To address these challenges, several innovative approaches have been proposed and tested. One promising solution is the development of photobioreactors, closed-loop systems that optimize light exposure, nutrient supply, and growth conditions for microalgae. These reactors can be used in conjunction with wastewater treatment plants to enhance the efficiency and consistency of the phyco-application process. Additionally, genetic engineering of microalgae strains has the potential to improve their resistance to toxins and increase their pollutant uptake rates, making them more effective in treating industrial wastewater. Hybrid treatment systems that combine microalgae with other biological or chemical processes have also shown promise. For example, coupling microalgae with anaerobic digestion can enhance the removal of organic pollutants and provide a source of renewable energy through biogas production. Furthermore, advances in automation and sensor technologies can help monitor and optimize microalgae cultivation and wastewater treatment in real-time, reducing operational costs and improving scalability.

Wastewater contains a high concentration of nutrients, and the presence of human waste can result in excessive nutrient levels in nearby water bodies, leading to eutrophication or algae blooms. Combining algae cultivation with existing industrial or wastewater treatment processes, as well as addressing flue gas emissions, could be an effective strategy for reducing costs and energy consumption in biofuel production. For instance, co-locating algal operations with power plants or wastewater treatment facilities could enable the use of feedstocks, waste streams, and industrial flue gases. This kind of industrial symbiosis can provide beneficial remedial services and significantly reduce economic and environmental costs. Numerous research have been conducted that demonstrate the numerous waste-drawn alternatives that can be used in place of traditional feedstock for algae development (Singh et al. 2016). Phyco-remediation, the term for the use of microalgae in wastewater treatment, is one great way to put the circular economy idea to use. Among other pollutants, this procedure uses photosynthetic microalgae to remove pathogens, heavy metals, and organic pollutants from wastewater. This entire process is carried out concurrently with the development of biomass or high-value products, such biofuels. Consequently, it achieves the goal of a circular economy and produces a profit. Nitrogen and phosphorus are just two of the many substances that microalgae absorb. These components are then utilized in the synthesis of ATP, nucleic acids, and proteins. Furthermore, heavy metals can be accumulated by microalgae by a variety of processes, including surface precipitation, ion exchange, adsorption, and absorption (López-Pacheco et al. 2021).

Despite the many advantages of using microalgae for wastewater treatment, several challenges hinder the widespread adoption of this technology. One of the primary challenges is the cost and scalability of microalgae cultivation systems. Large-scale cultivation requires significant space, infrastructure, and energy inputs, particularly in temperate climates where year-round sunlight is limited. Additionally, harvesting microalgae biomass from the treated water can be labor-intensive and costly, as it involves processes such as sedimentation, centrifugation, or filtration (Afkar et al. 2010). Another challenge is the variability of wastewater composition, which can impact the efficiency of microalgae treatment systems. Industrial effluents often contain high concentrations of toxic compounds, which can inhibit algal growth or reduce their pollutant removal capabilities. Developing strains of microalgae that are resistant to these conditions or optimizing treatment processes for specific types of wastewater remains an area of active research (Ahmad 2016). Furthermore, the integration of microalgae-based treatment systems into existing industrial infrastructure can be complex. Many industries are accustomed to using chemical treatment processes that provide rapid results, whereas phyco-application often requires longer treatment times due to the slower growth rates of algae. Overcoming these operational and logistical challenges is essential for scaling up microalgae-based technologies for widespread use in industrial wastewater treatment (Al-Homaidan et al. 2015).

A significant issue in the current situation is the lack of a suitable water supply for domestic, industrial, and drinking needs all throughout the planet. To overcome this obstacle, it is important to create a suitable wastewater treatment system. Drainage comprises two broad types of used water, such as wastewater from homes and effluent from industry. Utilizing treated wastewater can help with gardening and cleaning so that clean, fresh city water can be utilized for bathing, cooking, and drinking. The method applied to water affordable and environmentally sustainable treatment is a prerequisite (Chairat and Bremner 2016; Alqadami et al. 2018). Academics and companies are working to develop an inexpensive and environmentally friendly method due to the aforementioned shortcomings of the currently used procedures. The aforementioned shortcomings of the currently employed techniques have prompted researchers and industry players to devise an economical and ecologically sustainable solution for wastewater treatment. Bioremediation is the process of using naturally occurring organisms or techniques to get rid of environmental contaminants at a certain location. Water has been cleaned via bioremediation of soil, the atmosphere, or any other environmentally damaged area. Biological agents are frequently employed in environmental remediation and may originate from fungi, algae, or microbes (Bansal et al. 2018).

Phycoremediation and biodiesel production can be both cost-effective and environmentally friendly by integrating native algae cultivation with wastewater and industrial exhaust gases. Using wastewater as a nutrient source and flue gas as a carbon dioxide source significantly lowers the costs of producing biodiesel and treating wastewater and flue gases. To achieve large-scale, cost-effective algae farming, it is essential to make efficient use of all available resources (Singh et al. 2016). The future of phyco-application in wastewater treatment lies in the continued development of cost-effective and scalable systems that can be integrated into various industrial sectors. Research into improving microalgae strains through biotechnology, as well as optimizing photobioreactor designs, will be key to enhancing the efficiency of these systems. Additionally, the co-production of valuable bioproducts from algal biomass, such as biofuels, pharmaceuticals, and bioplastics, will help offset the costs of wastewater treatment and promote the circular economy. There is also a growing interest in the potential for phyco-application to address emerging contaminants, such as microplastics, pharmaceuticals, and endocrine-disrupting compounds, which are increasingly found in industrial wastewater. Developing microalgae-based technologies to target these pollutants could significantly contribute to reducing their environmental impact, the use of microalgae for industrial wastewater treatment offers a sustainable, eco-friendly alternative to conventional methods. While challenges remain, ongoing research and technological advancements hold great promise for expanding the application of this technology in the future (Dao and Beardall 2016; Chang 2019).

This chapter emphasizes the potential of phycoremediation as an alternative method for treating swine wastewater generated from hog meat production, reflecting the importance of the circular economy concept and the current prevalence of pork production. We also explored how the cultivation of microalgae during this process can yield high-value products or serve as feedstock for producing biodiesel, biohydrogen, biogas, and electricity. Previous reviews on swine wastewater treatment with microalgae overlooked these aspects. Phycoremediation, the use of microalgae in wastewater treatment, is an effective way to apply the circular economy concept. This method leverages photosynthetic microalgae to remove pathogens, heavy metals, and organic pollutants from wastewater.

7.2 Composition of Wastewater

Watercourses are polluted by numerous sources that differ in both intensity and volume. The composition of wastewater reflects the lifestyle and technologies used in the society that generates it. It consists of a complex blend of natural organic and inorganic materials, along with synthetic compounds. Pollutants originate from various sources, such as sewage from towns and villages, discharges from industries, agricultural runoff, and leachates from solid waste disposal sites. These pollution sources present challenges that require solutions. Water scarcity, along with the growing demand for energy and food, is pushing us to explore the potential of wastewater recycling and resource recovery (Ansari et al. 2019; Darda et al. 2019). Wastewater serves as an optimal environment for a variety of microorganisms, particularly bacteria, viruses, and protozoa. Although many of these microorganisms are harmless and play a beneficial role in biological sewage treatment, wastewater also harbors pathogenic microbes that are excreted in significant quantities by both infected individuals and asymptomatic carriers. Sewage can harbor bacteria responsible for diseases like cholera, typhoid, and tuberculosis; viruses causing infectious hepatitis; protozoa leading to dysentery; and eggs of parasitic worms. The effectiveness of sewage disinfection is typically measured by the reduction in total coliform organisms (Apandi et al. 2019; Furuhashi et al. 2019).

7.3 Conventional Method for Wastewater Treatment

In wastewater treatment, the primary objectives are to remove biochemical oxygen demand (BOD), suspended solids, nutrients, coliform bacteria, and toxins to achieve purified water. BOD reduction relies on microorganisms that oxidize organic matter into CO_2 and water, using oxygen as the oxidizing agent. Suspended solids are mainly removed through physical sedimentation. Nutrient removal, particularly dissolved nitrogen and phosphorus, has become an important step in treatment systems, as discharging these nutrients into sensitive water bodies can cause eutrophication by promoting the excessive growth of plants like algae and aquatic macrophytes. Currently, no single process can effectively achieve all these goals, so a combination of treatment methods is necessary (Khan et al. 2017; Mantzorou et al. 2018).

7.3.1 Preliminary Treatment

Preliminary treatment in wastewater plants removes large solid materials that could block the flow. These materials include floating objects like rags, wood, fecal matter, and heavier grit particles. Large floating debris is removed by passing the sewage through bars spaced 20–60 mm apart, with the trapped materials raked off regularly. Grit is separated by slowing the flow to a velocity where grit and silt settle, while organic matter remains suspended, typically at a speed of 0.2–0.4 m/s (Lahiri et al. 2017; Kwaansa-Ansah et al. 2019).

7.3.2 Primary Treatment

Once the coarse materials are removed, the sewage flows into sedimentation tanks, where gravity is used to separate settleable solids. A properly designed sedimentation tank can eliminate up to 40% of the biochemical oxygen demand by removing these solids. The effectiveness of pathogen removal during primary treatment varies, with different organisms showing different removal rates (Nowicka et al. 2016).

7.3.3 Secondary Treatment

The goal of secondary treatment is to lower the BOD by breaking down organic matter. This process is mainly carried out by a diverse group of heterotrophic bacteria that use the organic material for energy and growth. Various biological methods are available to achieve aerobic oxidation of BOD, and these can be categorized into two types: fixed film or dispersed growth processes. In fixed film reactors, biofilms develop on a stationary surface, where organic matter is absorbed and broken down aerobically. In contrast, suspended growth reactors involve microorganisms that are freely mixed with wastewater, kept suspended by mechanical stirring or air diffusers. Studies suggest that biological oxidation systems can remove over 90% of pathogenic bacteria from sewage, though viral removal varies significantly. The primary method for virus removal is believed to be adsorption. In suspended growth reactors, the close interaction between solid flocs and sewage achieves 90% viral removal, while the smaller surface area in fixed film reactors leads to more variable results (Poo et al. 2018; Priyadarshini et al. 2019; Yakamercan et al. 2021).

7.3.4 Tertiary Treatment

Tertiary treatment focuses on removing all remaining organic ions, using either biological or chemical methods. Biological treatment tends to be more efficient compared to chemical processes, which are generally more expensive and can result in secondary pollution. Each additional treatment stage significantly increases the overall cost of wastewater management. Advanced treatments, such as chemical precipitation, ozonation, reverse osmosis, and carbon adsorption, are technologically complex and often designed to target-specific nutrients. For smaller-scale improvements in effluent quality, removing fine particles through methods like lagoon storage, land application, or filtration using sand or gravel can ensure compliance with discharge standards. In many areas, primary and secondary treatments are used to remove easily settled materials and oxidize organic matter, producing a clear, seemingly clean effluent that is released into natural waters. However, this secondary effluent still contains inorganic nitrogen, phosphorus, and other long-lasting pollutants like heavy metals, contributing to eutrophication and other long-term environmental problems (Al Ketife et al. 2019; Gururani et al. 2021). Da Silvaet al. (2007) highlighted that natural purification processes, driven by chemoorganotrophic microorganisms, break down organic matter using both aerobic and anaerobic pathways. Modern biological wastewater treatment systems are based on these natural self-purification processes but are engineered to accelerate the treatment rate (Da Silvaet al. 2007).

7.3.5 Disinfection of Wastewater

Primary, secondary, and even tertiary wastewater treatments cannot eliminate 100% of the waste, leaving some microorganisms in the treated water. To reduce the risk of waterborne diseases and protect public health, regulatory authorities may mandate the elimination of harmful pathogens in wastewater. While many of these microorganisms are harmless, it is assumed that pathogens may be present. Therefore, reducing bacterial numbers is crucial, especially when treated wastewater is discharged into bodies of water used for drinking, swimming, or shellfishing. Disinfection refers to the process of destroying pathogens in the effluent, while sterilization is the destruction of all microorganisms. Wastewater treatment focuses on disinfection rather than sterilization, significantly reducing microbial levels to safe standards. Disinfection can be achieved through various physical or chemical methods, such as boiling, incineration, irradiation with X-rays or ultraviolet rays, or using strong acids, alcohols, and oxidizing agents (Sun et al. 2016; Azimi et al. 2017).

However, practical and cost-effective methods must be used for large-scale wastewater treatment. Chlorine has been the primary disinfectant used historically because of its effectiveness and relatively low cost. However, rising costs and concerns about chlorine's toxicity to fish and aquatic life, along with the potential formation of harmful chlorinated hydrocarbons, have made chlorination less favorable. As a result, ozone (ozonation) and ultraviolet (UV) light are becoming increasingly considered as alternatives for wastewater disinfection. Both ozone and UV light are effective, leave no toxic residue, and ozone also increases dissolved oxygen levels in water. Ozone generation, while more expensive initially, has recently become more economically competitive with chlorination. Similarly, studies are ongoing to assess UV light's effectiveness and cost-efficiency in large wastewater treatment facilities. Early findings suggest UV light is both effective and economically viable. Though both ozone and UV light show promise, chlorine remains widely used, and chlorination's principles and practices are still a central focus in wastewater disinfection (Cai et al. 2013; Leng et al. 2018; Bhatnagar et al. 2021).

7.4 Algae as an Alternative for Wastewater Treatment

Algae have proven to be an environmentally friendly and sustainable alternative to the energy-intensive and conventional biological treatment methods commonly used today. In addition to serving as a renewable source of biomass, the use of microalgae in wastewater treatment offers a cost-effective and practical approach for CO_2 bio-fixation. The rationale for using microalgae in wastewater treatment lies in their ability to utilize both organic and inorganic carbon, as well as inorganic nitrogen (N) and phosphorus (P), for growth, which leads to reductions in these substances in the water. A key advantage of integrating microalgae into wastewater treatment is the oxygen produced through photosynthesis, which is essential for heterotrophic bacteria to break down carbon-based materials. While the specific impact of algal culture on wastewater treatment can be difficult to measure, various studies have shown that algae can effectively support nutrient removal. Using algae granules in synthetic wastewater has proven highly efficient for phosphorus removal, with the added benefit of phosphorus recovery and reuse from the phosphorus-rich algae biomass. In addition to capturing CO_2 and removing nutrients, microalgae have shown potential for energy generation. They can use both organic and inorganic nitrogen, as well as nitrite and nitrate. Furthermore, wastewater treated with an algal-bacterial co-culture does not require multiple operational steps to remove inorganic nitrogen and phosphorus, simplifying the process and reducing energy use. This is because microalgae directly assimilate ammonia and phosphate for cell growth and metabolism. As a result, microalgae-based treatment processes emit fewer greenhouse gases (Du et al. 2018; Shahid et al. 2020; Song et al. 2022).

7.5 Algal-Based WasteWater Treatment

Algal cultures have been commercially utilized for approximately 75 years, mainly in wastewater treatment and large-scale production of species such as Chlorella and Dunaliella. They are particularly attractive for bioremediation because of their photosynthetic capacity to transform sunlight into biomass while absorbing nutrients that contribute to eutrophication. This concept was pioneered around 55 years ago in the United States by Oswald and Gotaas (1957) and has since been extensively studied worldwide (Oswald and Gotaas 1957). Palmer et al. (1981) conducted a survey of microalgal genera across various waste stabilization ponds, identifying Chlorella, Ankistrodesmus, Scenedesmus, Euglena, Chlamydomonas, Oscillatoria, Micractinium, and Golenkinia as the most common. Given that microalgal wastewater systems require significant land area, there is ongoing research into more space-efficient systems using hyperconcentrated algal cultures, which have shown remarkable effectiveness in rapidly removing nitrogen and phosphorus (Palmer et al. 1981). Traditionally used as a tertiary treatment, algae have also been proposed for secondary treatment. Algae can be employed in wastewater treatment for various purposes, including the removal of coliform bacteria, reduction of chemical and biochemical oxygen demand, removal of nitrogen and phosphorus, and elimination of heavy metals (Chu et al. 2021).

7.5.1 Removal of Coliform Bacteria

Moawad (2013) noted that environmental conditions favorable for algal growth tend to be unfavorable for the survival of coliform bacteria. Pathogens of concern in wastewater include bacteria such as *Salmonella* and *Shigella*, along with viruses and protozoa. Among these, bacteria make up the largest portion of the microbial community in all biological wastewater treatment processes. Research indicates that pathogenic bacteria typically have shorter survival times in the environment than coliforms, whereas viruses tend to persist longer (Moawad 2013). The success of sewage disinfection is commonly assessed by the degree of total coliform removal. In this context, sewage stabilization ponds are considered more effective than traditional treatment methods. Studies have demonstrated that stabilization ponds can significantly reduce coliform bacteria levels (Li et al. 2019).

7.5.2 Reduction of COD and BOD

As previously mentioned, wastewater contains various compounds and microorganisms that can pollute water bodies. Wastewater pollution can be classified into three main categories: organic materials, inorganic materials, and microbial content. Organic compounds in wastewater consist of a wide range of substances, all containing at least one carbon atom, which can be chemically or biologically oxidized to produce carbon dioxide. When biological oxidation is used, it is measured by the biochemical oxygen demand (BOD) test, while chemical oxidation is measured by the chemical oxygen demand (COD) test. In essence, BOD relies on microorganisms to oxidize organic matter into carbon dioxide and water, using molecular oxygen as the oxidizing agent. BOD is, therefore, a measure of the oxygen consumed by bacteria during the metabolism of organic matter in wastewater. Excessive BOD can reduce dissolved oxygen levels in receiving waters, leading to fish deaths and anaerobic conditions, making its removal a key goal in wastewater treatment (Hu et al. 2019; Lee et al. 2019; Saravanan et al. 2021). Colak and Kaya (1988) explored the potential of algae for biological wastewater treatment and found that domestic wastewater treatment achieved BOD and COD reductions of 68.4 and 67.2%, respectively (Colak and Kaya 1988).

7.5.3 Removal of N and/or P

The idea of using algae for wastewater treatment to remove nutrients like nitrogen and phosphorus while supplying oxygen for aerobic bacteria was first proposed by Oswald and Gotaas (1957) over 50 years ago. Since then, many laboratory and pilot studies have been conducted, and several wastewater treatment plants using different versions of this system have been built (Oswald and Gotaas 1957). Nitrogen in sewage effluent primarily comes from the metabolic conversion of externally derived compounds, while over 50% of phosphorus originates from synthetic detergents. The removal of these nutrients is referred to as nutrient stripping (Horan 2019). While wastewater is mainly treated through aerobic or anaerobic biological degradation, the treated water often still contains inorganic compounds which can lead to eutrophication in lakes and harmful algal blooms (Solimeno and García

2019). Sahu and Chaudhari (2013) identified phosphorus and nitrogen as key contributors to eutrophication. Therefore, additional treatment is needed to prevent eutrophication in sensitive water bodies, which can promote the overgrowth of undesirable plants like algae and aquatic macrophytes. Microalgal cultures offer a cost-effective solution for nutrient removal from wastewater. Microalgae have a strong capacity for inorganic nutrient uptake and can be cultivated in large-scale outdoor solar bioreactors (Sahu and Chaudhari 2013). Biological processes perform better compared to chemical and physical methods, which are often too expensive and may result in secondary pollution. Microalgal systems are particularly effective for tertiary and quaternary treatments, as they utilize inorganic nitrogen and phosphorus for growth and also remove heavy metals (Sutherland et al. 2018). Lau et al. (1996) studied the nutrient removal ability of *Chlorella vulgaris* in wastewater treatment and reported removal efficiencies of 86% for inorganic nitrogen and 78% for inorganic phosphorus (Lau et al. 1996). Interest in microalgal cultures arises from the limitations of conventional treatment methods, which include inconsistent nutrient removal efficiency, high operational costs, the risk of secondary pollution from chemical processes, and the loss of valuable nutrients like nitrogen and phosphorus. The final drawback is especially important, as conventional treatment methods lead to the incomplete use of natural resources. Numerous studies have shown that algal cultures are effective in removing nitrogen and phosphorus compounds from nutrient-rich wastewater (Foladori et al. 2018). Mohamed and Osman (2014) highlighted that *Scenedesmus* species are commonly found in various freshwater bodies and play a key role as primary producers, contributing to the purification of eutrophic waters. The presence or absence of specific *Scenedesmus* species can serve as an indicator of water quality (Mohamed and Osman 2014). To prevent nutrient recycling in receiving waters and recover the produced biomass, it is essential to harvest or physically retrieve the algal cells, though this poses significant technical and economic challenges. So far, most research has focused on planktonic and unicellular microalgae, which are challenging to collect. Using plants or microalgae for the removal of inorganic compounds offers the benefits of renewability and solar energy utilization. Under favorable conditions, cyanobacteria can grow faster than higher plants, making nutrient removal systems based on cyanobacteria a promising option (Zeraatkar et al. 2016; Arias et al. 2019).

Biological nitrogen removal is generally considered a viable option with several advantages over tertiary chemical and physicochemical treatments. De la Nouë and Basseres (1989) utilized cultures of *Phormidium bohneri* to remove nitrates from effluents produced after the anaerobic digestion of swine manure. Additionally, nitrogen removal from polluted waters using bench-top bioreactors with the thermophilic cyanobacterium *Phormidium laminosum* has been reported. The use of thermophilic cyanobacteria in wastewater treatment has benefits, as it avoids contamination due to the organism's tolerance for high temperatures, allowing treatment at elevated temperatures. *Phormidium* species cells were attached to chitosan flakes and used to remove nitrogen and orthophosphate from urban secondary effluents (De la Noüe and Basseres 1989). Over the past three decades, biological phosphorus removal processes have gained increasing attention. It is well known that in nitrogen-sufficient cyanobacterial cells, the uptake and reduction of nitrate is driven by photosynthesis, likely linked to the incorporation of ammonium into carbon skeletons. Phosphate uptake by cyanobacteria, characterized in several strains, follows a hyperbolic function in relation to external phosphate concentration. After the cyanobacteria absorb nutrients from the effluent, the cleansed water can be drained, and the cyanobacteria can be readily harvested. The harvested biomass offers potential commercial uses, such as extracting valuable pigments (Fang et al. 2017; Mishra et al. 2022).

7.5.4 Removal of Heavy Metals

Algae are well known for their ability to sequester heavy metals. As industrialization has progressed, the discharge of toxic pollutants into wastewater systems has increased. High concentrations of heavy metals and toxic organic compounds have been found in municipal wastewater, highlighting the importance of treatment systems' ability to manage and eliminate these harmful substances (Sousa et al. 2018). Microalgae are particularly effective at absorbing heavy metals, and their bioaccumulation of these metals presents a potential method for remediating metal-contaminated wastewater. A key advantage of algae is that they can be cultivated in ponds with minimal nutrient input and maintenance. While heavy metal concentrations in certain drainage systems may not be as high as those found in industrial effluents, particularly from metal processing industries, their presence remains a public concern, especially in densely populated regions. It is well established that several marine and freshwater algae can selectively absorb various heavy metals from water and accumulate them within their cells. Researchers have concluded that using algae to remove heavy metals from wastewater, along with separating the metal-saturated algae from the medium, is an economical method that produces high-quality reusable effluent water. Many algal species can sequester significant quantities of toxic heavy metal ions from water. Algal metal sequestration occurs through different mechanisms depending on the species of algae, the type of metal ion, solution conditions, and whether the algal cells are alive or dead. In living algal cells, trace metals like Co, Mo, Ca, Mg, Cu, Zn, Cr, Pb, and Se are accumulated intracellularly through active biological transport. Studies have also reported the accumulation of metals such as Cu^{2+}, Pb^{2+}, Cr^{3+}, Ni^{2+}, Cd^{2+}, Co^{2+}, Fe^{2+}, and Mn^{2+} by algae. For instance, algae in experimental rice paddies were found to concentrate Cd^{2+} up to 1000 times the ambient levels. Algae are also efficient at accumulating compounds such as organochlorides and tributyl tin and have been shown to degrade some of these compounds. Furthermore, algae have demonstrated the ability to break down hydrocarbons, such as those found in oily wastes (Santos et al. 2019; Zhang et al. 2019; Urrutia et al. 2019; Pradhan et al. 2019; Leong and Chang 2020; Shen et al. 2020).

7.5.5 Factors Affecting Algal Growth and Nutrient Removal

Algal growth and nutrient uptake are influenced not only by nutrient availability but also by complex interactions with physical factors like pH, light intensity, and temperature, as well as biotic factors. One key biotic factor affecting algal growth is the initial cell density. Generally, a higher algal density is expected to promote better growth and improve nutrient removal efficiency. However, excessively high algal density can lead to self-shading, the buildup of autoinhibitory substances, and a decline in photosynthetic efficiency (Di Pippo et al. 2012; Fica and Sims 2016; Mohsenpour et al. 2021).

7.6 Some Case Studies

Over the past decade, several companies, primarily from the United States, United Kingdom, and Australia, have begun exploring algae biomass production using wastewater as a resource. Although limited details are available about the specific microalgal species utilized by these companies, a general overview of their technological approaches can be provided. Algal Enterprises in Australia, for example, has developed a wastewater treatment plant that can handle various wastewater types, including municipal, industrial, and agricultural sources. Local algae species in a closed photobioreactor (PBR) system use photosynthetically active radiation as their primary energy source. The resulting algal biomass is then anaerobically co-digested to produce methane-rich biogas, which is subsequently converted into electricity (Sivasubramanian and Prasad 2024).

Microbio Engineering's RNEW® technology in the United States utilizes mechanically mixed, CO_2-gassed open raceway ponds to treat nitrogen- and phosphorus-rich municipal wastewater, producing biomass that serves as feedstock for biofuel production. According to predictions by Solimeno et al. (2017) algal biomass accounts for 58–68% of the total suspended solids in these ponds, while bacterial biomass constitutes 23–30%. In a subsequent study, the dominant species within the algae microbiome of these high-rate algal ponds were identified as *Micractinium*, *Scenedesmus*, *Chlorella*, and pennate diatoms (Solimeno et al. 2017).

Oswald Green Technologies has created the Advanced Integrated Wastewater Pond System (AIWPS®), also referred to as Energy Ponds™. This system utilizes a symbiotic consortium of bacteria and algae to address both organic and inorganic pollutants in municipal, agricultural, and industrial wastewater (Priyadharshini et al. 2021).

7.7 Limitations and Future Prospectives

The wastewater treatment process must effectively remove nutrients and produce biomass while ensuring the treated water is free from bacterial contamination. Laboratory tests often assess the effectiveness of the treatment process by evaluating the sanitary quality of the water. Given the limitations of conventional chemical disinfection methods—such as chlorination, ozonation, peracetic acid, and UV treatment—there is a growing need for more sustainable and cost-effective alternatives. Many studies have investigated the use of algae-based advanced treatment methods to reduce pathogens and coliforms in wastewater. Research has shown that abiotic factors, including pH, dissolved oxygen (DO), temperature, and sunlight, are crucial in removing pathogenic organisms from wastewater. High pH and DO levels can inhibit the growth of pathogens. Various strategies have been employed to prevent biological contamination in algal cultures, such as acidification, ozone treatment, UV radiation, removal of particulate matter, high ammonia concentration, and diurnal anaerobic conditions. While some of these methods are effective, they are often energy-intensive, increasing treatment costs. Additionally, issues such as inadequate agitation, rapid algae settlement, and nutrient limitations can lead to dead microalgal cells, which in turn can increase organic matter in the water and promote bacterial growth. Some researchers have noted that algae might sometimes lead to an increase in fecal coliforms, although the mechanism behind this is not well understood. Seasonal changes can also impact microbial diversity in open systems, affecting algal growth and treatment efficiency. However, larger ponds are increasingly prone to contamination, making it challenging to maintain them as contamination-free (Kaloudas et al. 2021; Krishnamoorthy and Manickam 2021; Ahmed et al. 2022).

Phycoremediation has emerged as a promising and cost-effective alternative to traditional wastewater treatment methods. However, further research is necessary to scale this technology for large-scale wastewater treatment. Inoculation involves using a suitable algal strain selected through screening methods or allowing an indigenous consortium to develop. The estimated biodiversity of algae ranges from 1 to ten million species. High-throughput screening techniques can be employed to identify various algal strains for potential use in treatment. Results from such screenings can inform the design of artificial consortia for testing and evaluating their wastewater treatment efficiency. One challenge is to create a consortium that can naturally evolve and adapt to changes in wastewater characteristics and environmental conditions. In recent decades, pollutants like nanoparticles, personal care products, pharmaceuticals, and illicit drugs have become increasingly concentrated in different types of wastewater. Although several advanced treatment methods have been employed to address these contaminants, they often remain in the treated effluent. Phycoremediation offers a cost-effective and environmentally friendly approach to target these specific micropollutants (Moondra et al. 2021; Koul et al. 2022).

7.8 Conclusion

Phycoremediation, utilizing both microalgae and macroalgae, offers a promising approach for treating industrial wastewater. It effectively addresses pollutants such as nutrients, heavy metals, and organic compounds, while also facilitating resource recovery. Macroalgae are advantageous due to their high biomass yield and ease of harvesting, making them ideal for large-scale applications and helping to mitigate eutrophication risks. Microalgae, on the other hand, grow faster and are adaptable to various wastewater compositions, allowing them to remove a wider range of pollutants, including pharmaceutical residues and heavy metals. However, challenges such as high operational costs and integration with existing infrastructure persist. Advances in biotechnology and system design, such as enhanced photobioreactors and genetic engineering, offer potential solutions to improve the scalability and economic feasibility of phycoremediation systems. With continued research and innovation, phycoremediation could become a key component in sustainable industrial wastewater treatment and resource recovery.

References

Afkar E, Ababna H, Fathi AA (2010) Toxicological response of the green alga Chlorella vulgaris, to some heavy metals. Am J Environ Sci 6(3):230

Ahmad P (2016) Plant metal interaction: emerging remediation techniques. Elsevier

Ahmed SF, Mofijur M, Parisa TA, Islam N, Kusumo F, Inayat A et al (2022) Progress and challenges of contaminate removal from wastewater using microalgae biomass. Chemosphere 286:131656

Al-Homaidan AA, Alabdullatif JA, Al-Hazzani AA, Al-Ghanayem AA, Alabbad AF (2015) Adsorptive removal of cadmium ions by Spirulina platensis dry biomass. Saudi J Biol Sci 22(6):795–800

Alqadami AA, Khan MA, Otero M, Siddiqui MR, Jeon B-H, Batoo KM (2018) A magnetic nanocomposite produced from camel bones for an efficient adsorption of toxic metals from water. J Clean Prod 178:293–304

Al Ketife AM, Almomani F, Muftah EN, Judd S (2019) A technoeconomic assessment of microalgal culture technology implementation for combined wastewater treatment and CO2 mitigation in the Arabian Gulf. Process Saf Environ Prot 127:90–102

Ansari FA, Ravindran B, Gupta SK, Nasr M, Rawat I, Bux F (2019) Techno-economic estimation of wastewater phycoremediation and environmental benefits using Scenedesmus obliquus microalgae. J Environ Manag 240:293–302

Apandi NM, Mohamed R, Al-Gheethi A, Kassim AHM (2019) Microalgal biomass production through phycoremediation of fresh market wastewater and potential applications as aquaculture feeds. Environ Sci Pollut Res Int 26:3226–3242

Arias DM, Rueda E, García-Galán MJ, Uggetti E, García J (2019) Selection of cyanobacteria over green algae in a photo-sequencing batch bioreactor fed with wastewater. Sci Total Environ 653:485–495

Azimi A, Azari A, Rezakazemi M, Ansarpour M (2017) Removal of heavy metals from industrial wastewaters: a review. Chem Bio Eng Reviews 4(1):37–59

Bansal A, Shinde O, Sarkar S (2018) Industrial wastewater treatment using phycoremediation technologies and co-production of value-added products. J Bioremed Biodegr 9(1):1–10

Bhatnagar P, Gururani P, Bisht B, Kumar V (2021) Algal biochar: an advance and sustainable method for wastewater treatment. Octa J Biosci 9(2)

Cai T, Park SY, Li Y (2013) Nutrient recovery from wastewater streams by microalgae: status and prospects. Renew Sust Energ Rev 19:360–369

Chairat M, Bremner JB (2016) Biosorption of lac dye by the red marine alga Gracilaria tenuistipitata: biosorption kinetics, isotherms, and thermodynamic parameters. Color Technol 132:472–480

Chang Y-C (2019) Microbial biodegradation of xenobiotic compounds. CRC Press, Boca Raton

Chu R, Li S, Zhu L, Yin Z, Hu D, Liu C, Mo F (2021) A review on co-cultivation of microalgae with filamentous fungi: efficient harvesting, wastewater treatment and biofuel production. Renew Sust Energ Rev 139:110689

Çolak Ö, Kaya Z (1988) A study on the possibilities of biological waste water treatment using algae

Dao LH, Beardall J (2016) Effects of lead on growth, photosynthetic characteristics and production of reactive oxygen species of two freshwater green algae. Chemosphere 147:420–429

Darda S, Papalas T, Zabaniotou A (2019) Biofuels journey in Europe: currently the way to low carbon economy sustainability is still a challenge. J Clean Prod 208:575–588

da Silva AK, Le Saux JC, Parnaudeau S, Pommepuy M, Elimelech M, Le Guyader FS (2007) Evaluation of removal of noroviruses during wastewater treatment, using real-time reverse transcription-PCR: different behaviors of genogroups I and II. Appl Environ Microbiol 73(24):7891–7897

De la Noüe J, Basseres A (1989) Biotreatment of anaerobically digested swine manure with microalgae. Biol Wastes 29(1):17–31

Di Pippo F, Ellwood NTW, Guzzon A, Siliato L, Micheletti E, De Philippis R, Albertano PB (2012) Effect of light and temperature on biomass, photosynthesis and capsular polysaccharides in cultured phototrophic biofilms. J Appl Phycol 24:211–220

Du ZY, Alvaro J, Hyden B, Zienkiewicz K, Benning N, Zienkiewicz A et al (2018) Enhancing oil production and harvest by combining the marine alga Nannochloropsisoceanica and the oleaginous fungus Mortierellaelongata. Biotechnol Biofuels 11:1–16

Fang Y, Hu Z, Zou Y, Zhang J, Zhu Z, Zhang J, Nie L (2017) Improving nitrogen utilization efficiency of aquaponics by introducing algal-bacterial consortia. Bioresour Technol 245:358–364

Fica ZT, Sims RC (2016) Algae-based biofilm productivity utilizing dairy wastewater: effects of temperature and organic carbon concentration. J Biol Eng 10:1–7

Foladori P, Petrini S, Nessenzia M, Andreottola G (2018) Enhanced nitrogen removal and energy saving in a microalgal–bacterial consortium treating real municipal wastewater. Water Sci Technol 78(1):174–182

Furuhashi Y, Honda R, Noguchi M, Hara-Yamamura H, Kobayashi S, Higashimine K, Hasegawa HJBEJ (2019) Optimum conditions of pH, temperature and preculture for biosorption of europium by microalgae Acutodesmusacuminatus. Biochem Eng J 143:58–64

Gururani P, Bhatnagar P, Bisht B, Kumar V, Joshi NC, Tomar MS, Pathak B (2021) Cold plasma technology: advanced and sustainable approach for wastewater treatment. Environ Sci Pollut Res 28:1–21

Horan NJ (2019) Potable water and wastewater treatment. In: Water resources. Routledge, pp 289–322

Hu X, Meneses YE, Stratton J, Wang B (2019) Acclimation of consortium of micro-algae help removal of organic pollutants from meat processing wastewater. J Clean Prod 214:95–102

Kaloudas D, Pavlova N, Penchovsky R (2021) Phycoremediation of wastewater by microalgae: a review. Environ Chem Lett 19:2905–2920

Khan S, Shamshad I, Waqas M, Nawab J, Ming L (2017) Remediating industrial wastewater containing potentially toxic elements with four freshwater algae. Ecol Eng 102:536–541

Koul B, Sharma K, Shah MP (2022) Phycoremediation: a sustainable alternative in wastewater treatment (WWT) regime. Environ Technol Innov 25:102040

Krishnamoorthy S, Manickam P (2021) Phycoremediation of industrial wastewater: challenges and prospects. Bioremediat Environ Sustain:99–123

Kwaansa-Ansah EE, Nti SO, Opoku F (2019) Heavy metals concentration and human health risk assessment in seven commercial fish species from Asafo market, Ghana. Food Sci Biotechnol 28:569–579

Lahiri S, Ghosh D, Bhakta JN (2017) Role of microbes in eco-remediation of perturbed aquatic ecosystem. In: Bhakta J (ed) Handbook of research on inventive bioremediation techniques. IGI Global, Hershey, pp 70–107

Lau PS, Tam NFY, Wong YS (1996) Wastewater nutrients removal by Chlorella vulgaris: optimization through acclimation. Environ Technol 17(2):183–189

Lee SA, Lee N, Oh HM, Ahn CY (2019) Enhanced and balanced microalgal wastewater treatment (COD, N, and P) by interval inoculation of activated sludge. J Microbiol Biotechnol 29:1434
Leng L, Li J, Wen Z, Zhou W (2018) Use of microalgae to recycle nutrients in aqueous phase derived from hydrothermal liquefaction process. Bioresour Technol 256:529–542
Leong YK, Chang JS (2020) Bioremediation of heavy metals using microalgae: recent advances and mechanisms. Bioresour Technol 303:122886
Li K, Liu Q, Fang F, Luo R, Lu Q, Zhou W et al (2019) Microalgae-based wastewater treatment for nutrients recovery: a review. Bioresour Technol 291:121934
López-Pacheco IY, Silva-Núñez A, García-Perez JS, Carrillo-Nieves D, Salinas-Salazar C, Castillo-Zacarías C et al (2021) Phyco-remediation of swine wastewater as a sustainable model based on circular economy. J Environ Manag 278:111534
Mantzorou A, Navakoudis E, Paschalidis K, Ververidis F (2018) Microalgae: a potential tool for remediating aquatic environments from toxic metals. Int J Environ Sci Technol 16:1815–1830
Mishra S, Singh V, Cheng L, Hussain A, Ormeci B (2022) Nitrogen removal from wastewater: a comprehensive review of biological nitrogen removal processes, critical operation parameters and bioreactor design. J Environ Chem Eng 10(3):107387
Moawad EY (2013) Purification of sewage water through the protection of the environment from radioactive contamination. Energy Environ Eng 1:55–61
Mohamed EHA, Osman AR (2014) Heavy metals concentration in water, muscles and gills of Oreochromisniloticus collected from the sewage-treated water and the White Nile. Int J Aquacult 4(6)
Mohsenpour SF, Hennige S, Willoughby N, Adeloye A, Gutierrez T (2021) Integrating micro-algae into wastewater treatment: a review. Sci Total Environ 752:142168
Moondra N, Jariwala ND, Christian RA (2021) Integrated approach of phycoremediation in waste-water treatment: an insight. WCM 5(1):8–12
Nowicka B, Pluciński B, Kuczyńska P, Kruk J (2016) Physiological characterization of Chlamydomonasreinhardtii acclimated to chronic stress induced by Ag, Cd, Cr, Cu and Hg ions. Ecotoxicol Environ Saf 130:133–145
Oswald WJ, Gotaas HB (1957) Photosynthesis in sewage treatment. Trans Am Soc Civ Eng 122(1):73–97
Palmer LL, Beyer AH, Stock J (1981) Biological oxidation of dissolved compounds in oilfield produced water by a field pilot biodisk. J Pet Technol 33(06):1136–1140
Poo K-M, Son E-B, Chang J-S, Ren X, Choi Y-J, Chae K-J (2018) Biochars derived from wasted marine macro-algae (Saccharina japonica and Sargassumfusiforme) and their potential for heavy metal removal in aqueous solution. J Environ Manag 206:364–372
Pradhan D, Sukla LB, Mishra BB, Devi N (2019) Biosorption for removal of hexavalent chromium using microalgae Scenedesmus sp. J Clean Prod 209:617–629
Priyadarshini E, Priyadarshini SS, Pradhan N (2019) Heavy metal resistance in algae and its application for metal nanoparticle synthesis. Appl Microbiol Biotechnol 7:1–20
Priyadharshini SD, Babu PS, Manikandan S, Subbaiya R, Govarthanan M, Karmegam N (2021) Phycoremediation of wastewater for pollutant removal: a green approach to environmental protection and long-term remediation. Environ Pollut 290:117989
Santos FM, Mazur LP, Mayer DA, Vilar VJ, Pires JC (2019) Inhibition effect of zinc, cadmium, and nickel ions in microalgal growth and nutrient uptake from water: an experimental approach. Chem Eng J 366:358–367
Sahu OP, Chaudhari PK (2013) Review on chemical treatment of industrial waste water. J Appl Sci Environ Manag 17(2):241–257
Saravanan A, Kumar PS, Varjani S, Jeevanantham S, Yaashikaa PR, Thamarai P et al (2021) A review on algal-bacterial symbiotic system for effective treatment of wastewater. Chemosphere 271:129540
Singh V, Tiwari A, Das M (2016) Phyco-remediation of industrial waste-water and flue gases with algal-diesel engenderment from micro-algae: a review. Fuel 173:90–97
Sivasubramanian V, Prasad MNV (2024) Phycoremediation and business prospects. In: Bioremediation and bioeconomy. Elsevier, pp 531–582

Shahid A, Malik S, Zhu H, Xu J, Nawaz MZ, Nawaz S et al (2020) Cultivating microalgae in wastewater for biomass production, pollutant removal, and atmospheric carbon mitigation; a review. Sci Total Environ 704:135303

Shen L, Wang J, Li Z, Fan L, Chen R, Wu X et al (2020) A high-efficiency Fe2O3@ microalgae composite for heavy metal removal from aqueous solution. J Water Proc Eng 33:101026

Solimeno A, García J (2019) Microalgae and bacteria dynamics in high rate algal ponds based on modelling results: long-term application of BIO_ALGAE model. Sci Total Environ 650:1818–1831

Song Y, Wang L, Qiang X, Gu W, Ma Z, Wang G (2022) The promising way to treat wastewater by microalgae: approaches, mechanisms, applications and challenges. J Water Proc Eng 49:103012

Sousa JC, Ribeiro AR, Barbosa MO, Pereira MFR, Silva AM (2018) A review on environmental monitoring of water organic pollutants identified by EU guidelines. J Hazard Mater 344:146–162

Solimeno A (2017) Numerical modelling of microalgae systems for wastewater treatment

Sun Y, Chen Z, Wu G, Wu Q, Zhang F, Niu Z, Hu HY (2016) Characteristics of water quality of municipal wastewater treatment plants in China: implications for resources utilization and management. J Clean Prod 131:1–9

Sutherland DL, Heubeck S, Park J, Turnbull MH, Craggs RJ (2018) Seasonal performance of a full-scale wastewater treatment enhanced pond system. Water Res 136:150–159

Urrutia C, Yañez-Mansilla E, Jeison D (2019) Bioremoval of heavy metals from metal mine tailings water using microalgae biomass. Algal Res 43:101659

Yakamercan E, Ari A, Aygün A (2021) Land application of municipal sewage sludge: human health risk assessment of heavy metals. J Clean Prod 319:128568

Zhang C, Tao Y, Li S, Tian J, Ke T, Wei S et al (2019) Simultaneous degradation of trichlorfon and removal of Cd (II) by Aspergillus sydowii strain PA F-2. Environ Sci Pollut Res 26:26844–26854

Zeraatkar AK, Ahmadzadeh H, Talebi AF, Moheimani NR, McHenry MP (2016) Potential use of algae for heavy metal bioremediation, a critical review. J Environ Manag 181:817–831

Wastewater Treatment by Soil Infiltration Process

8

Nuridayah Yusmaidi and Nurfadhila Nasya Ramlee

Abstract

Wastewater treatment is a critical part of safeguarding environmental health and sustainability because the discharge of untreated wastewater poses major hazards to ecosystems and human well-being. In the pursuit of more sustainable and nature-centric options, soil infiltration stands out as a possible method for wastewater treatment. In contrast to conventional treatment approaches, which frequently depends on intricate infrastructure and energy-demanding procedures, soil infiltration utilizes the innate capability of the soil to naturally filter and purify water. This method utilizes the physical, chemical, and biological characteristics of the soil, offering an environmentally friendly strategy for treating wastewater that aligns with the tenets of sustainable development. Generally, the soil infiltration process involves the introduction of wastewater into specially designed infiltration basins or trenches, initiating a journey through the soil layers. The complex interconnection among soil particles, microorganisms, and organic material works together to decompose pollutants, improving the overall quality of the treated water. Its versatility positions it as a practical choice for treating wastewater in both residential and industrial contexts, providing a decentralized and economical method. This chapter focuses on exploring the soil infiltration process, analyzing its advantages, limitations, and its implementation to treat wastewater. Lastly, this chapterdiscuss the integration of the soil infiltration method with Sustainable Development Goals (SGDs).

N. Yusmaidi (✉)
Department of Chemistry, Faculty of Science, UniversitiTeknologi Malaysia, Johor, Malaysia

N. N. Ramlee
School of Chemical and Energy Engineering, Faculty of Engineering, UniversitiTeknologi Malaysia, Johor, Malaysia

N. Dwivedi et al. (eds.), *Wastewater Treatment Through Nature-Based Solutions*,
https://doi.org/10.1007/978-981-96-6699-7_8

Keywords

Wastewater treatment · Infiltration process · Infiltration capacity · Soil texture · Soil infiltration systems

8.1 Introduction

Wastewater treatment is a critical part of safeguarding environmental health and sustainability, because the discharge of untreated wastewater poses major hazards to ecosystems and human well-being (Pratap et al. 2023). In the pursuit of more sustainable and nature-centric options, soil infiltration stands out as a possible method for wastewater treatment (Chen et al. 2023). Soil infiltration is the critical initial phase of the hydrologic cycle, governing the rate at which rainwater penetrates into the soil matrix. This process underpins ecosystem health by replenishing groundwater reserves, sustaining plant life, and fostering diverse soil biota. Soil texture significantly influences infiltration rates, with sandy soils exhibiting higher permeability than their clayey counterparts (Zou et al. 2023). Ultimately, the soil's infiltration capacity serves as a pivotal indicator of its hydrological function, acting as both a water filter and a temporary storage medium, which are important for various purposes such as drinking water supply, irrigation, and ecosystem health. The schematic diagram illustrating the soil infiltration process is shown in Fig. 8.1. The absorption rate of water at the initial stage of dry soil infiltration process is fast due to the presence of many empty pores. As these pores become filled with water, the rate of absorption slows down. This is because the remaining pores are smaller and more difficult for water to penetrate. This rate measures the volume of water that enters the soil per unit area over a specific time.

In contrast to conventional treatment approaches, which frequently depends on intricate infrastructure and energy-demanding procedures, soil infiltration utilizes the innate capability of soil to naturally filter and purify water (Kaur et al. 2022; Singh et al. 2023). This method utilizes the physical, chemical, and biological

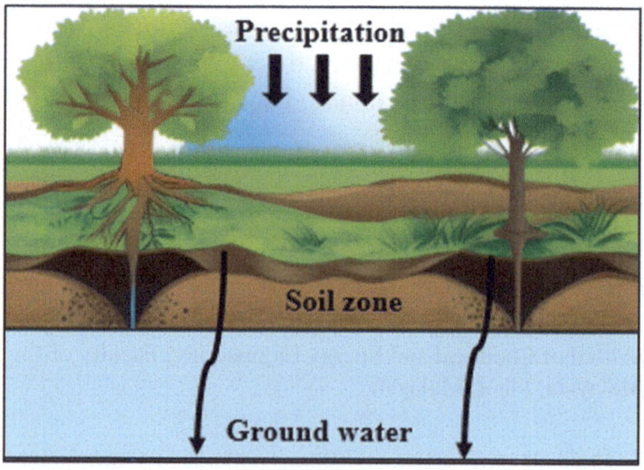

Fig. 8.1 The process of soil infiltration, where rainwater seeps through the soil zone and replenishes groundwater

characteristics of the soil, offering an environmentally friendly strategy for treating wastewater that aligns with the tenets of sustainable development. The soil infiltration process involves the introduction of wastewater into specially designed infiltration basins or trenches, initiating a journey through the soil layers (Zeleňáková et al. 2020; Maliva and Maliva 2020). The soil functions act as an inherent filter, eliminating contaminants, pathogens, and impurities through a blend of physical, chemical, and biological procedures (Rajendran et al. 2022). The complex interconnection among soil particles, microorganisms, and organic material works together to decompose pollutants, improving the overall quality of the treated water. This method not only reduces the requirement for energy-intensive interventions but also fosters the conservation and rejuvenation of natural ecosystems, rendering it an appealing option for communities in employing of eco-friendly wastewater management solutions.

Apart from its positive environmental impacts, the soil infiltration process exhibits its flexibility across diverse geographical and climatic settings. Its versatility positions it as a practical choice for treating wastewater in both residential and industrial contexts, providing a decentralized and economical method. As global communities work towards overcoming issues related to water pollution and scarcity, the soil infiltration process emerges as an appealing and sustainable remedy that aligns with the worldwide commitment to advocate responsible resource management and support the conservation of ecosystems.

8.2 Infiltration Capacity

Infiltration capacity is a critical hydrological parameter that quantifies the maximum rate at which water can penetrate the soil surface and infiltrate into the subsurface. This metric is essential for understanding soil water dynamics, groundwater recharge, and surface runoff processes (Zou et al. 2023). To quantify soil infiltration, two key parameters are often used: infiltrability and cumulative infiltration. Infiltrability measures the rate at which water can infiltrate the soil, typically expressed in mm/min or mm/h. On the other hand, cumulative infiltration, represents the total amount of water that has infiltrated the soil over a given period, usually measured in mm or cm (Lu et al. 2024). The rate at which water infiltrates the soil and the total amount of water that has infiltrated over time are interconnected. This relationship can be described by Eq. (8.1):

$$i = \frac{dI}{dt} \tag{8.1}$$

where i is the infiltration rate (mm/h), t is time (h), and I is the cumulative infiltration.

When water begins to infiltrate the soil, the initial infiltration rate is typically very high. However, as the soil becomes saturated, the infiltration rate decreases rapidly. Over time, the infiltration rate decreases more gradually, and the infiltration process eventually reaches a steady state. This final infiltration rate is closely related to the soil's saturated hydraulic conductivity, a measure of the efficiency of water flow through saturated soil. It generally takes several hours for the soil to reach this steady state (Fitriati and Rahman 2023).

The rate at which water can infiltrate the soil is determined by the matric potential gradient at the soil surface. A dry soil initially has a relatively high matric potential gradient, which contributes to the high initial infiltration rate. As the soil becomes wetter, the matric potential gradient decreases, leading to a slower infiltration rate (Guo et al. 2024). The rate at which water can infiltrate the soil is influenced by various factors, including the soil suction gradient, time, soil texture and structure, initial soil moisture content, topography, surface cover conditions, rainfall intensity, and soil conditions such as texture, aggregate stability, cracks, and crusts. As the soil becomes wetter, the soil suction gradient decreases, affecting infiltration rates. As soil moisture content increases, the hydraulic gradient within the soil diminishes. This reduction in hydraulic gradient leads to a corresponding decrease in the infiltration rates. The reason for this is that saturated soil conditions limit the available pore space for additional water to infiltrate.

Moreover, the decreased hydraulic gradient results in a reduced driving force for water movement into the soil, further impeding infiltration. Additionally, the characteristics of the soil itself, such as its texture, structure, and surface conditions, play a significant role in determining the efficiency of water to penetrate the soil (Gao et al. 2024). Sandy soils and soils with stable aggregates tend to have higher infiltration rates compared to silty soils. The swelling of clay soils can further impede water infiltration. In addition, the presence of macropores and high soil porosity, can promote the infiltration process (Peng et al. 2024).

Plants and an organic matter can significantly improve soil health by maintaining soil structure, reducing the impact of raindrops on the soil surface, and promoting water infiltration. On top of that, plant cover can help minimize runoff and soil erosion. This, in turn, contributes to healthier ecosystems and more productive lands (S. Wang et al. 2024b). Soils covered by plants generally have higher infiltration rates compared to bare surfaces. The topography of a watershed, including the steepness of the slopes and the characteristics of the soil surface, can also significantly impact soil infiltration processes. The combination of these factors can influence how water flow through the soil and contributes to runoff or erosion (Zhang et al. 2024).

While the previous descriptions assume a stable soil structure, it is important to note that the structure of the soil, particularly on the surface, is often unstable. This instability can significantly influence infiltration rates and other soil processes (Negrin et al. 2023). The infiltration rate of soil can fluctuate significantly over time due to changes in surface conditions, such as the disintegration of soil aggregates, the formation of surface seals, and the creation of air pockets within the soil. Darcy's equation, presented in Eq. (8.2), provides a conceptual framework for understanding this complex process.

$$i = K_s \frac{\Delta H}{\Delta z} \tag{8.2}$$

where K_s is the saturated hydraulic conductivity (mm/h) and $\Delta H/\Delta z$ is the hydraulic gradient (mm/mm), as determined by an increase or decrease in hydraulic pressure at the soil surface (ΔH) with soil depth (Δz).

The hydraulic pressure at the wetting front, the boundary between wet and dry soil, is primarily influenced by the forces of soil suction and gravity. These two forces work together to determine the movement of water through the soil (Ye et al. 2024). The force that drives water into the soil, known as the infiltration driving force, is influenced by the matric suction and elevation potential at the wetting front. These factors determine the overall pressure gradient that propels water into the soil (Lins and da Silva Filho 2024). Initially, the suction gradient within the soil is significant, leading to a high rate of water infiltration. As the soil becomes increasingly saturated, the suction gradient diminishes. Eventually, the infiltration process reaches a steady state, where the rate of water movement is primarily influenced by the elevation potential at the soil surface, and the suction gradient becomes negligible.

The initial high infiltration rate of dry soil is due to the strong adsorption forces exerted by the dry soil particles. As the soil becomes more saturated, a second stage of infiltration begins. In this stage, the water movement within the soil is primarily influenced by capillary forces and gravity (Li et al. 2024). As water infiltrates the soil, it displaces the air trapped within the soil pores. However, some air often remains trapped in the form of bubbles, which can hinder the movement of water. During this period, the soil's infiltration rate decreases rapidly until the soil becomes completely saturated. This rapid decline in infiltration is attributed to the short duration during which capillary forces are active within the soil, as these forces diminish as the pores fill with water (Lehmann and Or 2024). Infiltration can be divided into three distinct stages. The first two stages, where the soil is unsaturated, are often combined into a single leakage stage. During this stage, water movement is primarily influenced by gravity. Once the soil becomes fully saturated, the third stage, known as the steady infiltration period, begins. In this final stage, water moves through the saturated soil at a constant rate, controlled solely by gravitational forces.

While the two primary stages of soil infiltration are often discussed separately, they can occur simultaneously, making it difficult to clearly distinguish between them, especially in thicker soil layers. This overlap can complicate the understanding and prediction of soil infiltration processes (Bronstert et al. 2023). The saturated soil profile can be vertically partitioned into distinct zones. When adequate water is available, a thin layer of standing water may form on the soil surface (Nimmo and Shillito 2023). Beneath the surface water, the soil is fully saturated and is known as the saturated zone. Below this, the soil moisture content gradually decreases, forming a transitional zone. The water content in this layer ranges from approximately 60% to 80% of the saturated soil's moisture capacity (Liu et al. 2023). Within this zone, water primarily moves downward due to gravity. Below it lies the wetting front, which separates the saturated soil from the dry soil.

Several methods have been employed to measure soil infiltration, including direct measurement and watershed approaches (Mariani et al. 2024). Direct measurement techniques focus on small areas, allowing for the direct determination of infiltration curves at specific locations. The watershed method, as the name suggests, is applied to entire watersheds. By measuring rainfall intensity and total runoff during rain events, infiltration rates and cumulative infiltration can be estimated (Han et al. 2024). This method provides an average infiltration rate for the entire watershed.

Various methods have been developed to measure soil infiltration, including the double-ring infiltrometer method (Kiranaratri et al. 2024), the modified double-ring or Mariotte double-ring method (Vinatier et al. 2024), rainfall simulation, the runoff-on-ponding method (Jiang et al. 2024), the runoff-on-out method (Mongil-Manso et al. 2024), the disc permeameter method, and the linear source method (Al-Jburi et al. 2024).

8.3 Factors Affecting Soil Infiltration Process

The rate at which water can infiltrate the soil is influenced by various factors, especially characteristics of the soil as well as geographical factors (Garg and Goel 2019). For example, initial soil moisture content, the soil texture and structure, soil suction gradient, soil composition, aggregate stability, and surface cover conditions are among the important parameters in determining the best characteristics of soil for the infiltration process. These factors play a significant role in determining the efficiency of water to penetrate the soil. On top of that, it is crucial to analyze the geographical factors in choosing the perfect site for the infiltration process, such as the rainfall intensity, geology, and topography of the infiltration site. Lastly, infiltration time is important in determining the infiltration capacity of the soil.

8.3.1 Initial Moisture Content of the Soil

The initial moisture content of the soil can significantly impact its infiltration rate. Generally, drier soils tend to have higher initial infiltration rates compared to wetter soils (Jia et al. 2024). As water infiltrates the soil, it creates a distinct boundary between the wetted and dry portions of the soil, known as the wetting front. At this interface, the suction gradient decreases as the initial soil moisture content increases. This reduction in suction gradient ultimately reduces the overall infiltration rate (Ye et al. 2024). The soil infiltration rate is also influenced by the amount of water available at the soil surface. When the rate of the water supply to the soil is lower than the soil's ability to absorb water, the infiltration process is limited by the inflow rate. In other words, the infiltration rate will be equal to the rate of water is being supplied. However, if the rate of the water is being supplied exceeds the soil's infiltration capacity, some of the water will not be absorbed and it may run off or pool on the surface (Zhu et al. 2024). When the rate at which water is supplied to the soil exceeds its infiltration capacity, overland flow or runoff occurs In this scenario, the soil infiltration process is primarily controlled by the soil's inherent ability to absorb water, a phase known as the profile control stage.

8.3.2 Soil Texture

Soil texture refers to the size distribution of soil particles that play a decisive role in soil porosity. Soil structure is the arrangement of several soil particles into porous compounds and serves as the principal factor towards the capability of soil to absorb and retain water during irrigation and rainfall occurrence (Abou Najm et al. 2021). On top of that, soil texture determines the formation of aggregates of specific sizes through stimulation of larger aggregate sizes and inter-aggregate pores in clay (Ge et al. 2019). Based on the systematic review study conducted by Basset et al. (2023), they analyzed that soil amendments and crop management improve water infiltration and soil structure, whereas tillage negatively impacts water infiltration while improving soil structure.

In addition, plant roots also affect soil properties, water flow paths, and water infiltration rates. Zhang et al. (2023) analyzed physical characteristics of soil and hydraulic properties of three different types of forest in a subtropical area. The results showed that both monsoon evergreen broadleaf forests (BF) and mixed pine and broadleaf forests (MF) were characterized by a smaller bulk density and larger porosity than pine forests (PF). In addition, the observed soil water exchange amount also showed that both BF and MF had higher water-holding capacity than PF. In addition, the effects of root traits on soil properties are associated with soil texture and ultimately influence soil infiltration rate. Wang et al. (2023a) reported that rooted soils had higher infiltration rates than bare soils, with nonsignificant interspecies differences under three soil texture types. The infiltration study was performed in a water scarcity region located at Loess Plateau. Furthermore, soil texture and root morphological traits explained 92.03% and 74.75% of the total variance of soil properties (soil organic matter, aggregate stability, and soil porosity), respectively. These findings suggest that soil texture was the greater contributor to the soil infiltrability as compared to the root morphological traits.

8.3.3 Soil Structure

Soil structure refers to the shape in which soil particles group together and form aggregates. Soil structure is important as it indicates the arrangement of soil particles and can change over time due to changing conditions or disturbance (Amber Anderson 2023). The soil structure is influenced by the shape of the soil, whether granular, platy, blocky, or prismatic. On top of that, the biotic or abiotic component in the soil, in concomitance with soil management, will greatly influence soil structure. For instance, soils with a high content of organic matter provide a good ecosystem for a biotic components like earthworms. The movement of earthworms in the soil helps open the pore space and create continuous pores linking surface soil layers to subsurface soil layers (United States Department of Agriculture 2024). Franzluebbers (2002) found that tillage has a negative impact on the surface structural stability of the soil, surface residue accumulation, and surface organic carbon, which eventually control water infiltration and water storage in the soil. In addition, the pore structure of the soil also determines the infiltration rate, mechanism of infiltration, and the transmission activity of water into the soil (Ju et al. 2024).

8.3.4 Soil Composition

Soil composition is a mix of various components that can be broadly categorized into biotic (living and once-living things) and abiotic (non-living things) factors (Adhikari et al. 2024). The biotic factor is grouped into (i) organic matter, which includes decomposed plants, animals, and microorganisms, and (ii) live microorganisms, which composed of the bacteria, fungi, and other microorganisms that play a crucial role in breaking down organic matter and recycling nutrients. Meanwhile, the abiotic factors are (i) minerals that are essential for plant growth, such as phosphorus, potassium, calcium, magnesium, and sulfur; (ii) water, which is vital for plant uptake and biochemical processes; and (iii) air, which is necessary for root respiration and microbial activity (National Geographic Society 2023). The organic matter in the soil affects soil infiltration by binding the soil particles together in stable aggregates, increasing porosity and infiltration rate. Other than soil's composition, the types of soil (sandy, loamy, and clay) will give varying infiltration capacities, with sandy soils generally allowing more infiltration compared to clay soils (Simelane et al. 2024).

8.3.5 Slope Gradient

Slope gradient refers to the angle of inclination of the soil surface from the horizontal. It is usually expressed as a percentage, which represents the vertical rise or fall over a horizontal distance of 100 units (Franzmeier and Steinhardt 2004). Slope gradient is important because it influences the rate at which runoff flows on the soil surface and erodes the soil. The infiltration rate tends to increase with increasing slope gradient. Ribolzi et al. (2011) conducted a comparison study on the infiltration rate in Northern Laos at the site with a similar clay loam texture and organic matter content but at two different gradients (30% and 75%). The findings presented that on steeper slopes (75%), the micro-relief tends to form micro-terraces much more pervious and less erodible than the ripple-like roughness that formed on gentler slopes (30%). Meanwhile, the 75% slopes exhibited higher infiltration rates and greater soil permeability as compared to 30% slopes. Another study performed by Simelane et al. (2024) revealed that steeper slopes (8°) slowed down infiltration rate and led to increased runoff and soil loss, while gentler slopes (5°) minimized soil loss and ultimately facilitated greater infiltration rate.

Sites with gentle slopes are typically preferred, as they are more stable, as they facilitate gradual infiltration and lower the risk of erosion (Kang et al. 2024; La et al. 2023; Osman and Osman 2018). Steep slopes can accelerate runoff and may necessitate extra measures, such as terraces or retaining walls, in order to control erosion and sedimentation, and serve as an interventions to maintain the stability of the infiltration system. Apart from slope gradient, slope shape (straight, concave, or convex) and slope length are also important properties of soil surfaces that may affect the infiltration rate.

8.3.6 Aggregate Stability

Aggregate stability is defined as the ability of soil aggregates, also known as clusters of soil particles, to resist disintegration when exposed to disruptive forces such as raindrops, water, and wind (Vinayak 2018). Aggregate formation is a complex process. Soil aggregates are formed through physical, chemical, and biological activity below ground. They are even influenced by human factors, like tilling, walking on the surface, or fertilization methods (Nall I. Moonilall 2019).

8.3.7 Surface Cover Conditions

In environmental science, surface cover condition refers to the state and type of materials covering the ground. This can include vegetation, soil, rocks, water, and man-made surfaces like pavement. Surface cover affects how water moves across and into the ground. Vegetated areas allow more water to infiltrate into the soil, reducing runoff and the risk of flooding. In contrast, impervious surfaces like concrete increase runoff and can lead to water quality issues (Tian et al. 2024). Surface cover conditions can impact soil infiltration in the following ways: vegetative cover reduces soil erosion by water and increases infiltration, soil moisture content affects infiltration rate, with higher moisture content leading to lower infiltration, and dry soils tend to have faster infiltration due to pores and cracks (Agassi 2023).

Peng et al., (2024) investigated the effect of long-term cultivation (20, 40, 60, and 100 years) on the pore structure and infiltration characteristics in northeast Mollisol region of China. The findings presented that long term cultivation resulted in a simpler pore network, which is caused by the alteration in the soil pore size, shape distribution and a decrease in the macroporosity (>500 μm). In addition, long-term cultivation took a shorter time to reach a stable state in the infiltration curve over time as compared to the control site (forest) (Peng et al. 2024). Infiltration rate is primarily affected by conditions near the soil surface and can change drastically according to soil management. Infiltration is a rapid process through large continuous pores at the soil surface and slows as pores become smaller. Stable-state infiltration rates typically occur when soil is nearly saturated, and the results may vary according to different soil textural classes.

8.3.8 Rainfall Intensity

Rainfall intensity refers to the amount of rainfall that occurs over a specific period of time, typically measured in millimeters per hour (mm/hr) or inches per hour (in/hr) (Weather and Environmental Monitoring Directorate 2013). There are three classifications of rain intensity: (i) Light rain: <2.5 mm/yr, (ii) Moderate rain: 2.5–7.6 mm/yr, and (iii) Heavy rain: >7.6 mm/yr (Jan Barani 2020). It helps in understanding the severity of rainfall events and it is crucial for various applications like flood forecasting, agricultural planning, infrastructure design, and soil infiltration. As compared to other factors, Simelane et al. (2024) found that the combination of rainfall intensity and slope gradient played a significant role in soil infiltration rate, with 70 mm/h/5° combinations promoting higher infiltration rates (48.14 mm/h) and 100 mm/h/8° resulting in lower rates (37.07 mm/h for sandy loam and silt loam, 26.09 mm/h for clay loam). Increased rainfall intensity can initially enhance infiltration rate due to the force of water breaking through soil surface heterogeneities.

However, if the intensity is too high, it can lead to soil crust formation, which reduces infiltration rates. The amount and duration of rainfall also play a crucial role. As rainfall continues, the soil absorbs water until it reaches its infiltration capacity. Once saturated, any additional rainfall will result in runoff rather than infiltration. When runoff occurs on bare or poorly vegetated soil, erosion occurs. Runoff carries away nutrients, chemicals, and soil, resulting in decreased of soil productivity, offsite sedimentation of water bodies, and diminished water quality. Lu et al. (2024) discussed the effect of rainfall intensity and duration on the role of tillage-induced soil microrelief in soil erosion. As compared to the control site (smooth surface), the two types of tillage-induced soil microrelief, which are known as (i) depressions and mounds and (ii) troughs and ridges, demonstrated a decrement in soil erosion proportional to the increment of rainfall intensity, rain duration, and slope gradient. Therefore, rainfall conditions and slope gradient are important factors influencing soil and water conservation as well as reducing soil erosion in conjunction with assessment of climate change.

8.3.9 Geology and Topography

The integration of geology and topography knowledge during site selection not only maximizes the benefits of soil infiltration but also protects our vital water resources from adverse impacts (Tom and Adama 2023). Placing infiltration systems near lakes, rivers, or streams poses potential risks to groundwater quality (Mishra 2023). These systems are designed to manage stormwater runoff, which can carry various pollutants such as pesticides, fertilizers, and heavy metals (Ouedraogo et al. 2023). If the runoff is not effectively filtered or treated before it infiltrates the ground, these contaminants can migrate into the groundwater, eventually reaching and polluting nearby water bodies.

Additionally, improper maintenance or system overloads during intense rainfall events can lead to direct overflow of pollutants into adjacent water sources. Such scenarios can cause significant ecological damage and undermine the health of aquatic ecosystems (Ahmed and El-Rawy 2024). Therefore, careful planning and maintenance are crucial to mitigate these risks and ensure the effective functioning of infiltration systems. To safeguard against contamination of groundwater and surface water, meticulous attention must be given to the placement of infiltration systems. It is crucial to position these systems at a safe distance from adjacent water bodies, considering local regulations and specific soil conditions, which may set precise requirements for their location (Richards et al. 2023).

Even though a study on the factors affecting the soil infiltration process has been conducted, most of the studies focused on the single factor instead of the combination of different factors. Therefore, research on the multiple factors that influence soil infiltration process is crucial for deepening the understanding of the underlying mechanisms and enhancing the infiltration rate while minimizing problems associated with the process. Table 8.1 presents previous research on the multiple factors that affect the soil infiltration process.

8.4 Soil Infiltration Process, Implementation and Design

Implementation and design of infiltration systems require careful consideration of various factors to ensure their effectiveness and longevity. These factors include site selection, soil characteristics, installation methods, and construction materials, as depicted in Fig. 8.2.

8.4.1 Site Selection and Preparation for Soil Infiltration Systems

The type of soil is crucial in deciding whether a location is suitable for allowing water to soak into the ground (Kindo et al. 2024). Soils with large pores, like sandy or gravelly soils, allow water to infiltrate easily. Soils with small pores, like clay, can block water from soaking in and may need extra treatment or changes. Soils with large pores are perfect for systems that allow water to soak into the ground because they can absorb and hold water effectively (Putri et al. 2023). Soils with small pores or that are tightly packed together can restrict water infiltration. These soils may need changes, like adding sand or gravel, to improve their ability to allow water to pass through. Soil structure refers to the way soil particles clump together into small groups, as shown in Fig. 8.3.

8.4.1.1 Determination of Soil Texture by Jar Test and Soil Texture Triangle

In order to determine the proportions of different soil particles, a soil sample is mixed with water. The following steps can be used (Nilsson 2023). Figure 8.4 shows a jar containing a soil sample with three distinct layers. The top layer is clay, the middle layer is silt, and the bottom layer is sand. These layers can be easily distinguished based on their color and texture:

Table 8.1 Previous studies on the factors affecting soil infiltration

Factors affecting soil infiltration	Infiltration sites	Problem statements	Goals	Methods	Results	References
Physical properties of soil and hydraulic properties	Dinghushan Biosphere Reserve, located in Zhaoqing City, south China	Lack of knowledge about the patterns of soil water flow in different stages of forest succession	Maintaining and improving water conservation capacity of forest ecosystems	Quantify the change tendency of soil infiltration and water flow paths in subtropical different forest types Analyze the effect of soil's physical properties on hydraulic properties	Monsoon evergreen broadleaf forests (BF) and mixed pine and broadleaf forests (MF) were characterized by a smaller bulk density and the larger porosity than pine forests (PF) The observed soil water exchange amount also showed that both BF and MF had higher water-holding capacity than PF	(Zhang et al. 2023)
The effects of plant roots and soil moisture	Arid grasslands in China	No study was reported on the study of combination effects of soil moisture and plant roots on soil infiltration	Determine the effects of soil moisture and plant roots on infiltration rate Quantify the changes of these effects during the infiltration process	Measuring infiltration rate by the point source device at different cultivated grasslands in an arid area	Soil infiltration rates were negatively correlated to soil moisture, but a positive correlation was observed with the below-ground biomass Soil moisture gave significant influence than plant roots. Higher soil infiltration capacity in the leguminous grasslands than in the gramineous and the mixed grasslands	(Liu et al. 2019)

Soil texture and root morphological traits	Water-scarcity region located at Loess Plateau	Limited study on the root–soil interactions and their effects on soil water infiltration over large spatial scales	Identify the unique characteristics of soil infiltration processes and different contributions of soil texture and root morphological traits to soil infiltrability	153 plots of natural revegetated land with two contrasting plant species (*Artemisia gmelinii* Web. with tap root systems and *Stipa bungeana* Trin. with fibrous root systems) and bare land were selected	Rooted soils had higher infiltration rates than bare soils	(P. Wang et al. 2023b)
Vegetation restoration type and topography	Zhifanggou watershed, situated in the central Loess Plateau of Northwest China	Limited research on the combination effects of on soil water and infiltration in the Loess Plateau of China	Investigated soil water content soil water storage, soil infiltration rates, and related soil properties in typical forestland, shrubland, and grassland across five slope positions on steep slopes, and compared them with cropland	Soil sampling and analysis were conducted Infiltration rate was evaluated using double-ring infiltrometer	Vegetation restoration decreased deep soil water storage in the Loess Plateau Revegetation increased steady infiltration but reduced initial and mean rates Soil infiltration rates exhibited gradual decline trends along the slope The effect of vegetation type on infiltration was higher than slope position Soil water and clay content were key factors influencing infiltration rates	(Qiu et al. 2024)

(continued)

Table 8.1 (continued)

Factors affecting soil infiltration	Infiltration sites	Problem statements	Goals	Methods	Results	References
External factors: elevation, frozen soil, slope gradient, vegetation, and soil depth Internal factors: soil porosity, soil organic carbon content, and other soil physicochemical properties	High Mountain Asia surrounding the Tibetan Plateau	Lack of systematic quantitative analysis regarding soil–water infiltration properties in this region	Investigate different factors that affect water infiltration process	Literature review and meta-analysis based on experimental data from the literature	As the slope gradient increase, the infiltration rates decreased Increase of initial soil moisture, increase infiltration rate The effect of different vegetation on soil infiltration rates was significantly different Soil organic carbon content increases in average infiltration rates reached Noncapillary porosity increases in average infiltration rates reached Soil–water infiltration studies are subject to spatial and quantitative limitations	(Jia et al. 2024)

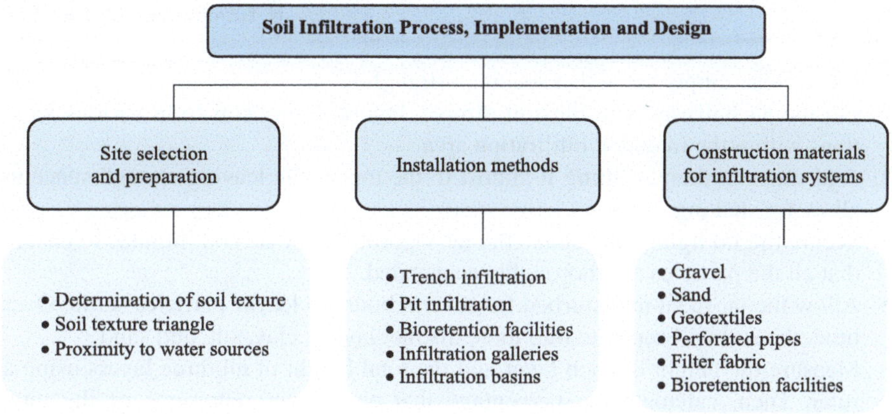

Fig. 8.2 Flowchart showing the steps of site selection, preparation, installation, and material selection for infiltration systems

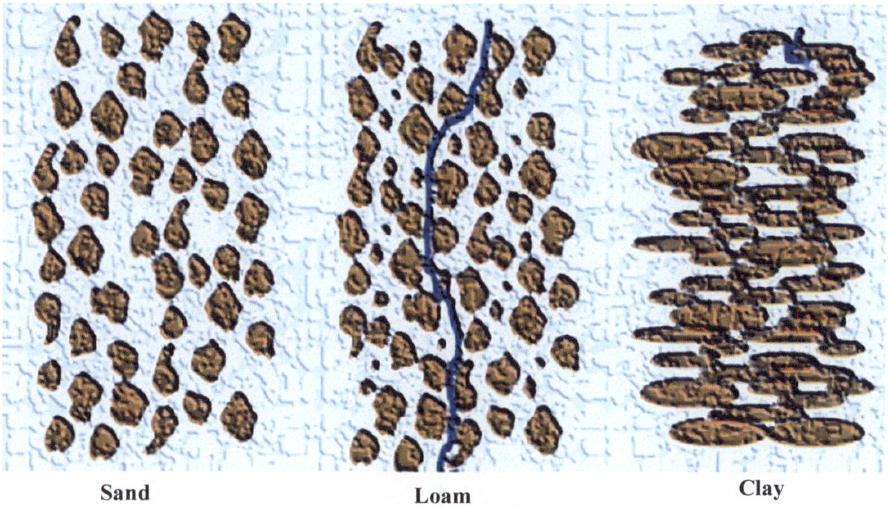

Fig. 8.3 The types of soil when it drains and absorbs water

Fig. 8.4 Soil sample composed of clay, silt, and sand layer

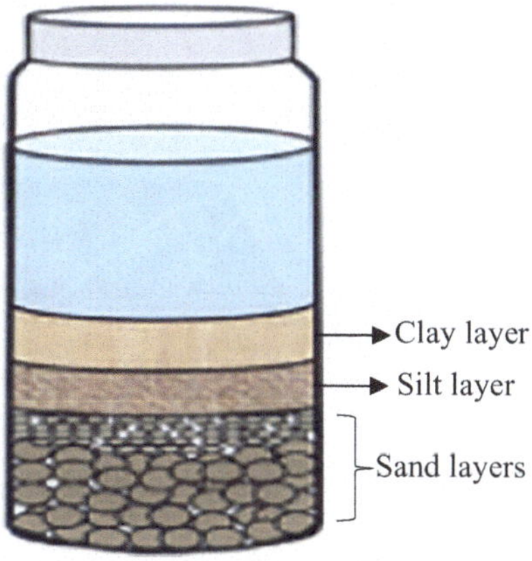

1. Use a clear, empty jar with a secure, airtight lid.
2. Fill the jar halfway with the soil sample, incorporating soil from various locations within the intended infiltration area.
3. Add water to the jar, filling it almost to the top, while leaving enough space to allow for shaking.
4. Secure the lid tightly and shake the jar vigorously for several minutes to ensure that all the particles are thoroughly suspended.
5. Allow the jar to sit undisturbed for several hours to let the particles settle. Over time, the soil will separate into three distinct layers: clay, silt, and sand.
6. Measure the height of each layer and the total height of all three layers using a ruler. Then, calculate the percentage that each layer represents of the total soil column.
7. Refer to the soil texture triangle to determine the soil's texture.

To interpret the results of the soil texture analysis, one can use the following guidelines:

1. The bottom layer will consist of the heavier particles, such as sand and rocks (as shown in Fig. 8.5).
2. Above this, you will find the silt particles.
3. The layer above the silt will contain the clay particles.
4. Organic material might be floating on top of the water.

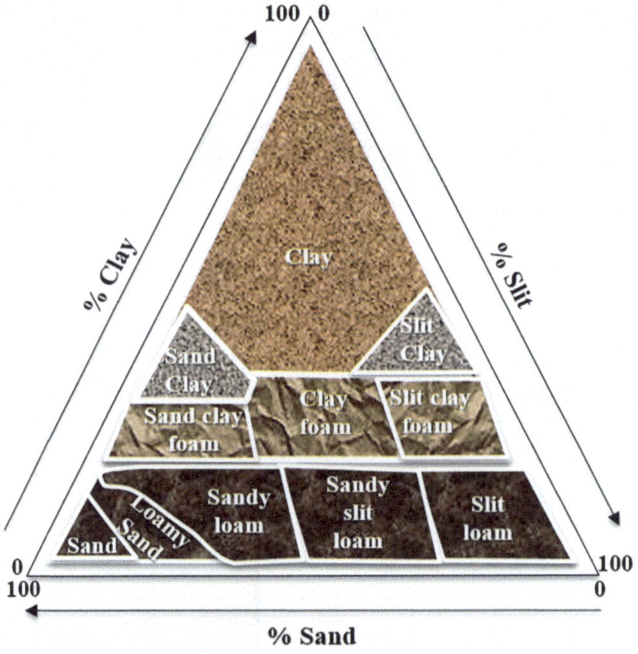

Fig. 8.5 Overview of the soil texture triangle (Dewangan et al. 2023)

8.4.1.2 Experimental Method to Determine the Soil Texture Using the Soil Texture Triangle

A jar test was conducted to determine the type of soil in a specific area where a trench for water infiltration is planned. Here are the results:

Measurement of the heights of the different layers	Calculation of the percentages
(a) Sand layer: 2 cm (b) Silt layer: 0.3 cm (c) Clay layer: 0.1 cm (d) Total height of all layers: 2.4 cm	(a) Sand layer: % sand = (2 cm/2.4 cm) × 100 = 83.3% (b) Silt layer: % silt = (0.3 cm/2.4 cm) × 100 = 12.5% (c) Clay layer: % clay = (0.1 cm/2.4 cm) × 100 = 4.2%

Based on the soil texture triangle classification, the soil sample is identified as loamy sand.

8.4.1.3 Proximity to Water Sources

Infiltration systems are indispensable for controlling stormwater runoff and enhancing water quality. To safeguard groundwater sources from potential contamination, it is essential to meticulously assess and strategically position these systems. Conducting comprehensive site assessments is essential to determine soil properties, groundwater levels, and proximity to water sources, ensuring the optimal positioning of the system. If situating the system near water bodies is unavoidable, incorporating supplementary treatment measures becomes imperative. Options such as advanced filtration devices or bioretention ponds can help remove pollutants from stormwater runoff before it enters the ground. Furthermore, regular monitoring is vital for maintaining system efficacy. This includes routine water quality testing and inspections to confirm that the system is functioning properly and to prevent any negative environmental effects. By adhering to these practices, the integrity of both groundwater and surface water can be effectively preserved (Vorstius 2023). Adhering to these guidelines enables the effective use of infiltration systems for managing stormwater while protecting both groundwater and surface water resources. By carefully considering site placement, implementing additional treatment measures when necessary, and maintaining rigorous monitoring practices, you can optimize the benefits of these systems while minimizing potential risks to water quality. This approach ensures that infiltration systems contribute positively to stormwater management without compromising the health of our vital water resources.

8.4.2 Installation Methods for Infiltration Systems

Infiltration systems can be installed using a range of techniques, each tailored to specific site conditions and requirements. The primary methods include trenches, pits, and bioretention facilities. Trenches are designed to handle and infiltrate runoff along elongated, narrow areas, making them suitable for linear spaces. Pits are employed to capture and manage runoff in distinct, localized spots, providing targeted infiltration. Bioretention facilities, which incorporate vegetation and specially engineered soils, offer the dual benefit of infiltrating water while also enhancing filtration through natural processes. By selecting the appropriate installation method, one can effectively address stormwater management needs while accommodating unique site characteristics and achieving environmental objectives (Weerawardhana 2024).

Fig. 8.6 shows various techniques are available for installing infiltration systems, each with its own set of advantages and limitations. The choice of the most appropriate method depends on several key factors, including the unique conditions of the site, the desired infiltration rate, and the spatial constraints. By thoroughly evaluating these elements, you can determine the installation approach that best suits the site's characteristics and effectively supports the objectives of stormwater management. Understanding these considerations ensures that the chosen technique will optimize performance and address specific site needs (J. Wang et al. 2024a).

1. Trench infiltration systems are constructed by excavating a trench and filling it with a permeable material like gravel or sand. Within this trench, perforated pipes are installed to evenly distribute stormwater across the permeable medium. This setup facilitates the effective infiltration of runoff by allowing water to flow through the pipes and into the surrounding material, where it gradually seeps into the ground (Avinash et al. 2023). Trench infiltration systems are particularly effective in regions with moderate slopes and soils that support adequate infiltration rates. The design is well suited for environments where these conditions are met, allowing for efficient management of stormwater runoff through the gradual absorption of water into the ground. This expanded design allows for greater storage capacity and enhanced infiltration of stormwater. By accommodating a larger volume of runoff, infiltration trenches effectively manage and direct water into the surrounding soil, improving overall stormwater management (J. Wang et al. 2024a).
2. Pit infiltration systems are established by excavating a pit and filling it with a permeable material. Stormwater is directed into the pit, where it percolates through the permeable medium and gradually infiltrates into the surrounding soil. This method effectively captures and manages runoff, promoting its absorption into the ground (Anschütz et al. 2003). Pit infiltration systems are ideal for locations where space is constrained or where trench systems are impractical. The compact design makes them a suitable alternative in areas with limited available space, providing an effective solution for managing stormwater runoff even in challenging conditions.
3. Bioretention facilities represent a specialized type of infiltration system that integrates vegetation with a permeable layer to both filter and infiltrate stormwater. This approach combines the natural cleansing power of plants with the effective drainage provided by the permeable medium, enhancing the system's ability to manage and process runoff while improving water quality (Técher and Berthier

2023). Bioretention facilities frequently feature a pond or swale populated with wetland plants that play a crucial role in removing pollutants from stormwater before it seeps into the soil. These systems offer multiple advantages beyond basic infiltration, including enhanced water quality, reduced stormwater runoff, and the creation of valuable wildlife habitats. By incorporating both natural and engineered elements, bioretention facilities contribute significantly to effective stormwater management and environmental health.

4. Infiltration galleries consist of horizontal tunnels filled with permeable material and linked to a stormwater pipe. This design facilitates the efficient distribution and infiltration of stormwater through the permeable medium within the galleries, allowing for effective management and absorption of runoff. The connection to a stormwater pipe ensures that the system can handle and process water effectively, making it a valuable component in stormwater management strategies (Pieri 2023).
5. Infiltration basins are expansive, shallow depressions filled with permeable material designed to facilitate the infiltration of stormwater. These basins provide a spacious area for stormwater to collect and percolate through the permeable layer, effectively managing runoff and promoting groundwater recharge. Their design enables efficient stormwater processing while reducing surface water runoff (Asry et al. 2023) Fig. 8.6.

8.4.3 Construction Materials for Infiltration Systems

Infiltration systems play a key role in managing stormwater runoff and improving water quality. To achieve effective infiltration and filtration, these systems utilize a variety of construction materials, including the following:

1. Gravel is an essential element in infiltration systems, serving as a permeable layer that enables stormwater to seep into the soil below. The effectiveness of the system depends on the size and gradation of the gravel. Larger gravel particles facilitate quicker infiltration, whereas smaller particles are better at filtering out pollutants (Bronstert et al. 2023).
2. Sand is another permeable material commonly utilized in infiltration systems. Typically used alongside gravel, sand forms a layered filtration setup. It aids in filtering out finer particles from stormwater, including silt and clay (Nimmo and Shillito 2023).
3. Geotextiles are synthetic fabrics positioned below the gravel or sand layers in infiltration systems (Indhanu et al. 2023). They serve several purposes, including the following:
   (a) Separation: Geotextiles act as a barrier between the permeable material and the surrounding soil, helping to prevent clogging and erosion.
   (b) Filtration: Geotextiles assist in removing pollutants from stormwater.
   (c) Reinforcement: Geotextiles offer structural support to the system, helping to prevent the collapse of the trench or pit.

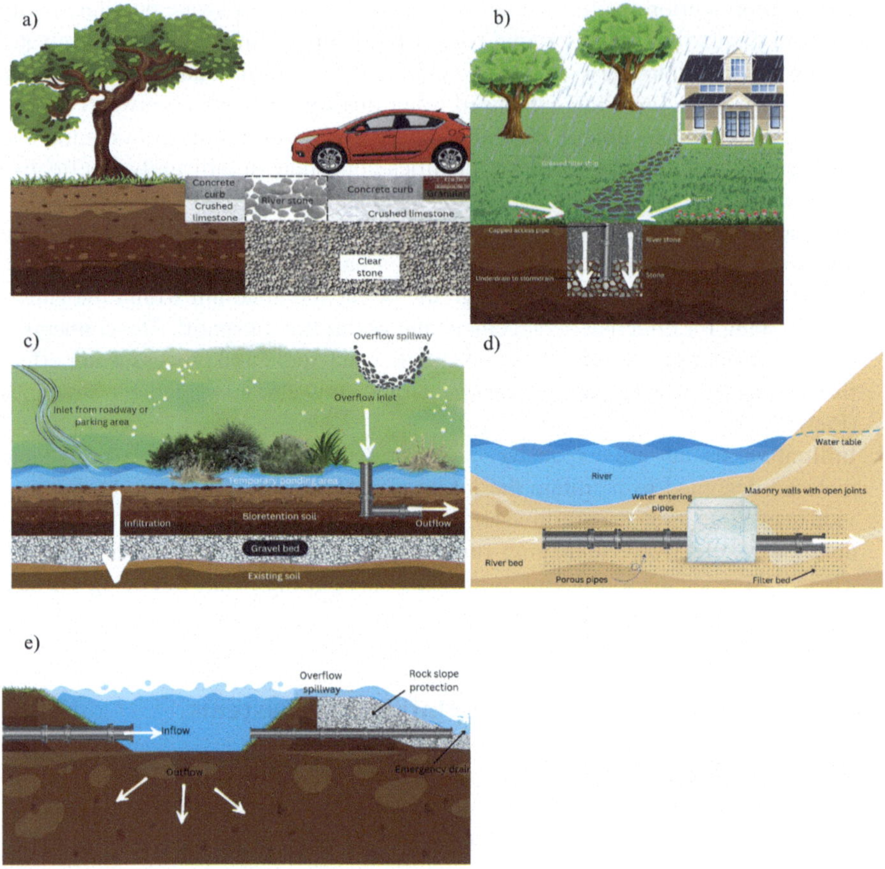

Fig. 8.6 Various installation techniques for infiltration system: (**a**) trench infiltration, (**b**) pit infiltration, (**c**) bioretention, (**d**) infiltration galleries, and (**e**) infiltration basins

4. Perforated pipes are used to distribute stormwater evenly throughout the permeable material in infiltration systems. The perforations allow stormwater to infiltrate into the surrounding soil. The type and size of the perforations can affect the infiltration rate (Raczková et al. 2024).
5. Filter fabric is a kind of geotextile designed to remove pollutants from stormwater. It is commonly used alongside other materials in infiltration systems (Esfandiar et al. 2024).
6. Bioretention facilities employ various types of media to filter and treat stormwater. This media can consist of soil, compost, wood chips, and other organic substances. These materials assist in pollutant removal from stormwater and create habitats for beneficial microorganisms (Wijesiri 2025).

Infiltration systems play a key role in managing stormwater runoff and improving water quality. To achieve effective infiltration and filtration, these systems utilize a variety of construction materials such as geotextiles, filter fabric, perforated pipes, gravel beds, sand layers, mulch layers, bioretention layers, and plantings.

8.5 Ensuring the Longevity of Infiltration Systems: Maintenance and Monitoring

Infiltration systems, which aim to replicate natural processes and improve water quality, play a vital role in sustainable urban development. Their effectiveness, however, depends on adequate maintenance and monitoring. This subsection explores the significance of these practices, highlighting their essential roles in extending system longevity and ensuring optimal performance. Maintenance practices for stormwater management systems involve conducting regular inspections to identify erosion, clogging, or structural damage while also ensuring proper water flow and monitoring changes in vegetation or soil conditions. Additionally, cleaning and debris removal require the removal of leaves, twigs, and other blockages from infiltration basins or trenches to keep water flow uninterrupted (Khabbaz Saberi 2009). Removing sediment is crucial to prevent clogging and maintain the system's capacity by regularly clearing accumulated sediment. Vegetation management involves managing invasive species, ensuring adequate vegetation cover to control erosion, and pruning or removing excess plants that could obstruct water infiltration. Furthermore, structural repairs are needed to fix any damage, such as cracks or subsidence, to maintain the system's overall integrity (Lamberti et al. 2023).

Monitoring techniques for stormwater management systems involve assessing water quality by measuring parameters such as pH, dissolved oxygen, nutrients, and pollutants to evaluate system performance and compare the results against established standards (Razguliaev et al. 2023). Groundwater monitoring includes tracking the levels and quality of groundwater to evaluate the system's effects on the subsurface environment. Soil moisture monitoring involves measuring the moisture content to assess infiltration rates and identify potential problems (Zou et al. 2023). Additionally, measuring the flow rate evaluates the speed of water infiltration to determine system efficiency and pinpoint any potential bottlenecks. The importance of maintenance and monitoring for infiltration systems lies in several key areas. Regular upkeep helps prevent premature failure and extends the system's lifespan, ensuring it operates at peak efficiency to meet water quality and infiltration goals. This proactive approach also contributes to environmental protection by safeguarding water resources and minimizing negative impacts. Additionally, adhering to maintenance and monitoring practices ensures regulatory compliance, helping to avoid penalties and meet environmental regulations.

Maintenance and monitoring are vital for managing infiltration systems effectively. By adopting robust practices, stakeholders can protect the long-term functionality and environmental advantages of these important assets. Regular inspections, cleaning, monitoring, and prompt repairs are key to ensuring the ongoing success of infiltration systems in supporting sustainable water management.

8.6 Advantages of Soil Infiltration

The soil infiltration system is an integrated system that helps in maintaining our ecosystem by stabilizing the hydrological cycle, increasing resilience towards climate-driven disturbances, playing a major role in water conservation, and serving as a pollutant remover. In addition, soil infiltration provides assistance for land management practices, facilitating water cycling for agricultural activities, sustaining plant life, and fostering soil biota.

8.6.1 Stabilizing Hydrological Cycle

Soil infiltration is a fundamental hydrological process whereby water penetrates the soil surface. It serves as a cornerstone for various natural and anthropogenic systems (Fayinminnu et al. 2024). The infiltration rate quantifies the speed at which water penetrates the soil, typically expressed in mm/h or in/h (Alharbi and El-Rawy 2024). Soil infiltration is a cornerstone of the hydrological cycle. By absorbing rainwater, it replenishes underground water reserves, known as aquifers. Healthy infiltration rates are crucial for maintaining a steady supply of water for both human consumption and ecosystems. Moreover, water that seeps into the soil provides essential moisture for plant growth and development (Wang et al. 2023a). Improving water conservation and water purification capacity in the soil and land is a key to stabilizing hydrological cycle. This process play a significant role in facilitating groundwater recharge, and ensuring effective water uptake by the plant.

8.6.2 Increase Resilience to Climate-Driven Disturbances

Soil infiltration also serves as a crucial buffer against the detrimental effects of heavy rainfall (He et al. 2024). When soil can efficiently absorb water, it significantly diminishes the risk of runoff, erosion, and flooding. This enhanced water retention capacity that plays a crucial role in maintaining ecological balance and protecting against environmental degradation (Nimmo and Shillito 2023). The ability of soil to absorb water effectively is not only beneficial for the environment but also for property owners. By reducing runoff and erosion, well-absorbing soil helps to protect soil resources, maintain water quality, and safeguard property. Additionally, it helps to prevent flooding, which can have devastating consequences. Conversely, when soil has low infiltration rates, rainwater is more likely to run off the surface, carrying pollutants into nearby water bodies. This can result in water pollution, harming aquatic life and compromising the overall health of ecosystems (AbuQamar et al. 2024).

8.6.3 Water Conservation

Water conservation is one of the purposes of soil infiltration. It allows us to manage the natural resource of freshwater and protect Earth's hydrosphere. Soil infiltration allows the absorption of water into the soil, increases the amount of groundwater, and minimizes the runoff. Rapid development of housing area especially in urban regions reduces rain catchment area, eventually leading to an increment of the runoff. In addition, a high quantity of runoff may cause floods during the rainy seasons (Yang and Zhang 2011). Ikhsan et al. (2024) reported that construction of infiltration wells successfully reduced 56% of the existing runoff, becoming one of the best alternatives for water conservation.

8.6.4 Remove Harmful Pollutant

Apart from that, a soil infiltration system is employed to remove harmful pollutants. Li et al. (2021) developed integrated multi-soil-layering and subsurface wastewater infiltration for sustainable nitrogen removal. In this system, the multi-soil-layering section located in the upper part facilitates the removal of organic matter, provides sufficient nitrification, and prevents the system from becoming clogged by mitigating the accumulation of pollutants. However, the subsurface wastewater infiltration section further purifies the sewage by removing residual pollutants and acts as a denitrification reactor to improve nitrogen removal. Other land treatment technologies for decentralized wastewater treatment called wastewater ecological soil infiltration systems have been employed to remove nitrogen content and reduce greenhouse gas emissions (Pang et al. 2020). Intermittent aeration is crucial in the system to enhance the bacteria-degrading pollutant. In addition, an increment of the influent C/N ratio led to high total nitrogen (TN) removal and low greenhouse gases emissions, such as CH_4 and N_2O.

Moreover, the soil infiltration method is one of the on-site techniques for wastewater treatment that has been proven in long-term phosphorus removal. Reportedly, there are several terms that have been used to denote soil infiltration systems, such as soil absorption system, septic infiltration system, or soil treatment system. Moreover, it is very efficient in terms of energy and resource use compared to more advanced treatment techniques. Eveborn et al. (2012) reported that the soil infiltration technique presented low phosphorus removal (average of 12%) for four types of the pretreated wastewater sample, indicating low capacity of soil treatment. The wastewater was treated by septic tanks and ponds to remove suspended solids. There are two mechanisms involved in phosphorus removal: $AlPO_4$ precipitation and/or sorption to poorly ordered aluminum compounds.

8.6.5 Land Management Practices

Land management practices can also significantly affect soil infiltration rates. Tillage and cover cropping, for example, can influence soil structure and organic matter content, ultimately impacting its ability to absorb water. In agricultural settings, the ability of soil to absorb water is critical for effective irrigation and optimal crop growth. When soil has high infiltration rates, water can be applied more efficiently, reducing water waste and minimizing the risk of runoff. This, in turn, contributes to higher crop yields and healthier agricultural systems (Ndegwa et al. 2023). When soil has adequate infiltration rates, water can effectively reach plant roots, maximizing water use efficiency and promoting healthy plant growth. Additionally, well-infiltrating soil helps to prevent waterlogging, which can damage plants.

In urban areas, soil infiltration plays a vital role in managing stormwater runoff, reducing the burden on drainage systems, and mitigating the risk of flooding (Raimondi et al. 2023). Preserving and enhancing soil infiltration is fundamental to sustainable land management. Practices such as cover cropping, reduced tillage, and the incorporation of organic matter can significantly improve soil structure and increase its ability to absorb water. Moreover, infiltration helps in fostering soil biota as well as sustaining plant life. Protecting natural vegetation and avoiding soil compaction are equally important for maintaining healthy soil infiltration rates. By implementing these strategies, we can contribute to the long-term health and resilience of our land resources.

8.6.6 Water Cycling for Agriculture Activities

Furthermore, understanding soil infiltration provides valuable insights into water cycling, especially for farmers, as it influences crop yields and help them prepare for the possible ecosystem risks such as pollution, soil erosion, floods, and droughts (Basche and DeLonge 2019). Infiltration is a critical ecosystem function that can mitigate drought and flood risk by facilitating water entry into the soil and reducing water losses by runoff. A soil infiltration system in the presence of healthy soil will be able to deliver the maximum amount of ecosystem services by storing larger volumes of water for a longer period of time after heavy rainfall, associated with a high infiltration rate, resulting in fewer waterlogging problems downstream, while low-quality soil can be attributed to the deficit of overall ecosystem service supply, which can lead to runoff, floods and erosion (Keesstra et al. 2021).

8.6.7 Limitations and Solutions

Despite all the advantages regarding soil infiltration, it possesses several limitations that are crucial to be addressed. The limitations of soil infiltration are due to clogging, soil compaction, root restriction, waterlogging, erosion, and spatial variability. Table 8.2 presents the limitations and solutions to resolve these problems.

Table 8.2 The limitations and solutions to resolve the problem in soil infiltration process

Limitations	Solutions	References
Clogging and Prevention: When soil pores become blocked, it can significantly reduce how quickly water seeps in	Filter Fabrics: Using filter fabrics can help prevent soil pores from becoming blocked by particles and debris Regular Maintenance: Cleaning and maintaining infiltration systems can help remove accumulated debris and prevent clogging Proper Design: Designing infiltration systems to minimize clogging by considering factors like hydraulic loading and soil characteristics is important	(Markiewicz et al. 2024; Mishra et al. 2024)
Soil Compaction: When soil is pressed down too tightly, there may be less empty space for water to seep in	Deep Tillage: Tilling the soil deeply can help improve its structure and reduce compaction No-Till Practices: Using farming practices that minimize disturbing the soil can help prevent compaction Cover Crops: Planting cover crops can protect the soil from erosion and improve its structure	(Agbetoye and Ewetumo 2023; Edwards et al. 2023; Pott et al. 2023)
Root Restriction: Dense root systems can limit how water moves through the soil	Pruning: Carefully pruning root systems can help reduce their density and allow for better water infiltration Plant Selection: Choosing plant species with less aggressive root systems or that are known to improve soil structure can be beneficial Soil Amendments: Adding organic matter or other amendments to improve soil structure and reduce root restriction can be helpful	(Rather et al. 2023; Deng et al. 2024)
Waterlogging: When the soil is excessively saturated with water, it can create conditions with no oxygen and reduce how quickly water seeps in	Proper Drainage: Ensuring adequate drainage can help prevent waterlogging and maintain conditions with oxygen Pumping: Using pumps can help remove excess water from infiltration systems Soil Amendments: Adding materials like sand or perlite can help improve soil drainage	(Modgil and Talekar 2024; Priyadharshini and Babu 2023; Das et al. 2023)
Erosion: Erosion can remove the top layer of soil, affecting how quickly water seeps in and the soil's structure	Vegetation Cover: Maintaining vegetation cover can help protect the soil from erosion Terracing: Creating terraces on slopes can help reduce erosion and improve how quickly water seeps in Mulching: Using mulch to cover the soil can help prevent erosion	(Peng et al. 2023; Unger 2023; Fan et al. 2023)
Spatial Variability: Differences in soil properties across a specific area can make managing infiltration difficult	Site-Specific Design: Designing the infiltration system to match the specific soil conditions and characteristics of the location is important Monitoring and Adjustment: Continuously monitoring infiltration rates and making necessary adjustments to the system to address differences in soil properties is crucial Adaptive Management: Implementing a flexible approach to respond to changes in soil conditions and infiltration patterns over time is essential	(Seifu et al. 2023; Kondratenko et al. 2024; Herrick et al. 2023)

8.7 Integration with Sustainable Development Goals (SDGs)

The implementation of the green economy in the soil infiltration system should integrate with SDG with a main purpose to overcome the challenges of sustainable development, focusing on three main aspects: economic, social, and environmental. Soil infiltration systems are closely related to SDG 6 and SDG 15, aiming to provide clean water and sanitation for all, as well as to protect, restore, and promote sustainable use of terrestrial land. Water scarcity currently affects more than 40% of the global population, as there is a lack of drinkable water for immediate consumption and a lack of irrigation water for agriculture. The role of soil infiltration in achieving SDG 6 lies in its capability to overcome water crises and reduce the impact of climate change on the environment. The primary factor that contributes to this goal is healthy soil for the infiltration process. A healthy soil can absorb and contain more green water by exhibiting a higher infiltration rate and greater water-holding capacity as compared to low-quality soil (Keesstra et al. 2021). Eventually, soil with good characteristics helps in lowering the rainfall-runoff coefficient and erosion. Mekonnen and Hoekstra (2024) reported that an estimated 74% of all freshwater used by humans derives from the soil.

Moreover, the assessment and analysis of soil properties is crucial to determine factors that impact soil productivity and soil health such as water retention, nutrient content, pollution, soil biodiversity, and organic carbon content. These assessments are crucial in developing an integrated and efficient system in agricultural and water management, as well as achieving SGD 6 and 15. In order to achieve SGD goals related to soil infiltration systems, sustainable water management must be prioritized by enhancing the collection of freshwater, increasing the efficiency of water usage, and reducing the number of people affected by water scarcity (Tóth et al. 2018).

The incorporation of the soil-water system is important during the optimization of solutions for SDGs because both resources are geographically connected and influencing one another. Furthermore, an effective collaboration among government, policymakers, stakeholders, and the community is essential to build a sustainable business model that prioritizes environmental protection and integrates environmental, social, and economic sectors (Keesstra et al. 2018). The strengthening of law enforcement must be executed to limit the multifunctional use of land and ensure that the land use is within the boundaries. These strategies can prevent further land degradation, promote land restoration, and promote sustainable water consumption.

For example, the agricultural management of UB Forest, located on the slopes of Mount Arjuno in Malang Regency, has built a green economy model in the perspective of SDG 15 (Protect, restore, and promote sustainable use of terrestrial land). In order to meet SDG criteria, sustainable management is executed with regard to economic, social, and environmental aspects. The deforestation by individuals and community groups is terminated as a primary step to restore forest degradation. On top of that, the soil infiltration method has been employed in the UB Forest as a strategy for the rainwater reclamation process without the occurrence of surface runoff. Soil infiltration also helps in forest rejuvenation by providing water to dry land caused by forest fires or other disasters (Susanto et al. 2021). In order to fulfill the SDGs, incorporation in different areas, including governance, economy, landscape, and management planning, is needed. In addition, the awareness-raising program must be executed to achieve the sustainable world for our and upcoming generations.

8.8 Conclusions

Soil infiltration is an important process in maintaining our ecosystem by stabilizing the hydrological cycle and nurturing our land. A comprehensive study is a must for proper and efficient soil and water management. There are lots of factors that affect the soil infiltration process. But most of these factors were studied independently, limiting the understanding of the significance integration between different factors towards soil infiltration. Moreover, further research is needed to understand the synergies between the soil infiltration rate and other ecological factors in vast areas, such as in relation to soil biology, nutrient cycling, and different types of climate change. Lastly, building a green economy model in the management of soil infiltration methods associated with the perspective of SDGs is one of the best alternatives in overcoming the challenges of sustainable development. The cooperation and the integration with economic, social, and environmental perspectives must be included in designing the model that fits with SDGs.

References

Abou Najm MR, Stewart RD, Di Prima S, Lassabatere L (2021) A simple correction term to model infiltration in water-repellent soils. Water Resour Res 57(2):e2020WR028539. https://doi.org/10.1029/2020WR028539

AbuQamar SF et al (2024) Ecological impacts and management strategies of pesticide pollution on aquatic life and human beings. Mar Pollut Bull 206:116613

Adhikari K et al (2024) Chapter 1 - Introduction to soils: soil formation, composition, and its spatial distribution. In: Dharumarajan S et al (eds) Remote sensing of soils. Elsevier, pp 3–11

Agassi M (2023) Soil Erosion, conservation, and rehabilitation. CRC Press, Boca Raton

Agbetoye LA, Ewetumo T (2023) A review on compaction and loosening of agricultural soils. Int J Sci Res Arch 10(1):893–899

Ahmed AKA, El-Rawy M (2024) The impact of aquifer recharge on groundwater quality. In: Managed aquifer recharge in MENA countries: developments, applications, challenges, strategies, and sustainability. Springer, pp 207–222

Alharbi H, El-Rawy M (2024) Soil aquifer treatment (SAT) for managed aquifer recharge and water quality improvement in the MENA region. In: Managed aquifer recharge in MENA countries: developments, applications, challenges, strategies, and sustainability. Springer, pp 73–95

Al-Jburi AHH, Mahmood HF, Subhi KA (2024) Effects of ploughing techniques on soil hydraulic conductivity and infiltration rate. In: IOP Conference Series: Earth and Environmental Science. 2024 IOP Publishing, p 92005

Anderson A (2023) Soil structure. In: Introduction to soil science. Iowa State University

Anschütz J et al (2003) AD13E Water harvesting and soil moisture retention. Agromisa Foundation

Asry A et al (2023) Sets of infiltration models for water infiltration in sustainable urban drainage systems. J Hydrol 623:129477

Avinash LS, Mishra A, Kami VB, Ojha CSP (2023) A critical appraisal of leachate recirculation systems in bioreactor landfills. J Hazard Toxic Radioact Waste 27(3):3123002

Basche AD, DeLonge MS (2019) Comparing infiltration rates in soils managed with conventional and alternative farming methods: a meta-analysis. PLoS One 14(9):e0215702. https://doi.org/10.1371/journal.pone.0215702

Basset C et al (2023) How does soil structure affect water infiltration? A meta-data systematic review. Soil Tillage Res 226

Bronstert A, Niehoff D, Schiffler GR (2023) Modelling infiltration and infiltration excess: the importance of fast and local processes. Hydrol Process 37(4):e14875

Chen T et al (2023) Application of Internet of Things (IoT) Technologies in Green Stormwater Infrastructure (GSI): a bibliometric review. Sustain For 15(18):13317

Das TK et al (2023) A review of compaction effect on subsurface processes in soil: implications on stormwater treatment in roadside compacted soil. Sci Total Environ 858:160121

Deng X, Greet J, Jones CS (2024) Soil moisture influences the root characteristics of a herbaceous riparian plant along a regulated river. Plant Ecol 225(1):25–36

Dewangan SK et al (2023) Using the soil texture triangle to evaluate the effect of soil texture on water flow: a review. Int J Res Appl Sci Eng Technol 11(6):389–390

Franzmeier DP, Steinhardt GC (2004) Soil and landscape properties. In: Indiana soil evaluation and conservation. Purdue University

Edwards N, Medina N, Asker E (2023) Mixing cover crops suppresses weeds and roto-till reduces urban soil penetration resistance and improves infiltration. Soil Tillage Res 231:105708

Esfandiar N, Suri R, McKenzie ER (2024) Evaluation of sorbent amendments used with stormwater management practices to remove contaminants: impacts of rainfall intensity and antecedent dry periods. Sci Total Environ 906:167766

Eveborn D, Kong D, Gustafsson JP (2012) Wastewater treatment by soil infiltration: long-term phosphorus removal. J Contam Hydrol 140–141:24–33. Available at: https://www.sciencedirect.com/science/article/pii/S0169772212001064

Fan D, Jia G, Wang Y, Yu X (2023) The effectiveness of mulching practices on water erosion control: a global meta-analysis. Geoderma 438:116643

Fayinminnu OO, Onitayo FA, Ogunkunle FA, Daodu BJ (2024) Soil pollution and climate change. In: Environmental pollution and public health. Elsevier, pp 289–302

Fitriati U, Rahman KM (2023) Relationship between soil physical characteristics and infiltration rate of the practice area of SMK PPN banjarbaru. Civil Environ Sci 6(2):117–123

Franzluebbers AJ (2002) Water infiltration and soil structure related to organic matter and its stratification with depth. Soil Tillage Res 66(2):197–205. Available at: https://www.sciencedirect.com/science/article/pii/S0167198702000272

Gao Z et al (2024) The Infiltration characteristics of expansive soil subjected to drying-wetting cycles under surchage. Can Geotechn J

Garg S, Goel A (2019) Infiltration—a critical review. In: Lecture notes in civil engineering. Springer, pp 111–120

Ge N et al (2019) Soil texture determines the distribution of aggregate-associated carbon, nitrogen and phosphorous under two contrasting land use types in the Loess Plateau. Catena 172:148–157. Available at: https://www.sciencedirect.com/science/article/pii/S0341816218303461

Guo H, Ng CWW, Zhang Q (2024) Three-dimensional numerical analysis of plant-soil hydraulic interactions on pore water pressure of vegetated slope under different rainfall patterns. J Rock Mech Geotech Eng

Han Q, Qi T, Khanaum MM (2024) Evaluation of the offsets of artificial recharge on the extra run-off induced by urbanization and extreme storms based on an enhanced semi-distributed hydrologic model with an infiltration basin module. Water 16(7):1032

He Q et al (2024) Runoff and infiltration responses of revegetated slopes to clipping management on the northern Loess Plateau. Int Soil Water Conserv Res 12(1):171–183

Herrick JE et al (2023) Practical guidance for deciding whether to account for soil variability when managing for land health, agricultural production, and climate resilience. J Soil Water Conserv 78(6):125A–133A

Ikhsan MA, Pandjaitan NH, Waspodo RSB, Rau MI (2024) Infiltration well planning to improve water conservation and reduce runoff in residential area. In: Recent advances in environmental science from the Euro-Mediterranean and surrounding regions. 3rd ed, Springer, Cham, pp 621–624

Indhanu T, Srisakul W, Chompoorat T, Chub-uppakarn T (2023) Evaluation and mitigation of ground loss and shear failure in silty sand due to static liquefaction potential with geotextile filtration. Int J Geosynth Ground Eng 9(6):80

Barani J (2020) Rain rate intensity classification. In: Barani design: meteorological sensors & weather stations

Jiang J, Luo S, Chang J, Zhang R (2024) An improved analytical model for rainfall infiltration in embankment slope constructed with red-bed solid waste. Transp Geotech 45:101207

Jia Z et al (2024) The effects of different factors on soil water infiltration properties in High Mountain Asia: a meta-analysis. Catena 234:107583. Available at: https://www.sciencedirect.com/science/article/pii/S0341816223006744

Ju X et al (2024) Impacts of the soil pore structure on infiltration characteristics at the profile scale in the red soil region. Soil Tillage Res 236:105922. Available at: https://www.sciencedirect.com/science/article/pii/S0167198723002891

Kang H et al (2024) Soil erosion characteristics of the agricultural terrace induced by heavy rainfalls on Chinese loess plateau: a case study. Agronomy 14(8):1840

Kaur S, Diwan F, Reddersen B (2022) Mitigating the impacts of drought via wastewater conversion to energy, nutrients, raw materials, food, and potable water. In: Climate change and environmental sustainability. The Academic Research Community Publication, pp 45–60

Keesstra S et al (2018) Soil-related sustainable development goals: four concepts to make land degradation neutrality and restoration work. Land 7(4)

Keesstra S et al (2021) The role of soils in regulation and provision of blue and green water. Philos Trans R Soc B Biol Sci 376(1834):20200175. https://doi.org/10.1098/rstb.2020.0175

Khabbaz Saberi H (2009) Hydraulic characteristics and performance of stormwater pollutant trap respect to weir's height, flow gradients, pipe diameters and pollutant capture

Kindo S, Agrawal N, Shori A (2024) Evaluation of infiltration models in clay loam and laterite soils under field conditions. Environ Conserv J 25(1):22–32

Kiranaratri AH, Bisri M, Asmaranto R (2024) Correlational of soil permeability and infiltration rate at the ITERA infiltration drainage development plan. In: IOP conference series: earth and environmental science. 2024 IOP Publishing, p 12023

Kondratenko J, Boogaard FC, Rubulis J, Maļinovskis K (2024) Spatial and temporal variability in bioswale infiltration rate observed during full-scale infiltration tests: case study in Riga Latvia. Water 16(16):2219

Lamberti L et al (2023) Field investigation of clogging with different sediment types in a permeable pavement system with low-cost maintenance. Ciência e Natura 45(esp. 1):e84089–e84089

La N et al (2023) Agroforestry with contour planting of grass contributes to terrace formation and conservation of soil and nutrients on sloping land. Agric Ecosyst Environ 345:108323

Lehmann P, Or D (2024) Analytical model for bare soil evaporation dynamics following wetting with concurrent internal drainage. J Hydrol 631:130800

Li D et al (2021) Decentralized domestic sewage treatment using an integrated multi-soil-layering and subsurface wastewater infiltration system. Water 13(4) Available at: https://www.mdpi.com/2073-4441/13/4/431

Lins JHR, da Silva Filho FC (2024) A numerical investigation of the influence of hydraulic parameters on the stability of a residual soil slope. Soils Rocks 47(1):e2024012022

Liu M et al (2023) Influence of vegetation type and topographic position on volumetric soil water content dynamics and similarity among surface and deep soil layers. Int Soil Water Conserv Res 11(1):183–196

Liu Y et al (2019) Influence of soil moisture and plant roots on the soil infiltration capacity at different stages in arid grasslands of China. Catena 182:104147. Available at: https://www.sciencedirect.com/science/article/pii/S0341816219302899
Li Y et al (2024) Study on the pore structure and capillary water migration characteristics of microbially improved strongly weathered phyllite. Constr Build Mater 425:136028
Lu S et al (2024) Characteristics of soil infiltration response to rainfall in differently converted forest hillslopes in a southern subtropical area of China. J Soils Sediments 24(1):1–16. https://doi.org/10.1007/s11368-023-03599-4
Maliva RG, Maliva RG (2020) Surface spreading system—infiltration basins. In: Anthropogenic aquifer recharge: WSP methods in water resources evaluation series No 5, pp 469–515
Mariani PP et al (2024) Different infiltration methods for swat model seasonal calibration of flow and sediment production. Water Resour Manag 38(1):303–322
Markiewicz A et al (2024) Polymeric products in erosion control applications: a review. Polymers 16(17):2490
Mekonnen MM, Hoekstra AY (2024) Four billion people facing severe water scarcity. Sci Adv 2(2):e1500323. https://doi.org/10.1126/sciadv.1500323
Mishra RK (2023) Fresh water availability and its global challenge. Br J Multidiscipl Adv Stud 4(3):1–78
Mishra SK, Chowdhury SD, Bhunia P, Sarkar A (2024) Clogging in subsurface flow constructed wetlands: mechanisms, influencing factors, measurements, modelling, and remediation. Ecol Eng 208:107374
Modgil S, Talekar N (2024) Root adaptation traits under water logging conditions. Int J Environ Clim Chang 14(6):1–12
Mongil-Manso J, Jiménez-Ballesta R, Navarro-Hevia J, San Martín R (2024) Soil infiltration rates in mediterranean mountain areas: relationship with soil properties and different land-use. Eur J For Res
Moonilall NI (2019) Soils matter, get the scoop! What are soil aggregates? Soil Science Society of America (SSSA)
National Geographic Society (2023) Soil composition. National Geographic
Ndegwa JK et al (2023) Integrated soil fertility and water management practices for enhanced agricultural productivity. Int J Agron 2023(1):8890794
Negrin I, Kripka M, Yepes V (2023) Metamodel-assisted meta-heuristic design optimization of reinforced concrete frame structures considering soil-structure interaction. Eng Struct 293:116657
Nilsson E (2023) Surface runoff on green urban areas: a study on driving forces behind surface runoff generation
Nimmo J, Shillito R (2023) Infiltration of water into soil. In: Oxford research encyclopedia of environmental science
Osman KT, Osman KT (2018) Soils on steep slopes. In: Management of soil problems, pp 185–217
Ouedraogo AS, Yuzhu Fu G, Yunus AI (2023) Treatment of highway stormwater runoff using sustainable biochar: a review. J Environ Eng 149(2):3122005
Pang J et al (2020) Does influent C/N ratio affect pollutant removal and greenhouse gas emission in wastewater ecological soil infiltration systems with/without intermittent aeration? Water Sci Technol 81(4):668–678
Peng J et al (2023) Effects of different vegetation restoration types on soil infiltration characteristics in severely eroded subtropical regions of South China. J Soil Water Conserv 78(4):364–375
Peng J et al (2024) Soil pore dynamics and infiltration characteristics as affected by cultivation duration for Mollisol in Northeast China. Geoderma 449:117021. Available at: https://www.sciencedirect.com/science/article/pii/S0016706124002507
Pieri AC (2023) Managed aquifer recharge: installation and early evaluation of infiltration galleries in the Cache River critical groundwater area. Arkansas State University, Northeast Arkansas
Pott CA et al (2023) Mixture of winter cover crops improves soil physical properties under no-tillage system in a subtropical environment. Soil Tillage Res 234:105854

Pratap B et al (2023) Wastewater generation and treatment by various eco-friendly technologies: possible health hazards and further reuse for environmental safety. Chemosphere 313:137547
Priyadharshini MP, Babu S (2023) Sub-surface drainage system for management of waterlogged saline soils. In: Emerging trends in agriculture and allied sciences, p 44
Putri FK, Hidayah E, Ma'ruf, M.F. (2023) Enhancing stormwater management with low impact development (LID): a review of the rain barrel, bioretention, and permeable pavement applicability in Indonesia. Water Sci Technol 87(9):2345–2361
Qiu D, Xu R, Gao P, Mu X (2024) Effect of vegetation restoration type and topography on soil water storage and infiltration capacity in the Loess Plateau, China. Catena 241:108079. Available at: https://www.sciencedirect.com/science/article/pii/S0341816224002765
Raczková A et al (2024) The influence of infiltration devices on the urban area. Pollack Periodica
Raimondi A, Marrazzo G, Sanfilippo U, Becciu G (2023) A probabilistic approach to stormwater runoff control through permeable pavements beneath urban trees. Sci Total Environ 905:167196
Rajendran S et al (2022) A critical review on various remediation approaches for heavy metal contaminants removal from contaminated soils. Chemosphere 287:132369
Rather AMUD, Hajam MA, Bhat MSA, Malik MI (2023) Horticulture: principles and practices. Academic Guru Publishing House
Razguliaev N, Flanagan K, Muthanna T, Viklander M (2023) Urban stormwater quality: a review of methods for continuous field monitoring. Water Res:120929
Ribolzi O et al (2011) Impact of slope gradient on soil surface features and infiltration on steep slopes in northern Laos. Geomorphology 127(1):53–63. Available at: https://www.sciencedirect.com/science/article/pii/S0169555X10005258
Richards CE, Tzachor A, Avin S, Fenner R (2023) Rewards, risks and responsible deployment of artificial intelligence in water systems. Nat Water 1(5):422–432
Seifu W et al (2023) Characterization and mapping of soil-landscape for site-specific soil management in Ayiba Watershed, Northern Highlands of Ethiopia. Appl Environ Soil Sci 2023(1):8899956
Simelane MPZ, Soundy P, Maboko MM (2024) Effects of rainfall intensity and slope on infiltration rate, soil losses, runoff and nitrogen leaching from different nitrogen sources with a rainfall simulator. Sustain For 16(11) Available at: https://www.mdpi.com/2071-1050/16/11/4477
Singh K et al (2023) Microalgae from wastewaters to wastelands: leveraging microalgal research conducive to achieve the UN Sustainable Development Goals. Renew Sust Energ Rev 188:113773
Susanto H, Sukoharsono EG, Hendroyono B, Leksono AS (2021) Soil infiltration management on perspective of sustainable development goals (SDGs). Int J Biol Biomed Eng 15:219–227
Técher D, Berthier E (2023) Supporting evidences for vegetation-enhanced stormwater infiltration in bioretention systems: a comprehensive review. Environ Sci Pollut Res 30(8):19705–19724
Tian L et al (2024) Combined role of ground cover management in altering orchard surface–subsurface erosion and associated carbon–nitrogen-phosphorus loss. Environ Sci Pollut Res 31(4):5655–5667. https://doi.org/10.1007/s11356-023-31535-z
Tom CN, Adama J (2023) Surface water and groundwater in a changing environment. Int J Technol Eng Stud 9(1)
Tóth G, Hermann T, da Silva MR, Montanarella L (2018) Monitoring soil for sustainable development and land degradation neutrality. Environ Monit Assess 190(2)
Unger PW (2023) Common soil and water conservation practices. In: Soil erosion, conservation, and rehabilitation. CRC Press, pp 239–266
United States Department of Agriculture (2024) Soil infiltration
Vinatier F et al (2024) Dynamics of quasi-steady ponded infiltration under contrasting plant cover and management strategies. Soil Tillage Res 237:105985
Vinayak S (2018) Soil aggregate stability – a soil health physical indicator. Agronomic Crops Network, Ohio State University Extension
Vorstius C (2023) Assessing and managing risks from climate change in drinking water supply sources: safeguarding raw water quality through improving catchment resilience

Wang B et al (2023a) The contributions of root morphological characteristics and soil property to soil infiltration in a reseeded desert steppe. Catena 225:107020

Wang J et al (2024a) Towards the cost-effective design of stormwater infiltration trenches: a hybrid model integrating cost–benefit analysis and an analytical stochastic approach. Environ Sci Water Res Technol 10(5):1108–1121

Wang P et al (2023b) Differential effects of soil texture and root traits on the spatial variability of soil infiltrability under natural revegetation in the Loess Plateau of China. Catena 220:106693. Available at: https://www.sciencedirect.com/science/article/pii/S0341816222006798

Wang S et al (2024b) Runoff and sediment deposition characteristics of gravel-mulched land: an experimental study. Land 13(4):445

Weather and Environmental Monitoring Directorate (2013) Manual of surface weather observations, Canada

Weerawardhana I (2024) Stormwater treatment measures and their performance analysis for removing pollutants

Wijesiri B (2025) Alternatives for urban stormwater treatment. In: Circular economy applications for water security. CRC Press, pp 185–200

Yang J-L, Zhang G-L (2011) Water infiltration in urban soils and its effects on the quantity and quality of runoff. J Soils Sediments 11(5):751–761. https://doi.org/10.1007/s11368-011-0356-1

Ye J et al (2024) Experimental and numerical investigations on soil–water dynamics in Moistube irrigation under dynamically regulated working pressure heads: scenario simulations under maize cultivation. J Hydrol 634:131083

Zeleňáková M, Hudáková G, Stec A (2020) Rainwater infiltration in urban areas. Springer

Zhang S, Li L, Zhu Z, Zhang P (2024) Spatial distribution and relationship between slope microtopography changes and soil aggregate stability under rainfall conditions. Water 16(5):648

Zhang W et al (2023) Changes in soil infiltration and water flow paths: insights from subtropical forest succession sequence. Catena 221:106748. Available at: https://www.sciencedirect.com/science/article/pii/S0341816222007342

Zhu P et al (2024) Variation in soil infiltration properties under different land use/cover in the black soil region of Northeast China. Int Soil Water Conserv Res 12(2):379–387

Zou C, Zhang S, Jiang X, Chen F (2023) Monitoring and characterization of water infiltration in soil unsaturated zone through an integrated geophysical approach. Catena 230:107243

Bioinspired/Bio-Nanomaterials for Wastewater Treatment

9

Soukaina El Abbadi, Abdelmjid Bouazizi, Mounia Baidou, Hajar El Moustansiri, Latifa E. L. Bardiy, Mohamed Douma, Najib Tijani, and Mohamed Ouammou

Abstract

Water contaminants have adverse effects on ecosystems, prompting significant attention toward purifying them in recent years. The primary challenge lies in discovering economically efficient methods for developing scalable water purification technologies. Various bioinspired and bio-nanomaterials have gained widespread use in treating environmental pollutants due to their high specific surface area, size, and catalytic properties. However, some bio-nanomaterials encounter challenges in eliminating pollutants due to a shortage of active adsorption sites. Modifying the surface of bioinspired through innovative adaptations could enhance their adsorption capability. Additionally, incorporating advanced

S. El Abbadi · M. Baidou · H. El Moustansiri · M. Douma · N. Tijani
Materials, Membranes and Nanotechnology Research Team, Moulay Ismail University of Meknès, Faculty of Sciences, Zitoune, BP: 11201, Meknès, MOROCCO

A. Bouazizi (✉)
Materials, Membranes and Nanotechnology Research Team, Moulay Ismail University of Meknès, Faculty of Sciences, Zitoune, BP: 11201, Meknès, Morocco

Interdisciplinary Laboratory for Modeling simulation and scientific innovation (LIMSIS), École Normale superieure of Meknès, Moulay Ismail University of Meknès, Meknès, Morocco

Laboratory of Materials, Membranes and Environment, Faculty of Sciences and Technologies of Mohammedia, Hassan II University of Casablanca, Casablanca, Morocco
e-mail: a.bouazizi@umi.ac.ma

L. E. L. Bardiy
Research Team "Data Science, Innovation and Development"; Affiliated with the Faculty of Legal, Economic and Social Sciences of Tangier, Tanger, Morocco

M. Ouammou
Laboratory of Materials, Membranes and Environment, Faculty of Sciences and Technologies of Mohammedia, Hassan II University of Casablanca, Casablanca, Morocco

N. Dwivedi et al. (eds.), *Wastewater Treatment Through Nature-Based Solutions*,
https://doi.org/10.1007/978-981-96-6699-7_9

membranes could significantly improve the ability to treat pollutants. While chemical and physical approaches have been successful, their reliance on hazardous chemicals makes them energy-intensive, costly, and environmentally detrimental. Exploring sustainable membrane preparation techniques, such as incorporating eco-friendly polymers innovative deposition methods, could offer promising alternatives to address these challenges in pollutant removal and improve the overall efficiency of filtration processes. In contrast, nanobiotechnology-based biogenic synthesis of nanoparticles (NPs) has emerged as a nature-derived, eco-friendly, and cost-effective fabrication method, reducing the ecological impact of the nanomaterial (NM) industry. Over the last decade, various biological materials have been explored to determine their suitability for NM fabrication. This chapter explores the application of innovative bioinspired/bio-nanomaterials and their composites for wastewater treatment, with a primary focus on synthesizing bioinspired/bio-nanomaterials for adsorption and membrane fabrication. The aim is to achieve removal of hazardous pollutants, and this chapter incorporates the latest advancements and inventive strategies in the field into its discussion.

Keywords

Bioinspired materials · Bio-nanomaterials · Sustainable development · Adsorption · Membrane · Wastewater treatment

9.1 Introduction

The existence of water's presence in the natural world is an essential resource necessary for the continued survival of all life on earth, globally. The importance of water use for preserving life stresses its vital role in preserving the environmental equilibrium. Considering its crucial importance, drinkable water sources are extremely limited, rendering it difficult to satisfy the daily needs of communities worldwide (Bouazizi et al. 2022). Pollution of water represents an exceptionally serious problem since it compromises not only the accessibility this vital resource but also the health of aquatic ecosystems and biodiversity. The accumulation of pollutants in waterways is significantly influenced by human, industrial, and agricultural activities. Various companies are actively working to safeguard human survival by contributing to the substantial rejection of pollutants from aquatic ecosystems (Alraddadi et al. 2023). Heavy metals, micropollutants, inorganic anions, and organic chemicals such as dyes, detergents, and phenols are frequently found in the aquatic region and can lead to water contamination (Hokkanen et al. 2016). Several chemical, physical, and biological techniques are used to treat wastewater (Fig. 9.1), including adsorption and membrane processes. Adsorption involves the use of materials with a high specific surface area to capture pollutants in water. This method is considered efficient, selective, environmentally friendly, cost-effective, and versatile techniques for removing harmful pollutants from water and

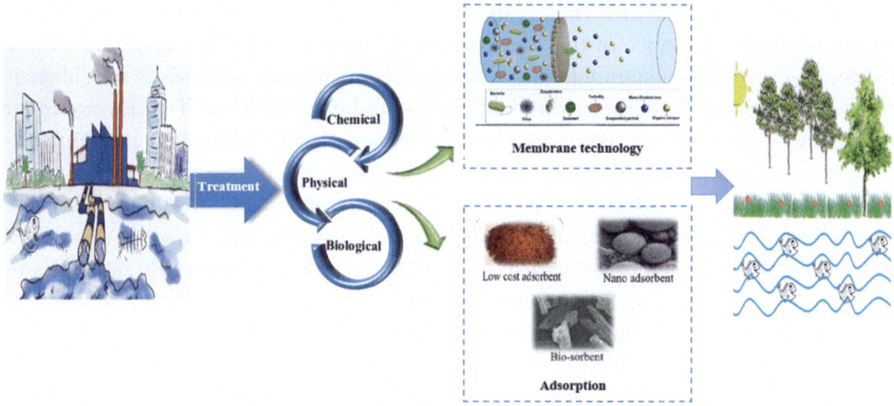

Fig. 9.1 Wastewater treatment by adsorption and membrane filtration process

wastewater (Sirajudheen et al. 2020). Traditional adsorbents such as bentonite, zeolites, activated carbon, and polymeric polymers, have been extensively researched but ate hindered by high cost (Chauhan et al. 2020). Bioinspired materials, reputed for their benefits in removing water contaminants, have captured the interest of the scientific community (Jamshaid et al. 2017; Sirajudheen et al. 2020; Du et al. 2023). These novel innovative bioinspired components play a significant role in environmentally friendly water pollutant elimination (Fan et al. 2019; Amusat et al. 2021; El-Fattah et al. 2024). The distinctive properties of bioinspired materials are attributed to derive from the surface morphological, chemical, and physical characteristics of the natural materials (Gao et al. 2017; Wang et al. 2018).

On the other hand, the membrane process relies on the use of specific membranes to separate contaminants, including solid particles, micro-organisms, dyes, and colloidal materials, from organic or aqueous solvents using microfiltration (MF), ultrafiltration (UF), and nanofiltration (NF). Filtration membranes are increasingly used in water treatment (Bouazizi et al. 2017), where they are made from ceramic or bioinspired materials. While polymeric materials dominate commercial membranes filters, ceramic membranes offer advantages such as high thermal and chemical stability, as well as extended service life (Bouazizi et al. 2016). However, the widespread use of commercial ceramic membranes is obstructed by their high cost; due to expensive industrial oxides (Al_2O_3, TiO_2, ZrO_2, SiO_2) and high sintering temperatures, efforts have been made to develop cost-effective ceramic membranes using geomaterials such as perlite, phosphate, bentonite, and clay (Bouazizi et al. 2022). Bioinspired membranes including aligned nanotube membranes, biomimetic protein–polymer hybrid membranes, and isoporous block copolymer membranes show promise for wastewater cleaning thanks to their hierarchical structures, controlled selective transport, stability, resistance, and self-cleaning properties (Zhao et al. 2014; Rabiee et al. 2023).

This chapter examines water treatment advancements, highlighting the application of biodegradable and bioinspired materials. The production of adsorbents and membranes from these materials is the first step in the comprehensive study, demonstrating the importance of this approach in the context of the environment. A key aspect that provides sustainable alternatives to conventional methods is the fabrication of membranes from bioinspired and biodegradable materials. The development of bio-nanomaterials is also covered in this chapter, along with instances of innovative uses for these materials in a variety of disciplines. A comprehensive discussion is provided on the role of these materials in synthesizing metallic bio-nanomaterials, highlighting the major elements that impact this process. This chapter concludes with an overview of the practical use of these environmentally friendly materials in the removal of pollutants from water, emphasizing their contribution to long-term solutions for water decontamination challenges.

9.2 Bioinspired and Biodegradable Materials for Adsorbent and Membrane Preparation

Materials that draw inspiration from nature and possess biodegradable qualities have captured significant attention in the realm of developing adsorbents, and membranes for wastewater treatment, primarily due to their sustainable nature and environmentally friendly attributes. These materials offer a multitude of advantages, including sustainability, cost-effectiveness, nontoxicity, and the ability to degrade naturally. Bioinspired and biodegradable adsorbents are characterized by their high specific surface area, enhanced selectivity, high adsorption capacity, and chemical and thermal stability. However, bioinspired membrane materials are generally conventional materials with excellent mechanical, adhesive, and hydrodynamic properties.

9.2.1 Chitosan

Chitosan a naturally derived biopolymeric adsorbent (Crini 2005) consists of two distinct monomers: one containing an acetamido group and the other an amino group. It is obtained from chitin using alkaline or enzymatic N-deacetylation processes (Crini and Badot 2008). The substitution of the acetamide group at position 2 is what separates the structures of chitin and chitosan. This has an immediate effect on the solubility characteristics of both substances, making chitosan soluble in mild acids and chitin insoluble and inert (Weska et al. 2007). Chitosan can be derived from waste materials within the fishing industry (Abdul Khalil et al. 2016). Its availability, biodegradability, lack of toxicity, and possession of reactive functional groups have spurred numerous researchers to utilize it in the elimination of various pollutants ranging from heavy metals (Zhang et al. 2016), dyes (Vakili et al. 2014), fluoride (Yu et al. 2013), to organic contaminants (Sirajudheen et al. 2020) present in wastewater. The presence of active adsorption sites, represented by reactive functional groups like hydroxyl (–OH) and primary amines (–NH_2) within its structure, contributes to its effectiveness as a bio-adsorbent for aquatic pollutants

(Crini 2005). Chitosan is also utilized as a bioinspired material in membrane fabrication. A previous study chitosan (CS) was employed to create a dual-layer nanofibrous membrane. This structure combines an active layer containing carbon nanotubes (CNT) with a porous nanofibrous support layer composed of chitosan (CS), polyvinylpyrrolidone (PVP), and polyvinyl alcohol (PVA) as components of the support layer. This design aims to ensure highly efficient removal of heavy metals and organic pollutants due to the high porosity and large specific surface area of the nanofibrous membranes (Wu et al. 2022).

9.2.2 Cellulose-Based Materials

Cellulose-based materials emerge as excellent environmentally friendly adsorbents owing to their prevalence in renewable sources like agricultural residues and wood. These materials exhibit distinctive surface properties characterized by the presence of hydroxyl and carboxyl groups. As a natural biopolymer, cellulose possesses notable attributes such as biodegradability, nontoxicity, and renewability, making it an appealing choice as a green adsorbent (Tanpichai et al. 2022). Cellulose primarily resides within lignocellulosic materials, boasting extensive availability, sustainability, and renewability, make it highly suitable for the adsorption of pollutants in wastewater (Liu et al. 2022). These innate attributes make cellulose-based adsorbents with remarkable efficiency in capturing pollutants, encompassing heavy metals (Jamshaid et al. 2017), dyes (Johari et al. 2016), and organic compounds (Toledo et al. 2019) from both aqueous and gaseous surroundings. This versatility positions them as an environmentally conscientious choice for environmental remediation efforts (Thomas et al. 2018).

Many chemicals, such as trimethyl ammonium chloride, citric anhydride, sodium metaperiodate, succinic anhydride, vinyl monomer glycidyl methacrylate, N,N-carbonyldiimidazole, aminopropyl tris ethoxysilane, maleic anhydride, polyamines, and acrylonitrile hydroxylamine, can be used to modify the abundant hydroxyl groups within cellulose molecular chains. Because of these possibilities for modification, cellulose has become a major focus of research initiatives (Hokkanen et al. 2016; Georgouvelas et al. 2021).

Cellulose modification strategies can be categorized into two primary approaches. The first involves direct modification, entailing the attachment of functional groups to the hydroxyl groups of cellulose molecules through various chemical pathways, including esterification, etherification, silylation, carbonylation, sulfation, and amination. The second strategy uses techniques including free radical polymerization, controlled/living radical polymerizations (CRPs), and ring-opening polymerization (ROP) to graft monomers onto the cellulose chain in homogeneous or heterogeneous solutions (Hokkanen et al. 2016). Particularly in water purification procedures and the elimination of hazardous metals and chemical compounds from water systems, these modification techniques enable the remarkable chelating properties of cellulose materials.

9.2.3 Lignin

Lignins extracted from lignocellulosic residues are potent alternatives for pollutant removal owing to their possession of phenolic, carboxylic, and hydroxyl functional groups. These groups can be attached to chemical pollutant structures, thereby facilitating the removal of these compounds from industrial effluents (Ge et al. 2014). Additionally, the polycyclic structure of lignin encompasses negatively charged functional groups that serve as effective chelating agents, aiding in the sequestration of heavy metals present in effluents (Du et al. 2023).

Due to its abundance resulting from the significant daily effluent generation in various sectors, lignin is advocated as a precursor material to enhance efficiency, reduce costs, and minimize toxic in industrial wastewater treatment (Halysh et al. 2018).

There exist multiple active groups within the molecular structure of lignin and its derivatives, offering effective pollutant adsorption capabilities in wastewater. Utilizing lignin as raw material facilitates the development of low-cost adsorbents. However, unmodified lignin confronts challenges due to its poor solubility, limited dispersion, and the inability of its unactivated structure and functional groups to yield sufficient pollutant adsorption efficiency (Okoronkwo and Olusegun 2013). The direct utilization of unmodified lignin for adsorption purposes exhibits drawbacks such as a restricted scope of application and low adsorption capacity. To enhance lignin's properties as an adsorbent and broaden its applicability, a series of chemical modifications have been undertaken. Given lignin's array of functional groups, it can undergo various reactions including oxidation, sulfonation, carboxylation, alkylation, amination, and copolymerization with other polymer monomers (Kai et al. 2016). However, despite advancements in lignin modification, limitations have arisen with singly modified lignin exhibiting drawbacks. Consequently, researchers have compounded lignin with other materials to create high-value lignin adsorbents endowed with specific functionalities (Wang et al. 2022). The resulting synthetic composites offer adjustable properties and enhanced adsorption performance, surpassing the capabilities of singularly modified lignin.

Lignin plays an essential role in the production of filtration membranes. As an example, the research team led by Wenge et al. succeeded in developing an environmentally friendly nanofiltration membrane based on lignin. To achieve this, they used alkaline lignin and sodium lignosulfonate as raw materials, and dopamine as a surface modifier (Sun et al. 2023). This breakthrough demonstrates the adaptability of lignin in the creation of sustainable solutions for diverse applications, notably in the field of environmentally-friendly filtration.

9.2.4 Alginate

Alginate, a natural polysaccharide from brown algae, is recognized for its high biocompatibility, biodegradability, and renewability. It stands as one of the extensively researched biopolymers for eliminating pollutants from aqueous solutions due to its affordability, nontoxic nature, and efficiency. Furthermore, its abundant hydroxyl and carboxyl groups exhibit a notable affinity for adsorbing heavy metal ions and dyes (Alraddadi et al. 2023). Hence, owing to the prevalence of numerous anionic groups, alginate emerges as a potential solution for tackling cationic pollutants present in wastewater, akin to cationic dyes.

However, alginate exhibits relatively low mechanical strength, stability, and heat resistance (Gao et al. 2017; Wang et al. 2018). Consequently, physical or chemical modifications are commonly employed to augment its suitability for heavy metal adsorption. Traditionally, modification methods for alginate-based adsorbents encompass surface grafting, cross-linking, and composite formation with other materials (Thakur et al. 2018). Surface grafting primarily enhances selectivity toward specific metal ions and elevates metal uptake capacities, whereas cross-linking can modify chemical resistance and mechanical strength (Jobby et al. 2018). Additionally, combining alginate with other materials further augments both adsorption capacities and the physical attributes of the composite materials (Wen et al. 2020).

9.2.5 Bacterial Cellulose

Bacterial cellulose, produced in vitro primarily by Acetobacter bacteria, shares a chemical structure similar to cellulose derived from higher plants, comprising glucose molecules linked by β-1, 4 glycosidic bonds (Cheng et al. 2017). However, in contrast to conventional plant-based cellulosic materials, bacterial cellulose boasts an unparalleled degree of purity, nearly reaching 100% cellulose content and devoid of common components found in plant fibers (Jin et al. 2017).

Distinguished by its cost-effectiveness, simple structure, high hydrophilicity, and biodegradability, bacterial cellulose offers unique chemical and physical properties that include exceptional water-holding capacity, heightened crystallinity, remarkable tensile strength, and moldability during fermentation traits that are notably advantageous for adsorption and filtration applications (Chen et al. 2016; Wang et al. 2023). Moreover, the micro-nanoscale fibers amplify contact area with pollutants, enhancing adsorption capabilities (Wu et al. 2016).

The extensive presence of hydroxyl groups along the backbone of bacterial cellulose allows for diverse chemical modifications through oxidation, acylation, silylation, and electrostatic interactions (Tamura et al. 2018). Previous literature has cited instances where bacterial cellulose has been utilized for adsorbing metal ions (Shen et al. 2009), dyes (Bai et al. 2023), and organic compounds (Ion et al. 2015).

It is also used as a base material for the fabrication of bioinspired adsorptive ultrafiltration membrane (BC/PDA14h/CNCs) for the removal of natural organic matter, lead ions, and organic dyes. The extracted bacterial cellulose is coated with polydopamine (PDA) as a functional layer and molecular bond to immobilize carboxylated cellulose nanocrystals (C-CNCs) to prepare the membrane. Bacterial cellulose serves as structural support for the membrane, providing a strong, three-dimensional structure crucial for the mechanical stability of the membrane

(Wang et al. 2023). The use of bacterial cellulose also extends to the creation of a bioinspired ultrafiltration membrane, aimed at effectively removing soluble dyes and insoluble oils (Hu et al. 2021). The development of this membrane, with exceptional separation capabilities and antifouling properties, is of particular importance in water treatment.

9.2.6 Polydopamine

Polydopamine (PDA) has obtained significant attention in environmental applications due to its remarkable biocompatibility, robust adhesion involving hydrogen bonding and metal chelation through catechol (Maier et al. 2015), and rich array of functional groups (Farnad et al. 2012). Its various functional groups, such as catechol, amine, and aromatic moieties, position PDA as a highly promising biosorbent with numerous active sites for the adsorption ~~of~~ dyes (Fu et al. 2016), heavy metals (Zhang et al. 2018b), and toxic organic pollutants by means of chelation and π–π stacking interactions (Lee et al. 2018). However, the potential of PDA microspheres is constrained by their propensity for easy aggregation and challenging separation.

Inspired by adhesive proteins in mussel feet, a nanofiltration membrane is developed using PDA as the base material (Liu et al. 2023). During this process, researchers modified the PDA deposition processes by incorporating a molecule called spiro-piperazine (SPIP), which contains two secondary amine groups and a quaternary ammonium salt. The resulting PDA-based nanofiltration membrane showed a significant improvement in water permeability while maintaining solute rejection capacity. On the other hand, PDA also plays a crucial role in the fabrication of ultrafiltration membranes for water treatment, such as the PDA@WS2 polyacrylonitrile membrane aimed at efficiently separating ~~of~~ oily saline wastewater and ~~the~~ removing soluble dyes (Ajibade et al. 2022). In another context, PDA is used in the fabrication of co-deposition-modified hydroxyapatite nanotubes, acting as environmentally friendly nanofillers in the preparation of hybrid ultrafiltration membranes. PDA is employed in the modification of hydroxyapatite nanotubes, enhancing the compatibility, structure, and hydrophilicity of the membrane, while contributing to synergy with other components for optimized performance. This membrane successfully filters water efficiently while retaining specific contaminants, such as oil and organic dyes, represented, respectively, by bovine serum albumin (BSA) and humic acid (HA) (Mu et al. 2020).

9.2.7 Biochar

Biochar, an enduring carbon-rich derivative resulting from the thermal decomposition of biomass, including forestry and agricultural wastes under oxygen-depleted conditions, manifests a composition marked by abundant phenolic hydroxylic groups, carboxylic groups, aromatic structures, hydroxylic groups, and aliphatic double bonds. Preserving its integrity, the original material's finely structured pores endow biochar with low solubility, substantial adsorption capacity, enduring chemical and physical stability, and an expansive pore structure, contributing to a considerable specific surface area (Yu et al. 2019). Various organic materials, such as sewage sludge, straw, rice husk, and animal manure, can serve as feedstocks for biochar production through a variety of encompassing slow pyrolysis, fast pyrolysis, thermochemical processes, gasification, flash carbonization, hydrothermal carbonization, and torrefaction (Tan et al. 2016). Biochar has emerged as a promising, cost-effective adsorbent for efficiently removing a variety of contaminants from water and wastewater (Paranavithana et al. 2016).

Bioinspired and biodegradable adsorbents and membranes offer distinctive and customized approaches to address different categories of contaminants, and significantly improve wastewater treatment technologies. The range of materials used illustrates the breadth of potential offered by bioinspired techniques, providing possibilities for continued development in the field of water treatment.

9.3 Synthesis of Bio-Nanomaterials Using Bioinspired/Biodegradable Materials: Step-by-Step Process

Bioinspired and biodegradable materials are integral components in the synthesis of bio-nanomaterials, influencing not only their structural composition but also imbuing them with environmentally friendly characteristics. This is particularly evident in the dynamic field of wastewater treatment, where the application of these materials is steadily evolving to provide sustainable and effective solutions. Here are several methods through which these materials are employed in the synthesis of bio-nanomaterials.

9.3.1 Cellulose Nanocrystals

Cellulose nanocrystals (CNCs) are the crystalline segments derived from cellulose, which exhibit exceptional physicochemical properties including low density, extensive surface area, significant pore structure, high crystallinity, remarkable stiffness, biocompatibility, high aspect ratio, and numerous functional groups. They also offer possibilities for surface modification, are nontoxic, hydrophilic, and exhibit high transparency (Aziz et al. 2020), making them valuable in various applications, particularly as adsorbents within wastewater treatment systems (El-Fattah et al. 2024; Fan et al. 2019). CNCs are nanorod structures with the same chemical composition as plant cellulose, sourced from materials such as wood, pulp, cotton, and straw. The production of CNCs involves preprocessing methods to remove lignin and hemicellulose, ensuring the obtainment of highly pure cellulose used in their preparation (Zhang et al. 2023). CNCs can be obtained through the disintegration of cellulose fibers into nanoscale particles using chemical, biological, or mechanical methods (Zhang et al. 2020). Figure 9.2 provides a summary of the various steps involved in the production of cellulose nanocrystals.

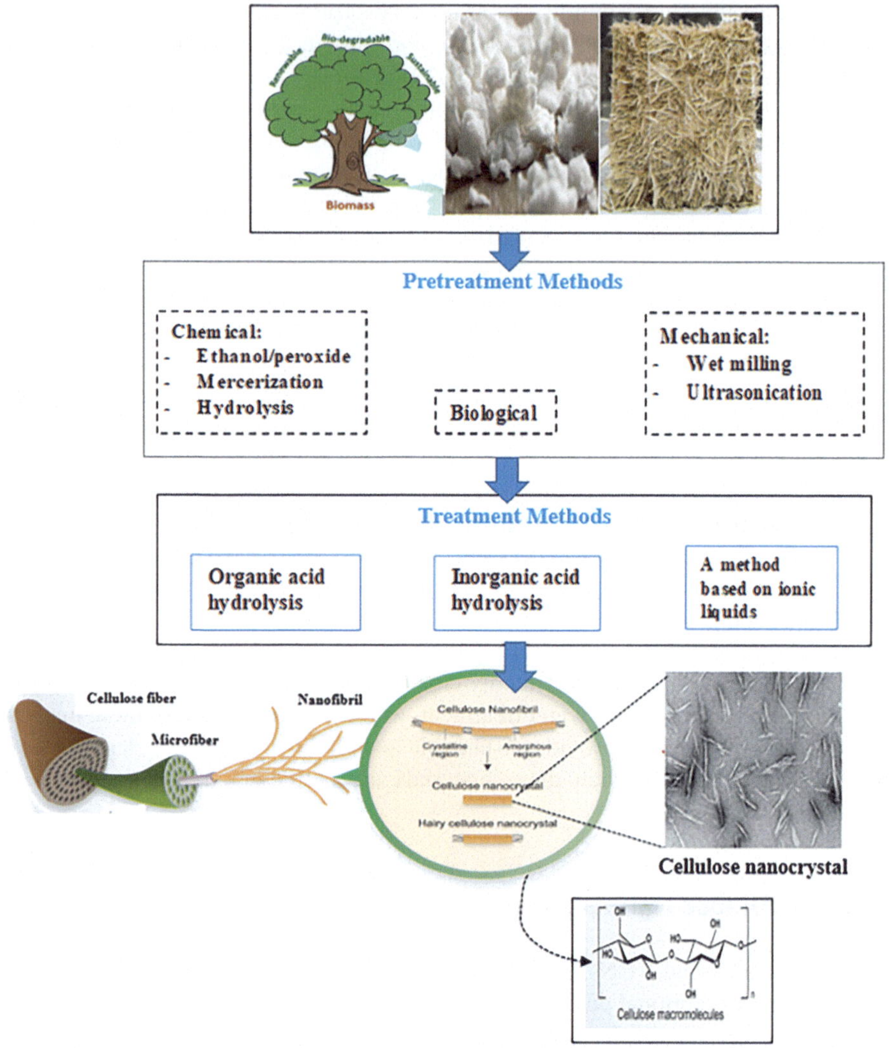

Fig. 9.2 Summary of the production steps of cellulose nanocrystals

Chemical pretreatment includes ethanol or peroxide treatment to decellulose lignocellulose and remove noncellulosic components. Involving NaOH to remove hemicelluloses and lignin, and acid hydrolysis, with H_2SO_4, as a catalyst. Ethanol and peroxide pretreatment, specifically in wood processing, involves heating in an ethanol/water solution with H_2SO_4, followed by NaOH and H_2O_2 treatment. Mercerization immerses cellulose fibers in a concentrated NaOH solution, and acid hydrolysis uses H_2SO_4 as a catalyst. While chemical pretreatment remains the most frequently employed method, the combination of traditional chemical methods with new physical and biological approaches is of particular interest, as this synergy significantly reduces the use of toxic chemicals such as H_2SO_4 (Zhang et al. 2023).

Two distinct approaches are used for mechanical pretreatment, namely the wet disk milling (WDM) method and ultrasonication (Teixeira et al. 2015; Qin et al. 2022). The WDM method involves reducing the particle size of a material by subjecting it to shear forces between two nonporous ceramic disk mills rotating at high speed to fibrillate the cell wall down to the nanoscale. This process, called WDM, generates long, twisted microfibers that, after hydrolysis, can be transformed into CNCs. Ultrasonication offers significant advantages in terms of high efficiency, increased crystallinity, and mild reaction conditions (Zhang et al. 2023). Biological pretreatment involves breaking down fibrous structures with cellulose enzymes, converting more carbonyl groups into carboxyl groups, which can be modified with sulfate groups in the subsequent acid hydrolysis process. This contributes to increasing the charge density of cellulose particles, thereby promoting stable dispersion of CNCs (Zhang et al. 2023).

After undergoing pretreatment, the acquired materials can be employed in the production of CNCs through various preparation approaches, including inorganic acid hydrolysis, organic acid hydrolysis, and a method based on ionic liquids. The most prevalent technique is inorganic acid hydrolysis, with sulfuric acid being extensively employed to conduct the process and remove amorphous cellulose (Zhang et al. 2023). Sulfate groups are introduced onto the surface of nanocrystals, promoting rapid dispersion of CNCs in aqueous and organic environments. After acid hydrolysis at 45 °C, the initial cellulose microfibers expand and separate into three forms of smaller crystalline cellulose products with rod-like, spherical (the most abundant form), and porous network morphologies (the least abundant) (Lu and Hsieh 2010). Manisha et al. prepared CNCs from rice straw using sulfuric acid treatment with a concentration of 75%, at a temperature of 30 °C for 5 h. The obtained yield is 90.28% (Thakur et al. 2020).

Organic acid hydrolysis is also employed in the preparation of CNCs. This method, both recyclable and mildly corrosive, relies on high-yield oxidation degradation. The organic acids used, such as maleic acid, can be easily recovered through low-temperature crystallization after the hydrolysis process. Following treatment with these dicarboxylic acids, the resulting surface of CNCs presents dispersed carboxyl groups, facilitating dispersion and allowing for subsequent functionalization (Zhang et al. 2023). Zhu et al. successfully synthesized phosphorylated cellulose nanocrystals (PCNCs) through ultrasound-assisted maleic acid hydrolysis with a 75% weight concentration and an acid hydrolysis time of 2.5 h at 120 °C. In this study, a higher yield of PCNCs was obtained (Zhu et al. 2023). On the other hand, a method based on ionic liquids has been employed for the preparation of functional CNCs. Using this ionic liquid, which is a salt melting at temperatures below 100 °C, the hydrolysis step can be carried out under mild conditions. This improves substrate accessibility while preserving its original crystalline regions (Y. Zhang et al. 2023). Abushamala et al. described the direct extraction of CNCs from wood using 1-ethyl-3-methylimidazolium acetate as an ionic liquid. They successfully recovered native cellulosic product with a yield of 44% relative to the wood cellulose content (Abushammala et al. 2015).

9.3.2 Polymer Nanocomposites

Polymer nanocomposites (PNCs) represent an advanced category of materials, combining the advantageous properties of both polymers and nanomaterials (Khodakarami and Bagheri 2021). This integration results in unique physical and chemical characteristics derived from the substantial surface area to volume ratios, exceptional mechanical properties within the polymer matrix, and the heightened interfacial reactivity of nanofillers (Pan et al. 2009). Furthermore, their compatibility and ability to be regenerated and reused further underscore their significance (Nassar 2010). These attributes position PNCs as a promising class of materials for wastewater treatment (Nassar 2010). Natural polymers, such as silk, wool, DNA, cellulose, and proteins (Ghorbani and Eisazadeh 2013), sourced from nature, are advantageous support material for PNCs fabricating due to their ease of formation, exceptional pore-forming attributes, selective chemical species transfer, simple functionalization, and comparatively economical nature (Adeola and Nomngongo 2022).

For the production of PNCs, many strategies are being used. They can be primarily classified into three processes: exfoliation/intercalation, in-situ polymerization, and sol–gel. Although other approaches, like model synthesis, phase separation, auto-assembly, and electrofiling, have been employed for laboratory research, they are not as prevalent at the industrial scale (Das et al. 2023). Intercalation or exfoliation method aims to manipulate bulk fillers at the nanoscale by means of intercalation and exfoliation. Intercalation, carried out chemically or mechanically, involves the insertion of organic polymers between layers of filler materials, weakening the Van der Waals forces between adjacent layers. The mechanical method, involving polymer melting under high shear conditions, induces exfoliation by dispersing polymers between layers. The hydrophilic layered materials are previously modified with hydrophobic materials to ensure their compatibility with the hydrophobic polymers, acting as bridges. Figure 9.3a illustrates the structure of nanocomposites, showing how nanofillers can be intercalated or exfoliated in the polymer matrix

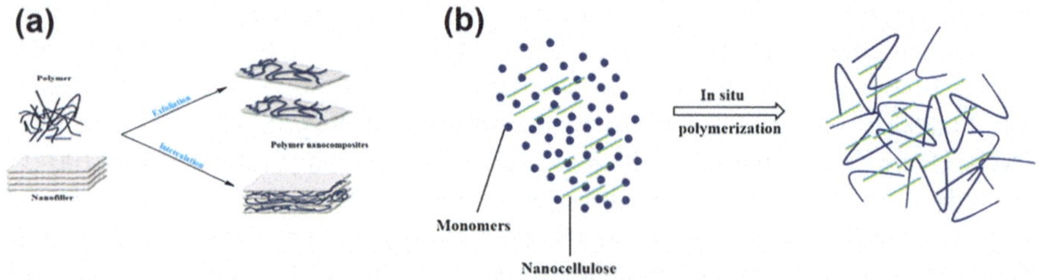

Fig. 9.3 (**a**) Exfoliation/intercalation, (**b**) in-situ polymerization processes for the production of PNCs

(Silva et al. 2019). The major distinction between intercalated and exfoliated nanocomposites lies in the Van der Waals forces between the layers, weakened or eliminated respectively, under the influence of external forces (Das et al. 2023).

In the in-situ polymerization method, initially synthesized nanomaterials are carefully mixed with monomers, emphasizing the importance of achieving a uniform distribution of these nanocharges (Fig. 9.3b). The nanocharges must be evenly dispersed throughout liquid monomers or monomers that contain solvants in order to achieve this homogeneity. Then, low-molecular-weight monomers settle in between the nanocharge gaps, creating conditions for polymerization, which can be accelerated by physical or chemical activation or by using particular initiators. The polymerization occurs in the spaces between the nanocharges, leading to the creation of PNCs, where nanocharges are incorporated into the newly formed polymer matrix. This method allows control over particle shape, strong interfacial adhesion of the nanocharges, high transparency, and the ability to use high nanocharge concentrations without agglomeration, improving product performance. However, there is a drawback, which is that unreactive precursors or by products of in-place processes may adversely affect the properties of the nanocomposite, emphasizing the need to carefully monitor this intricate process (Guo et al. 2014; Das et al. 2023).

The sol–gel method is an ascending approach that primarily involves two reaction steps. First, a colloidal suspension of solid particles in a liquid phase known as "sol" occurs, while the creation of an interconnected network between the phases, dubbed "gel," occurs during the second stage (Oliveira and Machado 2013). This method converts sol into gel. Employing this approach, both polymers and nanofillers are simultaneously formed through an in situ process, resulting in the creation of polymer nanocomposites (PNCs). The majority of the time, these PNCs organic matrix contains uniformly distributed inorganic nanocharges. The sol–gel method stands out as an environmentally friendly PNCs production method since it eliminates the need for washing when using raw materials that introduce no impurities into the finished product (Pomogailo 2005). This approach provides scientists and engineers with the opportunity to create a new generation of sophisticated PNCs with unique characteristics. Adopting the sol–gel method has the particularly intriguing benefit of being compatible with polymers and polymerization processes, allowing for the synthesis of nanoparticles in the presence of organic molecules (Oliveira and Machado 2013).

9.3.3 Carbon-Based Nanomaterials

Carbon-based nanomaterials extensively studied for their unique properties including small size, high surface area to volume ratio, elevated reactivity, and remarkable thermal and chemical stability have gained immense attention in water and wastewater remediation (Akharame et al. 2018). These nanomaterials have proven highly in removing a range of pollutants, including nitric oxide, hydrogen sulfide, heavy metals, dyes, and drug compounds, achieving notable adsorption efficiencies exceeding 80% and photocatalytic degradation efficiencies surpassing 98%

(Jayaraman et al. 2018). These materials, existing in diverse allotropic forms like carbon nanotubes, and graphene nanomaterials, offer extensive active sites for interaction with various chemical species found in wastewater.

Carbon nanotubes (CNTs) can be synthesized using a variety of methods, including hydrothermal, chemical vapor deposition (CVD), and arc-discharge synthesis methods (Paradise and Goswami 2007; Varshney 2017; Rao et al. 2021). In the hydrothermal process, a reaction mixture including metal catalysts and carbon precursors is first prepared in an aqueous solution typically utilizing nickel as the main catalyst other metal precursors such as polyethylene or ethylene glycol can also be employed. After that, this combination is placed inside a hydrothermal chamber that can sustain high pressures (60–100 MPa) and temperatures (700–800 °C). A complicated chemical process occurs when heat and pressure are applied, and the metal catalyst helps the carbon precursors separate, resulting in the formation of carbon nanotubes. The final size of nanotubes, shape, and other characteristics are determined by the particulars of the hydrothermal synthesis process (Rao et al. 2021). There are multiple advantages for this strategy over alternative methods. The hydrothermal process operates at a low temperature (150–180 °C) and doesn't require hydrocarbons or carrier, as the starting components are easily obtained and stable at room temperature (Varshney 2017).

The process of chemical vapor deposition (CVD) involves the decomposition of carbon-rich gaseous precursors over a metal catalyst. Essentially, supported transition metal catalysts are used in the catalytic decomposition of a hydrocarbon or carbon monoxide feedstock in CVD. The two phases of the CVD process include the deposition of metal catalysts (Fe, Ni, Co, etc.) on a substrate, followed by nucleation ~~is~~ achieved using heat annealing or chemical etching procedures, often utilizing ammonia as the etchant. The carbon source is added to the reaction chamber in the gas phase in the second step. Next, using plasma energy or a heated coil, the carbonaceous material such as methane, carbon monoxide, or acetylene, is converted to the atomic level. Subsequently, the carbon permeates to the substrate covered in catalyst. The interaction between the carbon and the substrate leads to the formation of CNTs. The temperature used for nanotube synthesis ranges from 650 to 900 °C, with a typical yield of around 30% (Paradise and Goswami 2007; Arunkumar et al. 2020; Rao et al. 2021).

Two carbon electrodes are submerged in a helium environment as part of the arc-discharge synthesis process. The graphite on the electrodes is vaporized and afterward condenses on the cathode and reactor walls by applying an intense electric current of around 50 amperes between these electrodes. The carbon nanotube-containing deposit is known as the cathode deposit. Depending on the parameters of the arc discharge, these CNTs can take on different shapes, such as single-walled nanotubes (SWNTs), which are made of a single layer of carbon atoms, or multi-walled nanotubes (MWCNTs), which are created from numerous concentrically connected nanotubes. Metals like Co and Ni in the anode can be used to create SWCNTs (Varshney 2017; Rao et al. 2021).

Graphen derivatives, such as graphene oxides, graphene quantum dots, graphene nanoplatelets, and graphene nanoribbons, are also classified among carbon-based

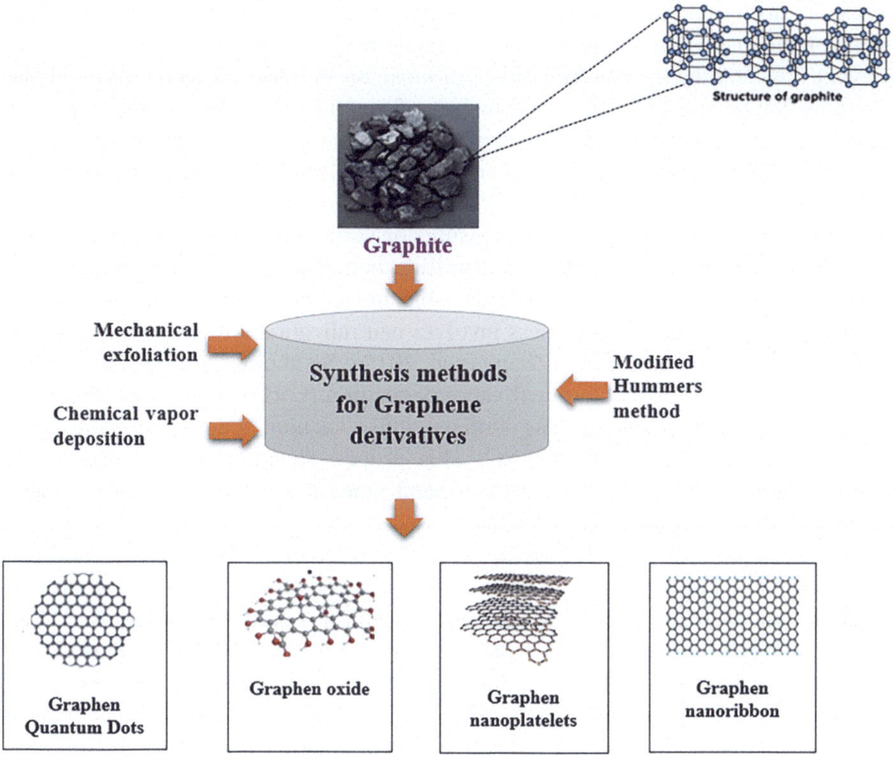

Fig. 9.4 Production steps of nanographene materials through various methods

nanomaterials. The synthesis of these various nanomaterials is achieved from graphite via a variety of methods (Saravanan et al. 2022). Figure 9.4 provides an overview of the steps involved in producing nanographene materials using different methods.

Several mechanical exfoliation methods, including micromechanical cleavage, sonication, and ball milling are employed for the synthesis of graphene-based nanomaterials (Saravanan et al. 2022). Graphite is processed using the micromechanical exfoliation method resulting in high-quality graphene thin films. This approach employs three primary techniques. The first involves graphene synthesis using scotch tape, where adhesive tape is applied to a graphite surface to peel off graphene layers, but this method is labor-intensive and yields are low. The second technique uses an ultra-sharp wedge to reduce graphite layers achieving a success rate of over 50%. The third method inspired by the scotch tape method employs a three-roll mill with rotating rollers to evenly distribute graphite using adhesive, followed by exfoliation, similar to the scotch tape method. Pure graphene with a single layer thickness is obtained by removing the adhesive materials after soaking in alcohol and heating to 500 °C (Bhuyan et al. 2016; Saravanan et al. 2022). Sonication, as another exfoliation method, involves dissolving graphite powder in solvents like N-methylpyrrolidone and N-dimethylformamide, followed by sonication to produce pure graphene. The resulting graphenes are separated by centrifugation, with potential issues of low yields due to graphene re-stacking during sonication. To prevent this, stabilizing materials like cellulose are introduced, creating a solvent based on NaOH and urea. The combination of graphite and cellulose facilitates the exfoliation of graphite into individual graphene layers during sonication, preventing re-stacking. Stable graphenes can then be isolated from the sonicated mixture by centrifugation (Yu et al. 2021a). Ball milling, the third method, exfoliates graphite into graphene flakes through shear pressures or vertical impacts. Both wet and dry ball milling can be utilized, with wet milling being more common (Zhang et al. 2018a). For example, graphite powder is subjected to ball milling, resulting in edge-carboxylated graphene. The process involves neutralization with strong hydrochloric acid and subsequent drying (Zhang et al. 2018a; Sivakumar et al. 2020).

Graphene synthesis by chemical vapor deposition (CVD) begins with the introduction of a precursor gas, such as methane, into a vacuum chamber containing the substrate heated to around 1000 °C. This process decomposes the precursors, adsorbs them onto the substrate surface, and generates a thin layer of graphene about 1 atom thick (Ashraf et al. 2022).

The fabrication of graphene oxides (GOs) is often carried out by the modified Hummer method, where graphite flakes undergo the Hummer process. After preparing a mixture of graphite, H_2SO_4, and KNO_3, $KMnO_4$ is gradually added. At 35 and 90 °C, the mixture forms a black paste. Hydrogen peroxide (H_2O_2) is released over and develops into a dark brown solution at 90 °C (Alkhouzaam et al. 2020). To produce a homogenous dispersion, the solid particles are separated and placed under sonication following neutralization. Centrifugation is used to separate the resultant graphene oxides (GOs) (Sujiono et al. 2020). An alternative approach synthesizes graphene oxide composites of ferrihydrite and goethite using $NaNO_3$. First, the standard Hummers process generates graphene oxide (GO) and ultrasonic irradiations applied to produce the required graphite oxide. The process of degrading graphene oxide (MGO) using $NaNO_3$ modifies its surface, and co-precipitation yields the necessary graphene oxide composites, enhancing the extraction of As(V) ions from wastewater. According to reports, GO composites have an adsorption capability of approximately 66.88 mg/g for As(V) ions (Vazquez-Jaime et al. 2020).

9.3.4 Chitosan Nanoparticles

Chitosan nanoparticles, derived from chitosan, combine the inherent characteristics of chitosan with distinctive nanoparticles properties. It stands out as exceptional nanoadsorbents owing to their substantial surface area, high adsorption capacity, and eco-friendly attributes (Gopinath et al. 2021). Their natural abundance coupled with the ease of modification using diverse chemical and biological agents allows for regeneration and multiple reuses across cycles. Meeting the crucial criteria for wastewater treatment, these nanoparticles exhibit chemical inertness and morphological resilience in complex conditions (Benettayeb et al. 2023). Their preparation not only enhances surface area and adsorption capacity but also endows them with selectivity due to the presence of functional groups (Yu et al. 2021b). Chitosan's biodegradability and antibacterial properties contribute its effectiveness as a water treatment adsorbent (Divya and Jisha 2018). The preparation of chitosan nanoparticles is carried out following two steps below. First, extraction and deacetylation of chitosan from chitin takes place in several stages, utilizing shrimp residues from the food industry. These residues are then ground into smaller particles, and the minerals, mainly calcium carbonate, are removed by demineralization and decalcification using dilute hydrochloric acid, while constantly stirring at room temperature (Hussain and Berry 2024). Then chitosan production is carried out by alkaline treatment (deproteinization), followed by deacetylation using a NaOH solution to form chitosan (Islam et al. 2017). Islam et al. developed chitin from shrimp head peels (Islam et al. 2023). These peels, previously dried, were mixed with 5% hydrochloric acid under stirring for 1 day at room temperature. Then the reaction mixture was filtered, and the residual skin was washed with distilled water, then sun-dried. The demineralized head skin was combined with a 5% aqueous sodium hydroxide solution and stirred at 70 °C for 3 h. The deproteinized skin was rinsed with distilled water and left to dry in the sun (30–33 °C) for 2 days. The resulting chitin was mixed with 70% aqueous sodium hydroxide and stirred at 80 °C for 4 h. After cooling, the mixture was filtered, and the residue was washed with distilled water before being sun-dried. The final product resulting from this process is chitosan (Islam et al. 2023).

Second, chitosan nanoparticles are produced from chitosan by the ionic gelation method, using sodium tripolyphosphate (STPP) anions (Dahmane et al. 2016). The process begins by dissolving chitosan in an acetic acid solution. Next, an aqueous solution of STPP is added dropwise to the chitosan solution, while stirring. This step leads to the instant formation of an opalescent suspension. The nanoparticles thus generated are recovered by centrifugation. After removal of the supernatant, the wet pellet of chitosan nanoparticles is collected, washed with ethanol and the material lyophilized and stored at 4 °C for further analysis (Chakraborty et al. 2018).

9.3.5 Biochar-Based Nanocomposites

The synergistic integration of nanomaterial advantages with functional groups presents in pyrolyzed biochar, such as hydroxyl, carboxyl, and amino acids, holds the potential for enhancing the characteristics of biochar nanocomposites. These functional groups, notably hydroxyl and carboxyl, play a pivotal role in water purification applications, improving contaminant removal (Amusat et al. 2021). Furthermore, the substantial specific surface area inherent to both nanomaterials and biochar contributes to the heightened efficiency of the resulting nanocomposites. Research has delved into the incorporation of nanomaterials into biochar, revealing significant improvements in its physicochemical properties. In pure biochar utilization for water and wastewater purification, (Rodriguez-Narvaez et al. 2019) certain limitations such as low surface area or limited functional groups may reduce its effectiveness compared to modified biochar. It is important to acknowledge that the synthesis of a biochar-based nanocomposite goes beyond simply improving physical or chemical properties; rather, it signifies the creation of a novel composite that combines the inherent advantages of biochar with those of other nanomaterials. Several techniques are used to produce biochar, including pyrolysis, gasification, hydrothermal processes, and microwave pyrolysis (Yu et al. 2024).

Biochar nanomaterials are synthesized from ordinary biochar through various techniques including, ball milling, hydrothermal reactor, and sonication. In ball milling, biochar particles are fed into a ball mill, where they are subjected to intense mechanical forces resulting from the rotation of the drum and the impact of the grinding balls. This process causes the biochar particles to shrink to nanometric size, generating biochar nanomaterials. This method is environmental-friendliness, and offers promising prospects for environmental sustainability and waste recovery, producing artificial biochar for advanced applications. Another method starts with producing initial biochar from agricultural residues and by-products through pyrolysis. This initial biochar is then subjected to acid digestion with concentrated nitric and sulfuric acids in a high-pressure hydrothermal reactor, to break it down into nanobiochar. The suspension obtained after acid digestion is isolated by vacuum filtration to remove larger particles. Finally, the filtered suspension is dried in an oven to obtain the final nanobiochars. This process, described as simple, fast and economical, offers an achievable method of producing biochar nanomaterials that could find use in several environmental areas, including water treatment. Sonication was also used to generate nano-sized particles. Oleszczuk et al. physically dispersed the biochars employing an ultrasonic vibrator. To disperse the biochars, a biochar mass was stirred in water and treated using a circulating water system to preserve a temperature below 20 °C. Then the aqueous suspension was filtered and centrifuged after sonication to retain the nanometric particles suspended in the supernatant. To extract the biochar nanoparticles, the suspension was centrifuged and then the method was repeated (Oleszczuk et al. 2016).

9.4 Key Factors Influencing the Synthesis of Bio-Nanometallic Materials

The synthesis of bio-nanometallic materials for wastewater treatment is influenced by various factors, significantly impacting their efficacy in eliminating pollutants from wastewater. Several pivotal factors influencing synthesis of bio-nanometallic materials for wastewater treatment include the choice of biomaterials, such as bacterial biomass, fungal biomass, and plant extracts, exhibit distinctive properties that impact the reduction and stabilization of metal ions. The selection of a specific biomaterial can shape the synthesis mechanism and the ultimate properties of the bio-nanometallic materials.

Metal precursors, usually in the form of metal salts, play a pivotal role in dictating the nature of metallic nanoparticles produced. Varieties such as silver nitrate, gold chloride, or copper sulfates are common choices. The reactivity and stability inherent in these precursors exert influence over the nucleation and growth of nanoparticles throughout the synthesis process.

Essential reaction parameters such as temperature, pH, pressure, and reaction time, play crucial roles with variations impacting the reduction of metal ions, nanoparticles characteristics, and distribution. On the other hand, the concentration of biomaterials and metal precursors affects the reduction rate and nanoparticles size. Elevated concentrations may accelerate the reduction process, albeit with the potential drawback of particle agglomeration. It becomes imperative to optimize these concentrations to effectively control both particle size and stability. Stabilizing agents like surfactants or capping agents prevent nanoparticle agglomeration by creating a protective layer on their surface, thus inhibiting undesired growth and preserving colloidal stability.

Effective agitation and mixing are essential for the uniform distribution of reactants, facilitating the nucleation and growth of nanoparticles. The selection of a mixing technique, such as magnetic stirring or ultrasonication, plays a crucial role in shaping the size and morphology of the nanoparticles. Understanding the composition of wastewater, including the varieties and concentrations of pollutants, is vital for customizing the characteristics of bio-nanometallic materials. This understanding informs the design of materials with a high affinity for particular pollutants found in the wastewater.

In conclusion, the synthesis of bio-nanometallic materials necessitates a meticulous examination of these factors to attain the specified properties and functionality required for efficient wastewater treatment. The strategic implementation of experimental design and optimization becomes imperative to precisely adjust these parameters, ensuring the reproducibility and reliability of the synthesized materials.

9.5 Elimination of Water Contaminants Using Bioinspired/Bio-Nanomaterials

9.5.1 Adsorption

Two essential elements for an effective assessment of adsorbents are high selectivity and rapid adsorption of the specified pollutants. This section explores bioinspired and bio-nanomaterials designed for the adsorptive removal of diverse water contaminants. The objective is to explore how these materials play a crucial role in eliminating water contaminants through adsorption, while also highlighting their general applicability and efficacy. Sirajudheen et al. focused on creating an efficient matrix comprising chitosan-reinforced graphene oxide-hydroxyapatite (CS@GO-Hap), using the co-precipitation technique under a nitrogen atmosphere (Fig. 9.5) (Sirajudheen et al. 2020). Chitosan played a pivotal role as a host during the preparation, facilitating the generation of the composite on its surface. Notably, the composite demonstrated significant adsorption efficiencies of 41.32, 40.03, and 43.06 mg/g for removing the Acid Red 1, Reactive Red 2, and Congo Red (CR) dyes from aqueous solution, respectively. Remarkably, heightened adsorption was

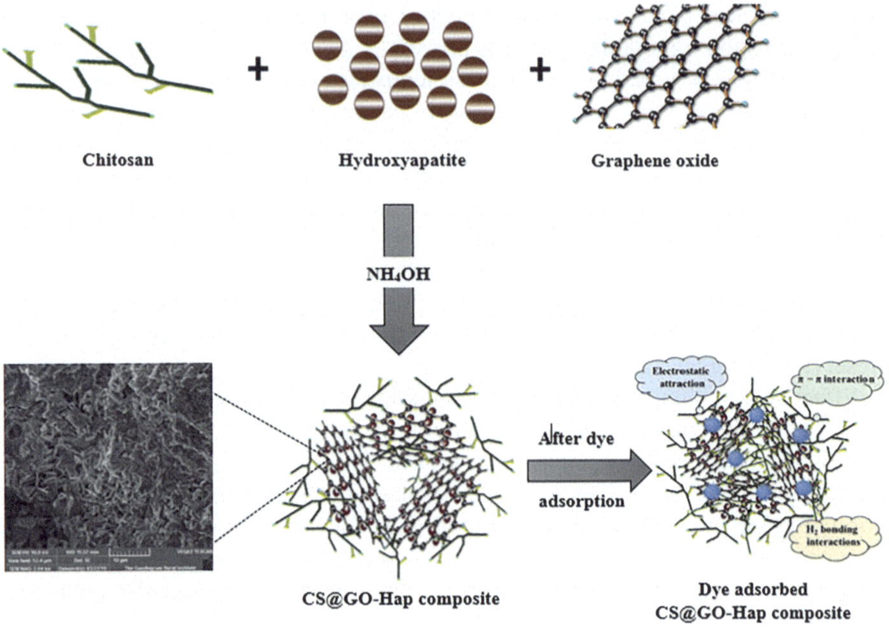

Fig. 9.5 Adsorption of dyes by CS@GO-Hap composite (Sirajudheen et al. 2020)

observed at a pH level of 2. However, the efficiency of removing substances by the CS@GO-Hap composite decreased as the pH increased. The adsorption mechanism on the composite involved hydrogen bonding, electrostatic forces, $\pi - \pi$ stacking interactions, and surface complexation. The study revealed that the adsorbent composite is highly effective in removing dyes from water solutions, highlighting its significant potential for future applications in water and wastewater treatment processes.

Furthermore, various cellulosic materials, including husks, fibers, shells, roots, leaves, seeds, stems, and barks, serve as adsorbents in diverse applications. In a study conducted by Johari et al., coconut husk (CH), coconut fiber (CF), and comprising coconut pith (CP) were employed as an adsorbent for eliminating elemental mercury (Johari et al. 2016). The researchers modified the surfaces of CP and CF using mercerization and bleaching techniques, leading to significant alterations in the structural characteristics and functional groups on the surface of these samples. These modifications resulted in distinct adsorption performances. The adsorption capacities for elemental mercury varied among the treated materials. CP-NaOH exhibited the peak adsorption capacity at 956.282 mg/g, succeeded by CP-Pristine, which recorded 730.250 mg/g, CF-NaOCl at 639.948 mg/g, CF-H_2O_2 at 634.347 mg/g, CF-NaOH at 611.678 mg/g, CF-H_2O_2 at 531.277 mg/g, CP-NaOCl at 501.126 mg/g, and CF at 431.773 mg/g. In another study, Boyu Du et al. developed a set of versatile lignin-derived magnetic adsorbents with recyclable properties (Fig. 9.6). These adsorbents were synthesized using carboxymethylated lignin (CL),

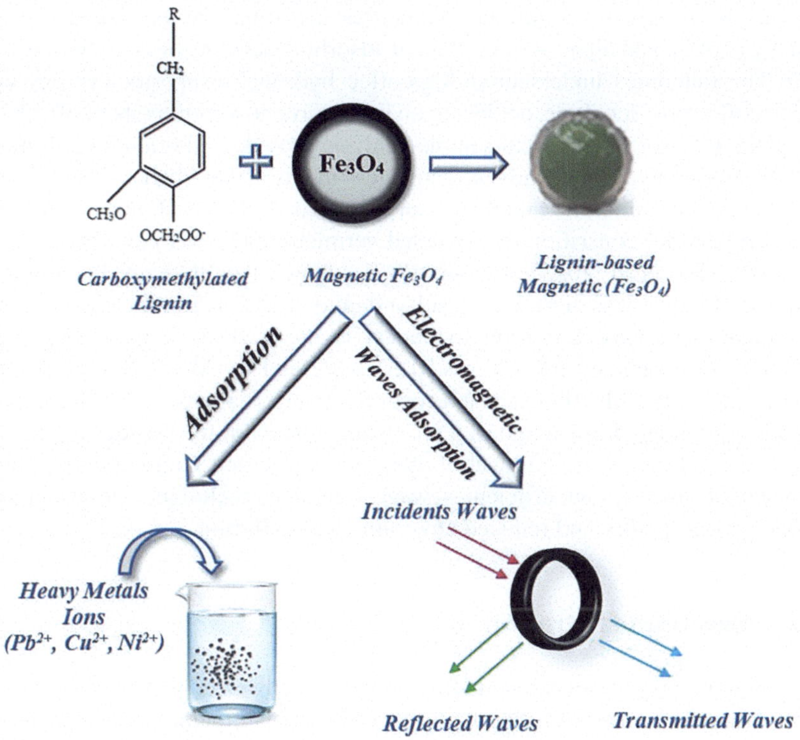

Fig. 9.6 Lignin-based adsorbent for efficient heavy metal ions removal and electromagnetic wave absorption (Du et al. 2023)

characterized by a high concentration of carboxyl groups (-COOH) (Du et al. 2023). Through the optimization of the mass ratio between CL and magnetite (Fe_3O_4), exceptional adsorption capabilities for heavy metal ions were achieved. The resulting adsorbent CL/Fe_3O_4 (3:1) demonstrated the maximum adsorption capacities of 106.97 mg/g for Ni^{2+}, 124.43 mg/g for Cu^{2+}, and 189.85 mg/g for Pb^{2+} ions. The prepared adsorbent showcased compelling microsphere structures and exceptional morphological distribution, providing a larger exposed specific surface area for enhanced adsorption site accessibility. Moreover, the composite CL/Fe_3O_4 (3:1) demonstrated remarkable capabilities in absorbing electromagnetic waves, displaying a reflection loss of −28.65 dB at 6.96 GHz with a thickness of 4.5 mm. This composite's effective absorption bandwidth covered a span of 2.24 GHz, extending from 6.08 to 8.32 GHz. Beyond its proficiency in heavy metal ion adsorption, the multifunctional CL/Fe_3O_4 (3:1) magnetic adsorbent exhibits outstanding electromagnetic waves characteristics, opening up diverse opportunities for utilizing lignin and lignin-based adsorbents.

On the other hand, the composite hydrogel known as Alg/coffee, incorporating sodium alginate and coffee waste, was developed by Haneen et al. (Alraddadi et al. 2023) serving as an effective adsorbent for eliminating both inorganic and organic pollutants from aqueous solutions. Numerous investigations on selectivity have shown that coffee and algae are excellent at adsorbing acridine orange dye (AO) and Pb(II). The outcomes underscored Alg/coffee hydrogel's superior efficacy compared to coffee powder alone, achieving an impressive adsorption rate of 80.53% for AO and 98.44% for Pb(II). Real sample analysis provided additional confirmation of the Alg/coffee hydrogel beads' effectiveness in adsorbing Pb(II).

For the goal of antibiotic and dye adsorption, Xiuzhi Bai et al. used an inventive composite material consisting of expanded vermiculite (EVMT) and bacterial cellulose (BC) (Bai et al. 2023). The study investigated the BC/EVMT composite's adsorption effectiveness in removing sulfanilamide (SA) and methylene blue (MB) from aqueous solutions. According to the results, the BC/EVMT composite's maximum adsorption capacity for SA was 71.53 mg/g and for MB it was 92.16 mg/g. However due to its high efficiency and cost-effectiveness, the BC/EVMT composite material is anticipated to emerge as a promising adsorbent for wastewater, offering a potential solution for the removal of dyes and antibiotics. Consequently, it holds significant promise as an efficient sewage treatment technique, contributing to improved water quality and reduced environmental pollution.

9.5.2 Membrane Filtration

Today, membranes play an extended functionality in advanced water treatment technology, crucially removing harmful substances such as organic and inorganic pollutants, heavy metals, and biomedical contaminants (Rabiee et al. 2023). Various types of membrane, including ceramic and polymer membranes, are used in wastewater treatment. Bioinspired membranes are also used in this context. Research focused on the treatment of water pollutants using bioinspired membranes represents an innovative field. These membranes, designed on the basis of biological processes, have unique for efficient removal of contaminants. Hu et al. (2021) prepared a bacterial cellulose-based ultrafiltration membrane for the efficient removal of soluble dyes and insoluble oils. In this study, researchers evaluated the performance of ultrafiltration (UF) membranes using Congo Red (CR) as a model dye. Results show that for pure water flux test, the membrane exhibited high permeability of 1149.3 $L\ m^{-2}\ h^{-1}$ under 0.1 MPa and a high rejection rate of 99.1% for Congo Red, outperforming other membranes dynamic oil repellency tests showcased the membrane's ability to rapidly repel oil droplets without apparent deformation on contact. BC/PDA/RGO composites showed high oil rejection rates (>99%) for epichlorohydrin, outperforming BC and BC/PDA membranes. Despite variations in rejection rates for different emulsions, the BC/PDA/RGO membrane maintained high rejection rates (>98%) for all organic solvents and water-immiscible oils (Fig. 9.7).

Similarly, Congo Red was removed by an ultrafiltration membrane designed by Liu et al. (2023). This membrane, developed on the basis of polydopamine (PDA) for the specific purpose of filtering dyes, exhibited a high permeability, approaching

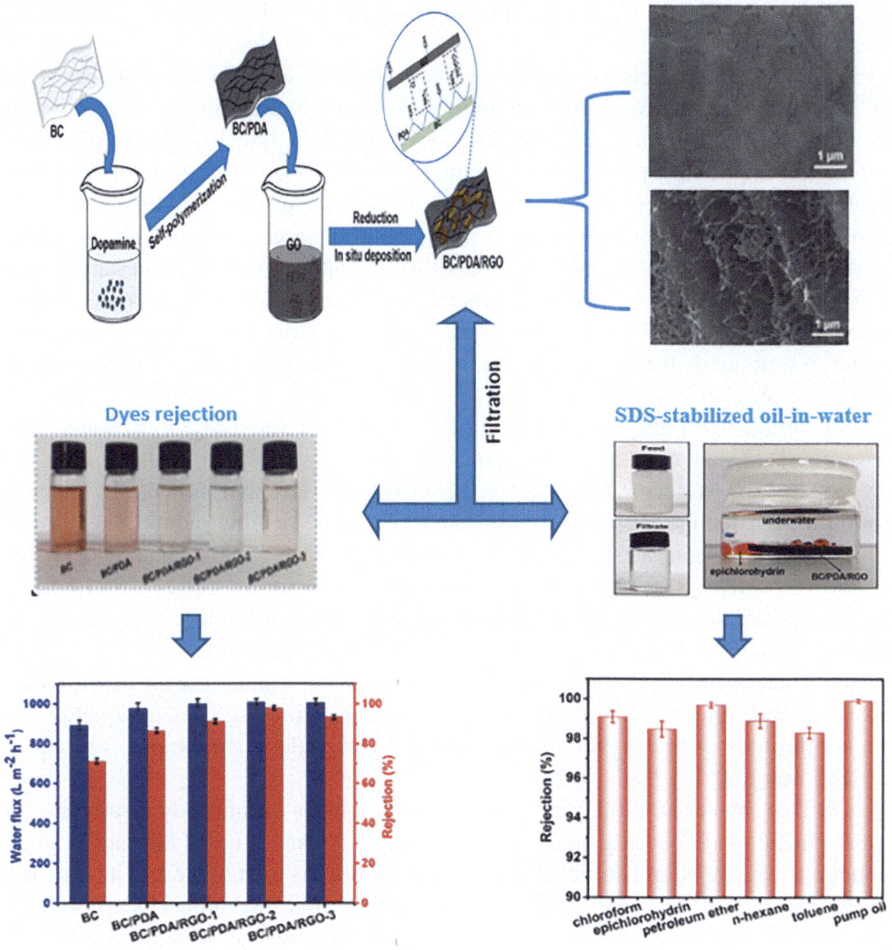

Fig. 9.7 Schematic representation of the bioinspired fabrication process for BC/PDA/RGO membrane (Hu et al. 2021)

73 L.m^{-2}.h^{-1}.bar. Results showed that this membrane was able to reject practically over 99% of Congo Red.

The coexistence of soluble dyes and oily saline wastewater in water can generate significant toxicity threat affecting both the aquatic ecosystem and human health. Oily saline wastewater, often resulting from industrial activities or production processes, frequently contains harmful compounds such as undissolved oils, heavy metals, and other toxic chemicals. To address these pollution problems, a study introduced a bioinspired PDA@WS2 polyacrylonitrile ultrafiltration membrane designed for efficient separation of oily saline wastewater and removal of soluble dyes. Results indicate that this membrane features high water permeability and demonstrates sufficient separation of oil and dye, up to 99.53% and 98.23%, respectively (Ajibade et al. 2022).

In alternative research, the authors evaluated the efficiency of ultrafiltration (UF) membranes using bovine serum albumin (BSA) and hyaluronic acid as contaminants in cyclic filtration experiments. At the same time, nature-inspired co-deposition-modified hydroxyapatite nanotubes (HANTs) were synthesized and exploited as environmentally friendly nanofillers in the design of hybrid ultrafiltration membranes. These membranes, designed for water treatment applications, were developed to offer favorable permeability, a high rejection rate and satisfactory antifouling capability. The resulting membrane demonstrated exceptional performance, rejecting 94.5% of bovine serum albumin and 97.8% of humic acid, two organic compounds frequently present in aquatic systems and likely to cause toxicity (Mu et al. 2020).

In a similar vein, the accumulation of salts like NaCl and $MgCl_2$ in water may pose a risk of danger to aquatic ecosystems along with human health if they are present in high concentration. Sun et al. developed an environmentally friendly nanofiltration membrane using lignin (Sun et al. 2023). This membrane demonstrated high rejection rates for NaCl and MgCl salts (85.1% for NaCl and 96.7% for $MgCl_2$, respectively), as well as a high rejection capacity for dyes (100% for Brilliant Blue R (BBR) and 96.1% for Congo Red (CR)). This outstanding performance is accompanied by ultra-high pure water permeability, measured at 65.2 L m^{-2}. h^{-1} bar^{-1} and positive dye separation results. These results underline the membrane's potential ability to mitigate the toxicity associated with these salts in aquatic systems, offering a promising solution for the removal of these compounds while preserving water quality (Sun et al. 2023).

Similarly, Wu et al. (2022) developed a double-layer nanofibrous membrane for water purification, focusing on the efficient removal of heavy metal ions and organic pollutants. Figure 9.8 illustrates the chitosan nanofibrous membrane and its function in the removal of dyes and heavy metals. Indeed, this membrane incorporates a nanofibrous support layer composed of chitosan (CS), polyvinylpyrrolidone (PVP), and polyvinyl alcohol (PVA), combined with an active layer containing carbon nanotubes (CNTs), and formed by the electrospinning–electrospraying technique. The incorporation of CNTs into the active layer creates additional nanochannels, significantly improving the permeation flux of pure water (1533, 26 Lm^{-2} h^{-1}) and the rejection of heavy metal ions as well as organic dyes (Cu^{2+} 95.68%, Ni^{2+} 93.86%, Cd^{2+} 88.52%, Pb^{2+} 80.41%, malachite green 87.20%, methylene blue 76.33%, and crystal violet 63.39%).

9.6 Prospects for Future Research in Bioinspired Material Science

Although the removal of waste from aquatic environments has been successful when bioinspired materials are used, it is imperative that future research address specific restrictions related to these materials. Extensive investigations employing

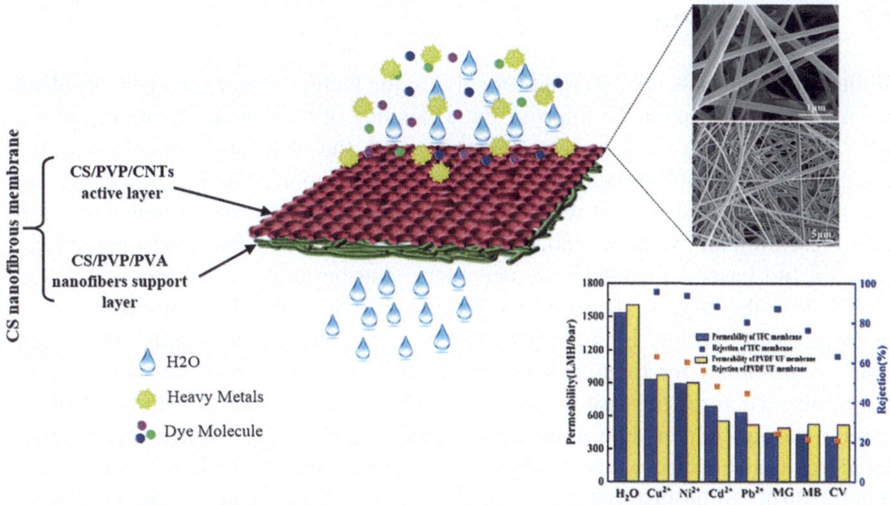

Fig. 9.8 Illustration of membrane performance in dye and heavy metal removal (Wu et al. 2022)

column-based bio-adsorption are necessary to augment the amount of bio-adsorbent and establish ideal conditions, guaranteeing effective removal of pollutants. Thorough research spanning a range of pollutant components in a particular setting are required to clarify and assess the competitive adsorption capabilities of materials inspired by biotechnology. This knowledge is essential for the actual implementation of bioinspired materials in real-world situations, highlighting the increasing significance of integrating membrane-based methodologies within the framework of research.

Moreover, research activities should go beyond isolated adsorption investigations and incorporate the creation and enhancement of bioinspired membranes for the purpose of purifying water. Evaluating these membranes' competitive adsorption properties will require examining how different pollutant components interact with them in a particular environment.

This multidisciplinary approach increases the study scope to incorporate membrane-based solutions while improving the basis for real-time bioinspired materials' applications. Furthermore, in order to address issues with the inadequate removal of pollutants within constrained reuse cycles, it is imperative to investigate methods for enhancing the reusability of bioinspired materials. This thorough research approach will make a major contribution to the development of effective and sustainable water purifying technologies.

9.7 Conclusion

Bioinspired materials are essential for solving the technical and scientific problems in wastewater treatment. To improve the efficiency of wastewater treatment procedures, a great deal of research has been done on the integration of different bio-based substances into synthetic processes. Bioinspired materials are being developed at a rapid pace in the field of water treatment. But due to advance science in bioinspired research, an extensive understanding of the structural and functional interactions of bio-based substances is necessary. Furthermore, reducing the budget allocated for the large-scale use of bio-based substances and the creation of materials inspired by biology is important. Although bio-based compounds are unquestionably important for treating wastewater, there are specific disadvantages when scaling up structures that use these compounds. In conclusion, the future of bioinspired materials depends on a thorough comprehension of the structural and functional relationships between bio-based molecules and the efficient commercial manufacturing of bioinspired materials. Opportunities for the development of novel technologies in efficient pollutant removal are created by the continuous advancement of bioinspired materials for the decontamination of water pollutants through adsorption and membrane filtration.

References

Abdul Khalil HPS, Saurabh CK, Adnan AS et al (2016) A review on chitosan-cellulose blends and nanocellulose reinforced chitosan biocomposites: properties and their applications. Carbohydr Polym 150:216–226. https://doi.org/10.1016/j.carbpol.2016.05.028

Abushammala H, Krossing I, Laborie MP (2015) Ionic liquid-mediated technology to produce cellulose nanocrystals directly from wood. Carbohydr Polym 134:609–616. https://doi.org/10.1016/j.carbpol.2015.07.079

Adeola AO, Nomngongo PN (2022) Advanced polymeric nanocomposites for water treatment applications: a holistic perspective. Polymers (Basel) 14:1–23. https://doi.org/10.3390/polym14122462

Ajibade TF, Tian H, Lasisi KH, Zhang K (2022) Bio-inspired PDA@WS2 polyacrylonitrile ultrafiltration membrane for the effective separation of saline oily wastewater and the removal of soluble dye. Sep Purif Technol 299:121711. https://doi.org/10.1016/j.seppur.2022.121711

Akharame MO, Fatoki OS, Opeolu BO et al (2018) Polymeric nanocomposites (Pncs) for wastewater remediation: an overview. Polym - Plast Technol Eng 57:1801–1827. https://doi.org/10.1080/03602559.2018.1434666

Alkhouzaam A, Qiblawey H, Khraisheh M et al (2020) Synthesis of graphene oxides particle of high oxidation degree using a modified Hummers method. Ceram Int 46:23997–24007. https://doi.org/10.1016/j.ceramint.2020.06.177

Alraddadi HM, Fagieh TM, Bakhsh EM et al (2023) Adsorptive removal of heavy metals and organic dyes by sodium alginate/coffee waste composite hydrogel. Int J Biol Macromol 247:125708. https://doi.org/10.1016/j.ijbiomac.2023.125708

Amusat SO, Kebede TG, Dube S, Nindi MM (2021) Ball-milling synthesis of biochar and biochar–based nanocomposites and prospects for removal of emerging contaminants: a review. J Water Process Eng 41:101993. https://doi.org/10.1016/j.jwpe.2021.101993

Arunkumar T, Karthikeyan R, Ram Subramani R et al (2020) Synthesis and characterisation of multi-walled carbon nanotubes (MWCNTs). Int J Ambient Energy 41:452–456. https://doi.org/10.1080/01430750.2018.1472657

Ashraf A, Liu G, Arif M et al (2022) Insights into the synthesis and application of biochar assisted graphene-based materials in antibiotic remediation. J Clean Prod 361:132211. https://doi.org/10.1016/j.jclepro.2022.132211

Aziz T, Fan H, Zhang X et al (2020) Advance study of cellulose nanocrystals properties and applications. J Polym Environ 28:1117–1128. https://doi.org/10.1007/s10924-020-01674-2

Bai X, Liu Z, Liu P et al (2023) An eco-friendly adsorbent based on bacterial cellulose and vermiculite composite for efficient removal of methylene blue and sulfanilamide. Polymers (Basel) 15. https://doi.org/10.3390/polym15102342

Benettayeb A, Seihoub FZ, Pal P et al (2023) Chitosan nanoparticles as potential nano-sorbent for removal of toxic environmental pollutants. Nano 13:1–25. https://doi.org/10.3390/nano13030447

Bhuyan MSA, Uddin MN, Islam MM et al (2016) Synthesis of graphene. Int Nano Lett 6:65–83. https://doi.org/10.1007/s40089-015-0176-1

Bouazizi A, Breida M, Achiou B et al (2017) Removal of dyes by a new nano–TiO2 ultrafiltration membrane deposited on low-cost support prepared from natural Moroccan bentonite. Appl Clay Sci 149:127–135. https://doi.org/10.1016/j.clay.2017.08.019

Bouazizi A, Ouammou M, Aaddane A et al (2022) Novel low-cost bentonite-based membranes for microfiltration, ultrafiltration, and nanofiltration applications. Elsevier Inc

Bouazizi A, Saja S, Achiou B et al (2016) Elaboration and characterization of a new flat ceramic MF membrane made from natural Moroccan bentonite. Application to treatment of industrial wastewater. Appl Clay Sci 132–133:33–40. https://doi.org/10.1016/j.clay.2016.05.009

Chakraborty P, Mustafa V, Abraham J (2018) Synthesis and characterization of Chitosan nanoparticles and their application in removal of wastewater contaminants. Nat Environ Pollut Technol 17:469–478

Chauhan M, Saini VK, Suthar S (2020) Ti-pillared montmorillonite clay for adsorptive removal of amoxicillin, imipramine, diclofenac-sodium, and paracetamol from water. J Hazard Mater 399:122832. https://doi.org/10.1016/j.jhazmat.2020.122832

Chen C, Zhang T, Zhang Q et al (2016) Biointerface by cell growth on graphene oxide doped bacterial cellulose/poly(3,4-ethylenedioxythiophene) nanofibers. ACS Appl Mater Interfaces 8:10183–10192. https://doi.org/10.1021/acsami.6b01243

Cheng Z, Yang R, Liu X et al (2017) Green synthesis of bacterial cellulose via acetic acid pre-hydrolysis liquor of agricultural corn stalk used as carbon source. Bioresour Technol 234:8–14. https://doi.org/10.1016/j.biortech.2017.02.131

Crini G (2005) Recent developments in polysaccharide-based materials used as adsorbents in wastewater treatment. Prog Polym Sci 30:38–70. https://doi.org/10.1016/j.progpolymsci.2004.11.002

Crini G, Badot PM (2008) Application of chitosan, a natural aminopolysaccharide, for dye removal from aqueous solutions by adsorption processes using batch studies: a review of recent literature. Prog Polym Sci 33:399–447. https://doi.org/10.1016/j.progpolymsci.2007.11.001

Dahmane EM, Taourirte M, Eladlani N, Rhazi M (2016) Preparation and characterization of α-chitin whiskers, chitosan nanoparticles and chitosan nanoscaffold from Parapenaeus longirostris. Mater Today Proc 3:2590–2598. https://doi.org/10.1016/j.matpr.2016.06.001

Das TK, Jesionek M, Çelik Y, Poater A (2023) Catalytic polymer nanocomposites for environmental remediation of wastewater. Sci Total Environ 901:165772. https://doi.org/10.1016/j.scitotenv.2023.165772

Divya K, Jisha MS (2018) Chitosan nanoparticles preparation and applications. Environ Chem Lett 16:101–112. https://doi.org/10.1007/s10311-017-0670-y

Du B, Chai L, Zheng Q et al (2023) Designed synthesis of multifunctional lignin-based adsorbent for efficient heavy metal ions removal and electromagnetic wave absorption. Int J Biol Macromol 234:123668. https://doi.org/10.1016/j.ijbiomac.2023.123668

El-Fattah WA, Guesmi A, Ben Hamadi N et al (2024) A green synthesis of cellulose nanocrystals biosorbent for remediation of wastewater containing industrial dye. Colloids Surfaces A Physicochem Eng Asp 681:132729. https://doi.org/10.1016/j.colsurfa.2023.132729

Fan XM, Yu HY, Wang DC et al (2019) Facile and green synthesis of carboxylated cellulose nanocrystals as efficient adsorbents in wastewater treatments. ACS Sustain Chem Eng 7:18067–18075. https://doi.org/10.1021/acssuschemeng.9b05081

Farnad N, Farhadi K, Voelcker NH (2012) Polydopamine nanoparticles as a new and highly selective biosorbent for the removal of copper (II) ions from aqueous solutions. Water Air Soil Pollut 223:3535–3544. https://doi.org/10.1007/s11270-012-1131-7

Fu J, Xin Q, Wu X et al (2016) Selective adsorption and separation of organic dyes from aqueous solution on polydopamine microspheres. J Colloid Interface Sci 461:292–304. https://doi.org/10.1016/j.jcis.2015.09.017

Gao X, Zhang Y, Zhao Y (2017) Biosorption and reduction of Au (III) to gold nanoparticles by thiourea modified alginate. Carbohydr Polym 159:108–115. https://doi.org/10.1016/j.carbpol.2016.11.095

Ge Y, Li Z, Kong Y et al (2014) Heavy metal ions retention by bi-functionalized lignin: synthesis, applications, and adsorption mechanisms. J Ind Eng Chem 20:4429–4436. https://doi.org/10.1016/j.jiec.2014.02.011

Georgouvelas D, Abdelhamid HN, Li J et al (2021) All-cellulose functional membranes for water treatment: adsorption of metal ions and catalytic decolorization of dyes. Carbohydr Polym 264:118044. https://doi.org/10.1016/j.carbpol.2021.118044

Ghorbani M, Eisazadeh H (2013) Removal of COD, color, anions and heavy metals from cotton textile wastewater by using polyaniline and polypyrrole nanocomposites coated on rice husk ash. Compos Part B Eng 45:1–7. https://doi.org/10.1016/j.compositesb.2012.09.035

Gopinath KP, Vo DVN, Gnana Prakash D et al (2021) Environmental applications of carbon-based materials: a review. Environ Chem Lett 19:557–582. https://doi.org/10.1007/s10311-020-01084-9

Guo Q, Ghadiri R, Weigel T et al (2014) Comparison of in situ and ex situ methods for synthesis of two-photon polymerization polymer nanocomposites. Polymers (Basel) 6:2037–2050. https://doi.org/10.3390/polym6072037

Halysh V, Sevastyanova O, Riazanova AV et al (2018) Walnut shells as a potential low-cost lignocellulosic sorbent for dyes and metal ions. Cellulose 25:4729–4742. https://doi.org/10.1007/s10570-018-1896-y

Hokkanen S, Bhatnagar A, Sillanpää M (2016) A review on modification methods to cellulose-based adsorbents to improve adsorption capacity. Water Res 91:156–173. https://doi.org/10.1016/j.watres.2016.01.008

Hu Y, Yue M, Yuan F et al (2021) Bio-inspired fabrication of highly permeable and anti-fouling ultrafiltration membranes based on bacterial cellulose for efficient removal of soluble dyes and insoluble oils. J Memb Sci 621:118982. https://doi.org/10.1016/j.memsci.2020.118982

Hussain S, Berry S (2024) A review study on green synthesis of chitosan derived schiff bases and their applications. Carbohydr Res 535:109002. https://doi.org/10.1016/j.carres.2023.109002

Ion VA, Pârvulescu OC, Dobre T (2015) Volatile organic compounds adsorption onto neat and hybrid bacterial cellulose. Appl Surf Sci 335:137–146. https://doi.org/10.1016/j.apsusc.2015.02.040

Islam MM, Islam R, Mahmudul Hassan SM et al (2023) Carboxymethyl chitin and chitosan derivatives: synthesis, characterization and antibacterial activity. Carbohydr Polym Technol Appl 5:100283. https://doi.org/10.1016/j.carpta.2023.100283

Islam S, Bhuiyan MAR, Islam MN (2017) Chitin and chitosan: structure, properties and applications in biomedical engineering. J Polym Environ 25:854–866. https://doi.org/10.1007/s10924-016-0865-5

Jamshaid A, Hamid A, Muhammad N et al (2017) Cellulose-based materials for the removal of heavy metals from wastewater—an overview. Chem Bio Eng Rev 4:240–256. https://doi.org/10.1002/cben.201700002

Jayaraman T, Murthy AP, Elakkiya V et al (2018) Recent development on carbon based heterostructures for their applications in energy and environment: a review. J Ind Eng Chem 64:16–59. https://doi.org/10.1016/j.jiec.2018.02.029

Jin X, Xiang Z, Liu Q et al (2017) Polyethyleneimine-bacterial cellulose bioadsorbent for effective removal of copper and lead ions from aqueous solution. Bioresour Technol 244:844–849. https://doi.org/10.1016/j.biortech.2017.08.072

Jobby R, Jha P, Yadav AK, Desai N (2018) Biosorption and biotransformation of hexavalent chromium [Cr(VI)]: a comprehensive review. Chemosphere 207:255–266. https://doi.org/10.1016/j.chemosphere.2018.05.050

Johari K, Saman N, Song ST et al (2016) Adsorption enhancement of elemental mercury by various surface modified coconut husk as eco-friendly low-cost adsorbents. Int Biodeterior Biodegrad 109:45–52. https://doi.org/10.1016/j.ibiod.2016.01.004

Kai D, Tan MJ, Chee PL et al (2016) Towards lignin-based functional materials in a sustainable world. Green Chem 18:1175–1200. https://doi.org/10.1039/c5gc02616d

Khodakarami M, Bagheri M (2021) Recent advances in synthesis and application of polymer nanocomposites for water and wastewater treatment. J Clean Prod 296:126404. https://doi.org/10.1016/j.jclepro.2021.126404

Lee H, Dellatore SM, Miller WM, Messersmith PB (2018) Mussel-inspired surface chemistry for multifunctional coatings. Science 318:426–430

Liu W, Liu K, Du H et al (2022) Cellulose nanopaper: fabrication, functionalization, and applications. Nano-Micro Lett 14:1–27. https://doi.org/10.1007/s40820-022-00849-x

Liu X, Peng H, Lu J et al (2023) Bio-inspired polydopamine nanofiltration membranes modulated by spiro-piperazine. Adv Membr 3:100059. https://doi.org/10.1016/j.advmem.2022.100059

Lu P, Hsieh YL (2010) Preparation and properties of cellulose nanocrystals: rods, spheres, and network. Carbohydr Polym 82:329–336. https://doi.org/10.1016/j.carbpol.2010.04.073

Maier GP, Rapp MV, Waite JH et al (2015) Adaptive synergy between catechol and lysine promotes wet adhesion by surface salt displacement. Science 349:628–632. https://doi.org/10.1126/science.aab0556

Mu Y, Feng H, Zhang S et al (2020) Development of highly permeable and antifouling ultrafiltration membranes based on the synergistic effect of carboxylated polysulfone and bio-inspired co-deposition modified hydroxyapatite nanotubes. J Colloid Interface Sci 572:48–61. https://doi.org/10.1016/j.jcis.2020.03.072

Nassar NN (2010) Rapid removal and recovery of Pb(II) from wastewater by magnetic nanoadsorbents. J Hazard Mater 184:538–546. https://doi.org/10.1016/j.jhazmat.2010.08.069

Okoronkwo AE, Olusegun SJ (2013) Biosorption of nickel using unmodified and modified lignin extracted from agricultural waste. Desalin Water Treat 51:1989–1997. https://doi.org/10.1080/19443994.2012.714896

Oleszczuk P, Ćwikła-Bundyra W, Bogusz A et al (2016) Characterization of nanoparticles of biochars from different biomass. J Anal Appl Pyrolysis 121:165–172. https://doi.org/10.1016/j.jaap.2016.07.017

Oliveira M, Machado AV (2013) Preparation of polymer-based nanocomposites by different routes. In: Nanocomposites: synthesis, characterization and applications, pp 1–22

Pan B, Pan B, Zhang W et al (2009) Development of polymeric and polymer-based hybrid adsorbents for pollutants removal from waters. Chem Eng J 151:19–29. https://doi.org/10.1016/j.cej.2009.02.036

Paradise M, Goswami T (2007) Carbon nanotubes – production and industrial applications. Mater Des 28:1477–1489. https://doi.org/10.1016/j.matdes.2006.03.008

Paranavithana GN, Kawamoto K, Inoue Y et al (2016) Adsorption of Cd2+ and Pb2+ onto coconut shell biochar and biochar-mixed soil. Environ Earth Sci 75. https://doi.org/10.1007/s12665-015-5167-z

Pomogailo AD (2005) Polymer sol-gel synthesis of hybrid nanocomposites. Colloid J 67:658–677. https://doi.org/10.1007/s10595-005-0148-7

Qin J, Li N, Jiang M et al (2022) Ultrasonication pretreatment assisted rapid co-assembly of cellulose nanocrystal and metal ion for multifunctional application. Carbohydr Polym 277:118829. https://doi.org/10.1016/j.carbpol.2021.118829

Rabiee N, Sharma R, Foorginezhad S et al (2023) Green and sustainable membranes: a review. Environ Res 231:116133. https://doi.org/10.1016/j.envres.2023.116133

Rao N, Singh R, Bashambu L (2021) Carbon-based nanomaterials: synthesis and prospective applications. Mater Today Proc 44:608–614. https://doi.org/10.1016/j.matpr.2020.10.593

Rodriguez-Narvaez OM, Peralta-Hernandez JM, Goonetilleke A, Bandala ER (2019) Biochar-supported nanomaterials for environmental applications. J Ind Eng Chem 78:21–33. https://doi.org/10.1016/j.jiec.2019.06.008

Saravanan A, Kumar PS, Srinivasan S et al (2022) Insights on synthesis and applications of graphene-based materials in wastewater treatment: a review. Chemosphere 298:134284. https://doi.org/10.1016/j.chemosphere.2022.134284

Shen W, Chen S, Shi S et al (2009) Adsorption of Cu(II) and Pb(II) onto diethylenetriamine-bacterial cellulose. Carbohydr Polym 75:110–114. https://doi.org/10.1016/j.carbpol.2008.07.006

Silva ADR, Stocco TD, Granato AEC et al (2019) Recent advances in nanostructured polymer composites for biomedical applications. Elsevier Inc

Sirajudheen P, Karthikeyan P, Ramkumar K, Meenakshi S (2020) Effective removal of organic pollutants by adsorption onto chitosan supported graphene oxide-hydroxyapatite composite: a novel reusable adsorbent. J Mol Liq 318:114200

Sivakumar M, Yadav S, Hung WS, Lai JY (2020) One-pot eco-friendly synthesis of edge-carboxylate graphene via dry ball milling for enhanced removal of acid and basic dyes from single or mixed aqueous solution. J Clean Prod 263:121498. https://doi.org/10.1016/j.jclepro.2020.121498

Sujiono EH, Zurnansyah ZD et al (2020) Graphene oxide based coconut shell waste: synthesis by modified Hummers method and characterization. Heliyon 6:e04568. https://doi.org/10.1016/j.heliyon.2020.e04568

Sun W, Zhang N, Li Q et al (2023) Bioinspired lignin-based loose nanofiltration membrane with excellent acid, fouling, and chlorine resistances toward dye/salt separation. J Memb Sci 670:121372. https://doi.org/10.1016/j.memsci.2023.121372

Tamura Y, Kanomata K, Kitaoka T (2018) Interfacial hydrolysis of Acetals on protonated TEMPO-oxidized cellulose nanofibers. Sci Rep 8:6–10. https://doi.org/10.1038/s41598-018-23381-8

Tan XF, Liu YG, Gu YL et al (2016) Biochar-based nano-composites for the decontamination of wastewater: a review. Bioresour Technol 212:318–333. https://doi.org/10.1016/j.biortech.2016.04.093

Tanpichai S, Boonmahitthisud A, Soykeabkaew N, Ongthip L (2022) Review of the recent developments in all-cellulose nanocomposites: properties and applications. Carbohydr Polym 286:119192. https://doi.org/10.1016/j.carbpol.2022.119192

Teixeira RSS, Da Silva ASA, Jang JH et al (2015) Combining biomass wet disk milling and endoglucanase/β-glucosidase hydrolysis for the production of cellulose nanocrystals. Carbohydr Polym 128:75–81. https://doi.org/10.1016/j.carbpol.2015.03.087

Thakur M, Sharma A, Ahlawat V et al (2020) Process optimization for the production of cellulose nanocrystals from rice straw derived α-cellulose. Mater Sci Energy Technol 3:328–334. https://doi.org/10.1016/j.mset.2019.12.005

Thakur S, Sharma B, Verma A et al (2018) Recent progress in sodium alginate based sustainable hydrogels for environmental applications. J Clean Prod 198:143–159. https://doi.org/10.1016/j.jclepro.2018.06.259

Thomas B, Raj MC, Athira BK et al (2018) Nanocellulose, a versatile green platform: from biosources to materials and their applications. Chem Rev 118:11575–11625. https://doi.org/10.1021/acs.chemrev.7b00627

Toledo PVO, Martins BF, Pirich CL et al (2019) Cellulose based cryogels as adsorbents for organic pollutants. Macromol Symp 383:1–12. https://doi.org/10.1002/masy.201800013

Vakili M, Rafatullah M, Salamatinia B et al (2014) Application of chitosan and its derivatives as adsorbents for dye removal from water and wastewater: a review. Carbohydr Polym 113:115–130. https://doi.org/10.1016/j.carbpol.2014.07.007

Varshney K (2017) Carbon nanotubes: a review on synthesis, electrical and mechanical properties and applications. Asian J Appl Sci Technol 1:15–20

Vazquez-Jaime M, Arcibar-Orozco JA, Damian-Ascencio CE et al (2020) Effective removal of arsenic from an aqueous solution by ferrihydrite/goethite graphene oxide composites using the modified Hummers method. J Environ Chem Eng 8:104416. https://doi.org/10.1016/j.jece.2020.104416

Wang S, Vincent T, Faur C, Guibal E (2018) A comparison of palladium sorption using polyethylenimine impregnated alginate-based and carrageenan-based algal beads. Appl Sci 8. https://doi.org/10.3390/app8020264

Wang T, Jiang M, Yu X et al (2022) Application of lignin adsorbent in wastewater treatment: a review. Sep Purif Technol 302:122116. https://doi.org/10.1016/j.seppur.2022.122116

Wang Y, Guo Y, Yang C et al (2023) Bio-inspired fabrication of adsorptive ultrafiltration membrane for water purification: simultaneous removal of natural organic matters, lead ion and organic dyes. J Environ Chem Eng 11:109798. https://doi.org/10.1016/j.jece.2023.109798

Wen R, Tu B, Guo X et al (2020) An ion release controlled Cr(VI) treatment agent: Nano zerovalent iron/carbon/alginate composite gel. Int J Biol Macromol 146:692–704. https://doi.org/10.1016/j.ijbiomac.2019.12.168

Weska RF, Moura JM, Batista LM et al (2007) Optimization of deacetylation in the production of chitosan from shrimp wastes: use of response surface methodology. J Food Eng 80:749–753. https://doi.org/10.1016/j.jfoodeng.2006.02.006

Wu S, Li K, Shi W, Cai J (2022) Chitosan/polyvinylpyrrolidone/polyvinyl alcohol/carbon nanotubes dual layers nanofibrous membrane constructed by electrospinning-electrospray for water purification. Carbohydr Polym 294:119756. https://doi.org/10.1016/j.carbpol.2022.119756

Wu ZY, Liang HW, Chen LF et al (2016) Bacterial cellulose: a robust platform for design of three dimensional carbon-based functional nanomaterials. Acc Chem Res 49:96–105. https://doi.org/10.1021/acs.accounts.5b00380

Yu P, Wang X, Zhang K et al (2021a) Aqueous cellulose solution assisted direct exfoliation of graphite to high concentration graphene dispersion. Mater Lett 285:129081. https://doi.org/10.1016/j.matlet.2020.129081

Yu S, Pang H, Huang S et al (2021b) Recent advances in metal-organic framework membranes for water treatment: a review. Sci Total Environ 800:149662. https://doi.org/10.1016/j.scitotenv.2021.149662

Yu S, Park J, Kim M et al (2019) Characterization of biochar and byproducts from slow pyrolysis of hinoki cypress. Bioresour Technol Rep 6:217–222. https://doi.org/10.1016/j.biteb.2019.03.009

Yu S, Zhang W, Dong X et al (2024) A review on recent advances of biochar from agricultural and forestry wastes: preparation, modification and applications in wastewater treatment. J Environ Chem Eng 12:111638. https://doi.org/10.1016/j.jece.2023.111638

Yu X, Tong S, Ge M, Zuo J (2013) Removal of fluoride from drinking water by cellulose@hydroxyapatite nanocomposites. Carbohydr Polym 92:269–275. https://doi.org/10.1016/j.carbpol.2012.09.045

Zhang H, Chen Y, Wang S et al (2020) Extraction and comparison of cellulose nanocrystals from lemon (Citrus limon) seeds using sulfuric acid hydrolysis and oxidation methods. Carbohydr Polym 238:116180. https://doi.org/10.1016/j.carbpol.2020.116180

Zhang J, Yang S, Chen Z et al (2018a) In Situ synthesis of SiC-graphene core-shell nanoparticles using wet ball milling. Ceram Int 44:8283–8289. https://doi.org/10.1016/j.ceramint.2018.02.012

Zhang L, Zeng Y, Cheng Z (2016) Removal of heavy metal ions using chitosan and modified chitosan: a review. J Mol Liq 214:175–191. https://doi.org/10.1016/j.molliq.2015.12.013

Zhang Q, Li Y, Yang Q et al (2018b) Distinguished Cr(VI) capture with rapid and superior capability using polydopamine microsphere: behavior and mechanism. J Hazard Mater 342:732–740. https://doi.org/10.1016/j.jhazmat.2017.08.061

Zhang Y, Zhang Y, Xu W et al (2023) Preparation methods of cellulose nanocrystal and its application in treatment of environmental pollution: a mini-review. Colloids Interface Sci Commun 53:100707. https://doi.org/10.1016/j.colcom.2023.100707

Zhao J, Zhao X, Jiang Z et al (2014) Biomimetic and bioinspired membranes: preparation and application. Prog Polym Sci 39:1668–1720. https://doi.org/10.1016/j.progpolymsci.2014.06.001

Zhu S, Sun H, Mu T et al (2023) Preparation of cellulose nanocrystals from purple sweet potato peels by ultrasound-assisted maleic acid hydrolysis. Food Chem 403:134496. https://doi.org/10.1016/j.foodchem.2022.134496

Designing and Technical Feasibility of Biohybrid System for Scaling Up Treatment Process

10

Anoushka Singh, Prajna Ghosh, Moksh Rajpal, and Naveen Dwivedi

Abstract

Biohybrid systems, integrating biological processes with artificial enhancements, present sustainable solutions for wastewater treatment, are the essential for achieving Sustainable Development Goal 6 (SDG 6). This chapter discusses the combination of natural elements and technologies to overcome limitations of conventional treatment, enhancing pollutant removal and resource recovery. It covers theoretical foundations, focusing on the synergy between microorganisms, plants, and components like sensors, nanomaterials, and automated controls. Design principles for scaling are outlined, emphasizing biological agent selection, reactor configurations, and modularity. Case studies illustrate technical feasibility across diverse settings, showcasing systems like aerated wetlands, membrane bioreactors, and algal photo-bioreactors with high efficiency. Optimization strategies aim at improving biological–artificial interactions, energy reduction, and incorporating real-time monitoring. Future prospects include advancements in biotechnology, materials science, and circular economy integration, while addressing challenges in stability, scalability, and economic viability. This chapter underscores biohybrid systems' potential in advancing sustainable wastewater management and supporting SDG 6.

Keywords

Biohybrid systems · Wastewater treatment · SDG 6 · Pollutant removal · Resource recovery · Nature-based solutions · Scalability

A. Singh (✉) · P. Ghosh · M. Rajpal · N. Dwivedi
Department of Biotechnology Engineering and Food Technology, University Institute of Engineering, Chandigarh University, Mohali, Punjab, India

N. Dwivedi et al. (eds.), *Wastewater Treatment Through Nature-Based Solutions*,
https://doi.org/10.1007/978-981-96-6699-7_10

10.1 Introduction: Relevance of Biohybrid Systems to SDG 6

10.1.1 Overview of SDG 6 and Its Importance

According to the UN Sustainable Development Goals Sustainable Development Goal 6 (SDG 6), the global community is to "Ensure availability and sustainable management of water and sanitation for all" by the year 2030. This goal is important because water remains a central theme and a major issue of concern in the current world. It is estimated that over two billion people have no access to proper drinking water, and over four billion people have no access to properly managed sanitation means, which are directly influencing health, livelihoods, and ecosystems negatively. Achieving SDG 6 is essential not only for human development but also for environmental sustainability, as clean water and well-managed sanitation systems are vital to preserving ecosystems, reducing pollution, and fostering resilience against climate change (Fonseca et al. 2020). There is a paramount importance of adapting water treatment technologies due to innovations that targets Sustainable Development Goal 6; it is also depicted in Fig. 10.1. While older methods of water treatment can work, they are not efficient in the long run, require energy and chemicals, and infrastructure, especially for larger populations, not to mention limited-resource areas. Consequently, an increasing number of researchers began to look for other approaches that combine native remediation processes with technological assistance treatments, as they are more environment friendly and can be scaled more easily.

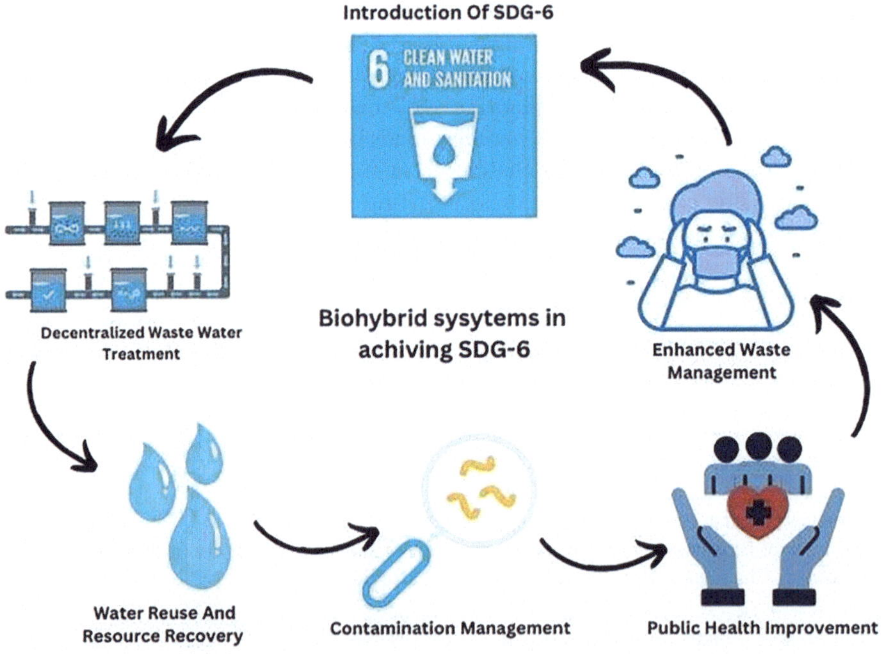

Fig. 10.1 Advantages of use of biohybrid system in achieving SDG 6

10.1.2 Understanding Nature-Based Solutions (NbS) in Water Management

Biohybrid systems are a relatively recent technological innovation where natural biological processes are integrated with artificial solutions for increased and sustainable efficiency of wastewater treatment. These systems are somewhere between the strictly "natural" NbS and traditional engineered systems using living organisms, such as plants, bacteria, and algae together with technology including sensors, nanomaterials, and a self-regulating control system. These systems often require lower energy inputs and produce fewer emissions compared to conventional treatment plants, making them attractive for sustainability (Gude 2018). Another outstanding feature of biohybrid systems is adaptive to the changes of wastewater characterization and treatment needs. For example, it can be employed to work selectively against selected contaminants including heavy metals, drugs, and nutrients which are highly challenging to eliminate by conventional procedures. Additional gadgets including aeration systems or filtration units can quicken the natural processes in biological treatment; some parts or units of the biohybrid system can be implemented on rural areas, other on the urban areas. This makes biohybrid systems a feasible solution to the multiple water management problems aligned to SDG 6 (Miksch et al. 2015).

10.1.3 The Role of Biohybrid Systems in Nature-Based Treatment

Biohybrid systems describe a relatively new class of technologies that can be applied in wastewater treatment, based on the synthesis of biological processes with engineering interventions. These systems are intermediate between NbS and traditional engineered systems and involve living organisms including plants, bacteria, and algae and technological materials including, sensors, structures made of nanomaterials, and actuating systems. Thus, biohybrid systems present a synergistic approach that improves technology by incorporating the advantages of both biological and artificial processes using pollutant removal, operational flexibility, and resource efficiency (Benjamin et al. 2019).

Another strong prospect of biohybrid systems is the flexibility of adjustments to the general composition of wastewater and the treatment processes needed. For example, they can be used to deploy targeted treatment for specific pollutants that include heavy metals, pharmaceuticals, and nutrients which cannot be eliminated effectively by conventional treatment processes. Bypass structures like aerators or filter units would facilitate these actions significantly and help biohybrid systems to work efficiently irrespective of whether they are tested on small rural farming regions or large metropolitan areas. Such flexibility makes biohybrid systems solution promising for various scratches under the water management within SDG 6 (Khan et al. 2019).

10.1.4 Current Gaps in Traditional Wastewater Treatment Systems

Although the use of activated sludge processes, membrane bioreactors (MBRs), and other chemical treatment methods has been employed worldwide, these contradicting methods have their drawbacks. These systems have many drawbacks because of the high operating costs involved in energy consumption, chemicals, and the high capital investment required for infrastructure. Moreover, they are not very amenable to changes in the composition of wastewater as well as the volumetric flow rates; this results in compromised efficiency, especially in situations where a decentralized or small scale mode of operation is employed.

Additionally, traditional systems are typically designed to address conventional pollutants, such as organic matter and suspended solids, but they struggle to effectively remove emerging contaminants, including pharmaceuticals, microplastics, and personal care products. As these pollutants become increasingly prevalent in wastewater streams, there is a pressing need for innovative treatment approaches capable of handling complex and evolving contaminant profiles. The growing recognition of these limitations has driven the search for more sustainable and versatile treatment technologies, paving the way for biohybrid systems to play a transformative role (Vallero and Gunsch 2020).

10.1.5 Relevance of Biohybrid Systems to Achieving SDG 6 Targets

Research on biohybrid systems shows promising prospects to advance efforts toward attaining SDG 6 since the systems can address water quality and access challenges. Specifically, biohybrid systems can contribute to the following:

- *SDG 6.1 (Access to Safe Water):* Biohybrid systems have the potential of enhancing the standards of functionality in wastewater treatment thus helping to supply safe water especially to areas where developing systems cannot be used due to their expense or inappropriateness.
- *SDG 6.3 (Reducing Water Pollution):* Biohybrid systems put them in a good position to eliminate a broad range of pollutants with water purification, especially those pollutants that are hard to be treated using conventional treatment processes with a view of combating and lowering cases of pollution of water.
- *SDG 6.4 (Water Efficiency):* By incorporating resource recovery processes and energy-efficient technologies, biohybrid systems can support improved and sustainable water use and recycling.
- *SDG 6.6 (Protecting Water-Related Ecosystems):* Such design ensures that operations of the developed biohybrid systems are eco-friendly and facilitate the recovery of aquatic habitats from the damaging effects of wastewater pollution (Tee et al. 2016).

The overview of the role and importance of biohybrid system in achieving SDG 6 is given in Table 10.1.

Table 10.1 Overview of biohybrid systems and their role in achieving SDG 6

Stage	Details	References
SDG 6: Water and sanitation for all by 2030	Goal to ensure sustainable water management and sanitation for everyone by 2030	Goal 6—Clean Water & Sanitation (2024)
Challenges	2+ billion people lack access to clean water 4+ billion people lack proper sanitation	Tortajada (2020)
Nature-based solutions (NbS) in water management	Using natural processes for water management, integrating biological and technological approaches	Johnston et al. (2011)
Biohybrid systems	Combine natural processes (plants, bacteria, algae) with technology (sensors, nanomaterials). Lower energy use and fewer emissions	Calicioglu and Bogdanski (2021)
Benefits of biohybrid systems	Adaptable to various pollutants (heavy metals, drugs, nutrients) Suitable for both rural and urban areas	Tee et al. (2016)
Limitations of traditional treatment	High costs, energy use, and infrastructure requirements Ineffective for emerging pollutants (microplastics, pharmaceuticals)	Mustafa and Hayder (2021)
Biohybrid systems' role in achieving SDG 6	SDG 6.1: Provide safe water access SDG 6.3: Reduce water pollution SDG 6.4: Enhance water efficiency SDG 6.6: Protect aquatic ecosystems	Calicioglu and Bogdanski (2021)

10.2 Theoretical Framework of Biohybrid Systems in Wastewater Treatment

10.2.1 Introduction to the Theoretical Underpinnings

The scientific concept of biohybrid system is developed from the biological and engineering practices which optimally employ naturally gathered biology along with artificial boost for effective wastewater treatment in large scale and in a sustainable way. In its simplest conception, biohybrid systems make use of biological entities that include microorganisms and plants with pollutant removal attributes, and alongside supporting technological improvements. This integration is designed to overcome two extremes of purely biological or purely artificial systems for designing a treatment method that can cater for the increasing sophistication of wastewater problems (Zhou and Smith 2002).

With a view to synthesizing the available information acquired, a subsequent aim is to provide a multi-perspective advanced overview of biological wastewater treatment as well as of how artificial parts augment these processes in biohybrid systems. This section examines the biological aspects of conventional natural treatment systems and the use of artificial augmentations, as well as the interactions that make biohybrids a prime weapon in wastewater treatment (Tantak et al. 2014).

10.2.2 Biological Components in Natural Treatment Systems

Ecosystems have been employed for wastewater treatment since antiquity because they avail natural cycles of pollutant decomposition. The microorganisms include bacteria, fungi, protozoa, and others, while plant and algae are other components integral in the biological systems used in minimization of organic wastes and pollutants.

- *Microbial Systems:* Microorganisms are primary dominants in the natural wastewater treatment processes that dominate most natural systems. In constructed wetlands, ponds as well as lagoons, the organic pollutants are transformed into nontoxic substances via the actions of bacteria and other microorganisms ranging from aerobic to anaerobic arrangements. Aerobic bacteria, for example, consume oxygen to break down organic matter into carbon dioxide, water, and biomass, while anaerobic bacteria work in oxygen-deprived environments to produce methane and other by-products. These microbial activities form the basis of biological oxygen demand (BOD) reduction in natural treatment systems (Terry 2010).
- *Plant-Based Systems:* Plants are very essential in natural wastewater treatment systems such as constructed wetlands. The technology of phytoremediation worked in such way that plant extracts nutrients like nitrogen and phosphorus, heavy metals, and any other pollutant found in the water. These systems offer suitable topographical substrate promoting microbial growth thus increasing pollutant degradation. Second, plants play a vital role in photosynthesis, supplemental oxygenation, and transpiration to allow for correct hydrogen balance in the treatment system.
- *Algae and Other Biological Agents:* Among the multiplication requirements for algae as wastewater treatment plants, the nutrients nitrogen and phosphorus are broadly predominant. Photosynthesis that occurs in algal organisms produce oxygen, according to which aerobic microbial action is produced hence making algae an important factor in natural systems like facultative ponds. In addition, algae can also be used for biomass production, which could be used in biofuel or biofertilizers making them circular economy products.

Despite practical utilization of biological components in the removal of pollutions, they are constrained by such environmentally sensitive factors as temperature, oxygen, and nutrient concentrations. These constraints increases mean that in comparison to biological systems that are purely constructed, systems with these structures are less effective in coping with high levels of pollutants or in cases where the compositions of the wastewaters differ. That is why the incorporation of artificial components in biological context within biohybrid systems is crucial.

10.2.3 Integration of Artificial Components

Prosthetics in biohybrid systems are enhancements to the biological activities accomplished by biological actors and interfaces, which provide enhanced

Table 10.2 Synergistic mechanisms in biohybrid systems

Mechanism	Description	References
Enhanced pollutant Removal	Combines biological processes (microorganisms, plants) with artificial components (nanomaterials, filtration) for improved removal of various pollutants, including micropollutants and heavy metals	Lu and Astruc (2018)
Optimization of biological processes	Artificial components (e.g., automated aeration, nutrient monitoring) provide optimal conditions for biological processes, increasing the efficiency of pollutant removal and system adaptability	Seco et al. (2018)
Resilience and flexibility	Can adjust to varying environmental conditions (temperature, pH, oxygen levels) and treatment needs using sensors, aerators, and control systems, suitable for different wastewater scenarios	Capps et al. (1995)
Resource recovery	Enables recovery of valuable resources (e.g., nutrients, biomass) from wastewater, which can be used as fertilizers, biofuels, or for energy production, contributing to a circular economy	Seco et al. (2018)

functionality, capacity, and manageability. Such technological components may as basic as aeration mechanisms, to as complex as sensors and automated controls that allow real-time control and optimization of treatment processes. It has also been discussed in Table 10.2.

Key artificial components integrated into biohybrid systems include the following:

- *Sensors and Real-Time Monitoring:* More sophisticated sensors have been developed to track some of the measurable quality indicators including water pH, dissolved oxygen (DO), temperature, and nutrient content. This is continuous data that help operators change system conditions and manage microbial and plant performance efficiently. For example, If the DO levels go low, one can use artificial aeration in order to maintain aerobic microbial action. Sensors also enable identification of new pollutants (e.g., pharmaceuticals, microplastics) thus making biohybrid systems better placed to combat the diverse pollutants.
- *Aeration and Oxygenation Systems:* Water aeration is very essential in improving the efficiency of aerobic bioprocesses. In artificial aeration, systems of purification of oxygen are introduced in the treatment medium to enhance the rate of degradation by aerobic micro flora. This is especially applicable in high strength wastewater or in situations where bacterial action of oxidization is inadequate. Due to the management of the rate as well as the time taken in aeration, biohybrid systems are able to offer optimal performance even when the environmental conditions are unstable.
- *Nanomaterials and Advanced Filtration Technologies:* Self-assembled biohybrid systems involving carbon nanotubes and metal oxide nanoparticles have been included in the removal and degradation of pollutants including heavy metals, organic compounds, and pathogens. These materials provide a numerous surface area for adsorption which makes them ideal for use in improving the removal

efficiency of biohybrid systems. Likewise, the other technologies like membrane filtration and nanofiltration are used for washing out the contaminants from the water along with biological intake of the water (Sarkar et al. 2017).

- *Automated Control Systems:* To increase an efficiency and reliability of the biohybrid systems, the process of automation is one of the most significant factors. By connecting through Internet of things (IoT) and programmable logic controller (PLC), biohybrids can control altering operating conditions through the acquired data from the onboard sensors. For example, if nutrients concentration increases to a certain level, the system will be able to regulate the biological or artificial systems to reduce nutrient concentration. This level of automation helps to minimize the raw reliance on manual control of the biohybrid systems, thus making the solutions more viably expansive.

10.2.4 Synergistic Mechanisms in Biohybrids

The interaction of biological and artificial elements in a single system characterizes biohybrid systems. Nevertheless, such separation of functions enables optimization of pollutant removal, operational stability and the versatility of the system for dealing with various types of waste waters.

- *Enhanced Pollutant Removal:* Thanks to biological and artificial mechanisms, the performance of biohybrids is rather versatile in terms of pollutants they can address. For example, microorganisms and plants decompose organic materials and assimilate nutrients and nanomaterials, and filter technologies retain micropollutants, heavy metals, and pathogen. This multiple-step approach guarantees biohybrid systems outperform biological or conventional systems through higher pollutant removal efficiencies.
- *Optimization of Biological Processes:* In biohybrid systems, artificial parts act to bring out the best in biological entities by offering them the most suitable conditions. For example, automated aeration systems make sure that aerobic bacteria have correct oxygen supply, while sensors and regulating systems keep the correct nutrient concentrations for plant and microbial development. Thus, improving these processes enables biohybrids to adapt to changes in wastewater composition and strength of pollutants.
- *Resilience and Flexibility:* Biohybrid systems should be capable of changing the operational mode as well as the load in response to varying environmental circumstances and treatment requirements. Natural systems are very receptive to changes in temperature, pH, and oxygen content but artificial parts like the sensor, aerator, on/off control systems are always there to regulate and balance the system. This makes biohybrids to be appropriate for both centralized and decentralized wastewater treatment especially where the local climate or wastewater content is not constant (Meuler et al. 2008).
- *Resource Recovery:* A further positive feature of biohybrid systems is the possible recovery of valuable materials. The nutrients from wastewater can be utilized by algae and plants, which in turns can be harvested as fuel, fertilizer or con-

verted to biomass for the production of energy. Additional features like filtration and separation features improve the recovery of wanted resources, making contribution to circular economy as well as decreasing the impact of wastewater treatment on the environment (Akyol et al. 2019).

10.2.5 Summary of the Theoretical Framework

In conclusion, biohybrid systems establish a cooperation and integration between inherent biological functions and external technological interventions as an optimum, effective, as well as versatile and sustainable, wastewater management approach. The biological elements which include microorganisms, plants, and algae form a bed for pollutant removal, while the artificial augmentation includes sensors, aeration system, and nanomaterials that enhance and increase the performance and versatility of the biological parts. This combined integration of biosystems and technologies enables accomplishments of higher pollutant removal, operational reliability, and resource recovery than the conventional treatment systems in biohybrids, which makes them an aplausible solution to global wastewater problems, especially toward the implementation of SDG 6 (Review on hybrid energy systems for wastewater treatment and bioenergy production 2016).

10.3 Design Principles for Scaling Biohybrid Systems

10.3.1 Introduction to Design Considerations for Scaling

This study found five key areas where biohybrid systems must handle design parameters to support large-scale wastewater treatment while maintaining efficiency, reliability, and flexibility. Since biohybrid systems have been defined as ensembles of living organisms and synthetic components, special consideration of the interactions between bio- and techno-proxies (such as, bacteria or vegetation and sensors, air delivery networks, and nanomaterials) is required. The aspects described above play a crucial role in achieving and sustaining the best treatment efficacy as the system is developed from the pilot one to the full-scale, ranging from municipal to industrial ones.

To maintain the benefits of the biohybrid systems, inherent issues like low power requirement, flexibility for treating a broad range of wastewater, and improved efficiency of pollutants removal, the design must follow the principles of scale-up. However, it is still an area of research that is best addressed using an interdisciplinary approach. This includes choosing suitable biological agents, adjusting the reactors in terms of biological and artificial systems, as well as providing modularity and system redundancy to support wide applications. Furthermore, the systems need to be capable to operate under varying environmental and operating conditions, for example, in flows of wastewater, concentrations of pollutants, and treatment goals (Alleman et al. 1988).

10.3.2 Selection Criteria for Biological Agents

The applicability of the biohybrid systems depends widely on the choice of biological entities that can effectively capture these pollutants within the wastewater treatment plants (WWTPs). Organisms such as bacteria and fungi, plants, and enzymes are, therefore, central in the breakdown of pollutants, nutrient, and water assimilation as discussed in Table 10.3. The following criteria should guide the selection process:

- *Pollutant-Specific Degradation Capacity:* It is agreed that microorganisms and plants accurately have to be selected according to their capability to utilize and transform the target pollutants which are in the wastewater stream. For instance, some genera of bacteria such as Pseudomonas and Bacillus are effective in degrading organic pollutants and hydrocarbons and other genera such as Nitrosomonas and Nitrobacter are profuse in nitrification and denitrification processes for nitrogen elimination. The preferred species for high nutrient uptake (e.g., *Typha latifolia* and *Phragmites australis*) are used in systems for phosphorus and nitrogen reduction (Choi et al. 2013).
- *Environmental Tolerance:* Living agents have to be durable under environmental conditions; changes in pH, temperature, DO levels, and load of contaminants. For example, certain microbial species have increased resistance to heavy metal concentrations or unfavorable environmental conditions (including salinity or low oxygen levels) and are thus best suited for industrial wastewater treatment or for environments that are inhospitable to other microorganisms.
- *Growth Rate and Stability:* The biological agents should have high growth rates to make certain of efficient pollutant elimination. For scaling purposes, these selected agents must have high metabolic activity, and, in particular, their population must be stable during the time, specifically, in conditions where the flow rates and concentrations of pollutants may fluctuate. Larger and more developed plants with roots are preferred since they posses great areas where microbes and pollutants can attach to (Sato et al. 2002).

Table 10.3 Selection criteria and solutions for biological agents in biohybrid systems

Criteria	Description	Solution	References
Pollutant-specific capacity	Target pollutant degradation (e.g., organics, nitrogen)	Use *Pseudomonas* for organics; *Typha* for nitrogen	Rittmann (2010)
Environmental tolerance	Resilience to pH, temperature, DO levels, and contaminant loads	Select resistant microbes; adaptive plants for wetlands	Beeman and Reitberger (2003)
Growth rate and stability	High growth and stable populations in varying conditions	Use fast-growing species; plants with extensive root areas	Bueno González (2020)
Compatibility with enhancements	Integration with aeration, filtration, and nanomaterials	Match agents with compatible technologies (e.g., aeration)	Cheng et al. (2020)

- *Compatibility with Artificial Enhancements:* Biological agents need to be fully compatible with engineered hardware like the aeration systems, filtration units, and nanomaterials. For example, aerobic microorganisms should be companions with oxygenation technologies capable of sustaining suitable DO levels; plants located in constructed wetlands should adopt nanomaterial-based adsorption systems to improve the elimination of heavy metals (Dinesha et al. 2017).

10.3.3 Engineering Design of Hybrid Systems

In particular, the engineering design of biohybrid systems indicates the key aspects of the design in order to achieve the most effective integration of the biological and artificial elements. One or two types of reactor systems are employed in biohybrid systems, although these kinds of systems are designed for various goals and tank sizes in wastewater treatment. Major types are biofilm reactors, MBR, and constructed wetlands which are all systems that use artificial additions for treatment processes.

- *Biofilm Reactors:* Biofilm-driven reactors are designed in such a manner that encourages the development of microorganisms that directly attach themselves to the stable surfaces in the reactor where biofilm is formed and pollutant breakdown happens. And these reactors can be designed in some types of reactor such as fixed-bed reactor, moving-bed reactor, and fluidized-bed reactor with differential pollutant removal rates and system scale-up. In biohybrid systems, however, devices such as aeration units or the incorporation of nanomaterials may be incorporated into the biofilm reactors in order to boost the rate of oxygen diffusion, microbial adaption to the biofilm structure, and commonly the overall degradation capacity of the system. In such systems, control in delivery of oxygen and thickness of the biofilm for the purpose of degradation of pollutants is necessary (González-Brambila and López-Isunza 2008).
- *Membrane Bioreactors (MBRs):* MBRs integrate the biological treatment process with membrane filtration and boasts a small footprint and high level of efficiency in wastewater treatment. Biological agents break down organic toxins in the wastewater stream in biohybrid MBRs, while membrane technology retains particulate matter and pathogenic organisms to yield high quality effluent. Some extras include the use of computerized control systems and the use of nanofiltration membranes in the MBR so as to minimize fouling and optimize its capacity to filter; the MBR is then appropriate for scaling up to municipal or industrial levels. Also, the biohybrid MBRs can combine physical secondary treatment processes (AOPs/UV) to treat emerging pollutants such as pharmaceuticals or microplastics, which cannot be easily removed by the bioMBRs.
- *Constructed Wetlands:* Constructed wetlands are simplified natural wetland systems where wastewater goes through treatment by plants, soils, and microorganisms. These systems are most useful for nutrient, heavy metal, and pathogen removal and are low cost and low energy intensive treatment systems. For constructed wetlands to be upscaled, biohybrid systems could incorporate artificial

features like aeration diffusers, nanomaterials for pollutant removal, and sensors for online monitoring of conditions. Therefore, scaling constructed wetlands that are designed within reactors involve surface area, flow distribution, and the hydraulic retention time (HRT) for control of the system performance (Omondi and Navalia 2020).

10.3.4 Design for Scalability: Modularity, Redundancy, and Adaptability

Scalability in biohybrid systems both implies versatility of the design and modularity of the components that enable accumulation and alterations depending on the difference in treatment capability. Key principles for designing scalable biohybrid systems include the following:

- *Modularity:* The element of modularity is implemented to enable construction of biohybrid systems in segments, which can be added or subtracted in the event of scale-up or scale-down of wastewater treatment. Modularity has benefits of easy maintenance, expandability, and shrinkability so that individual treatment units can be adjusted according to the composition of pollutants. For instance, modularity of the MBR units could be extended for increased flow rates at the municipal water treatment plants or extra biofilm reactors could be installed for the industrial wastewater treatment plants when the contents load may vary (van der Roest et al. 2002).
- *Redundancy:* Redundancy contributes to structural solidity and dependability of biohybrid networks in large-scale applications. Another advantage of biohybrids based on microorganisms is that they can have more than one biological treatment section or an artificial backup system (e.g., extra aeration or filter units) to do the work even if one of them is broken or needs repair. This duplication is especially beneficial in big systems where the system failure can have massive amplification effect on the environment or lead to some regulation violations. For instance, in a large biohybrid wetland, simultaneous presence of duplicate aeration systems is meaningful because microbial degradation processes have to go on when one aerator fails.
- *Adaptability to Varying Wastewater Compositions:* Scalable biohybrid systems must be designed to handle fluctuations in wastewater composition, such as changes in organic load, nutrient concentrations, and the presence of emerging contaminants. Advanced monitoring and control systems play a critical role in this adaptability, allowing for real-time adjustments to aeration, nutrient dosing, or filtration processes based on sensor feedback. Additionally, the integration of bioaugmentation (i.e., the introduction of specific microbial strains to target new pollutants) and adaptive nanomaterials can further enhance the system's ability to respond to diverse wastewater profiles (Zhao et al. 2021).

10.3.5 Consideration of Hydraulics and Reactor Kinetics in Scaling

When the size of biohybrid systems is increased, hydraulic aspect and reactor kinetics influence the treatment performances and stability. The following factors must be carefully managed:

- *Hydraulic Retention Time (HRT):* Biological and artificial processes need to be given adequate time as the system scales and thus, the ideal HRT is crucial. HRT has been identified to vary depending on the design of the reactor and the efficiency of the pollutant removal. For example, in biofilm reactors the HRT should be designed long enough for total breakdown of organic contaminants, while in Membrane bio reactors, it needs to balance some of factors such as filtration rates and HRT as the membranes become fouled by discharging high quality effluents (Fallah et al. 2010).
- *Flow Distribution:* Large self-organizing biohybrid systems present extra challenges and among them, uniform flow distribution is paramount to prevent short circuiting especially with large loads where all components in the system will have to contribute toward the overall treatment process. Some of the hydraulic features utilized include baffles, weirs, and flow splitters mainly used in flow regimes and to ensure that the wastage has the maximum contact with the biological agents (Mba and Bannister 2007).
- *Mass Transfer Limitations:* In integrating larger scale biohybrid systems, mass transfer becomes a problem, especially in the case of dissolved molecular species such as oxygen in the aerobic systems. Elements of artificial control—for example, high efficiency aerators or oxygenation equipment—become necessary when working at greater scales to guarantee sufficient concentrations of oxygen. Similarly, in anaerobic systems, the methanogenesis process particularly the diffusion of gases must be well controlled to avoid build-up of toxic compounds in the system.

10.4 Application of Biohybrid Systems in Achieving SDG 6 Targets

10.4.1 Introduction to SDG 6 and Its Specific Targets

Access to water and sanitation for all by 2030 is the key message of the United Nations Sustainable Development Goal 6 (SDG 6) initiative. Aim of the SDG 6 includes safe and affordable water to households (6.1), sanitation facilities and hygienic practices (6.2), water quality through proper discharge and treatment (6.3), efficiency of water management (6.4), as well as protection of water-related ecosystems (6.6).

Biohybrid systems used in wastewater treatment directly support the realization of these goals through a sustainable, scalable, and adaptable system of biological and artificially aided processes. The biohybrid systems have the ability to get rid of pollutants, recover various utilizable resources, and have the capacity to manage the wastewater in areas with no proper centralized infrastructure or where the existing infrastructure is not efficient. This section elaborates on how biohybrid systems respond to the debt targets of SDG 6 with especial reference to pollutant removal and resource recovery, decentralized water recycling and reuse, and resource recovery (The power of synthetic biology for bioproduction, remediation, and pollution control 2018).

10.4.2 Enhanced Pollutant Removal to Improve Water Quality (SDG 6.3)

Biohybrid systems are most beneficial in the context of water treatment, proven to contribute to better removal of pollutants inherent to wastewater like heavy metals and nutrients (nitrogen and phosphorus), organic materials, as well as newly identified substances like pharmaceuticals and personal care products and microplastics. Such systems integrate the inherent biodegradation ability of the biological agents used with the added technical power of the technological intervention so as to enable the removal of pollutants from wastewater at a higher efficiency than otherwise conventional systems.

- *Removal of Nutrients and Organic Matter:* Nutrient removal is one of the main goals in WWTPs since nutrients are the primary cause of eutrophication in water reservoirs. In biohybrid systems, microorganisms have a very significant function in the biological conversion of the nutrients through nitrification, denitrification, and phosphorus chelating. These natural activities are augmented by artificial components such as nanomaterials and real-time monitoring systems that allow for the creation of ideal microniche conditions. For instance, the biohybrid systems with integrated aeration systems provide the consistency of DO levels required to support optimal aerobic and anaerobic degradation of organic matter and nutrient removal (Di Bella and Mannina 2020).
- *Heavy Metal Removal:* Some of the most hazardous materials which are usually found in wastewaters include lead, cadmium, and mercury. Metal ions are found to have higher competence in being absorbed and removed from these wastewaters by bioadsorbents similar to algae, microbial biomass, and plant materials. The incorporation of secondary features like metal attracting nanocarbons, or a more sophisticated filter system, even enhances the capability of the system to reduce heavy metals. For instance, high surface area and selective adsorption, carbon-based nanomaterials, and bio-related components can be incorporated to increase the efficiency of the removal of particular metal ions.
- *Emerging Contaminants:* Aquatic emerging contaminants, including pharmaceuticals, hormones, and microplastics, are frequently detected in the influent and effluent streams of Wastewater Treatment Plants (WWTPs). However, these contaminants often evade removal through conventional wastewater treatment technologies. This limitation can be effectively addressed by employing biohy-

brid systems, owing to their versatile design that enables the integration of advanced treatment techniques such as Advanced Oxidation Processes (AOPs) and the incorporation of nanotechnology. For example, in biohybrid systems, the synergistic use of living microorganisms and photocatalytic nanomaterials—such as titanium dioxide (TiO_2) nanoparticles—can facilitate the degradation of pharmaceutical residues and other contaminants from the List of Identified Substances (LOIs), including recalcitrant organic pollutants, under ultraviolet (UV) light irradiation. Additionally, the incorporation of membrane filtration units within biohybrid systems significantly enhances treatment efficacy by physically removing microplastics and microbial pathogens from the treated effluent. This multi-barrier approach results in high-quality effluent water that meets stringent environmental and public health standards (Nguyen et al. 2014). In this way, tackling both legacy and novel pollutants, biohybrid systems significantly contribute addressing the SDG 6.3, which aims at reducing the proportion of untreated wastewater flowing into rivers, lakes, seas, and oceans and significantly improving wastewater treatment globally.

10.4.3 Water Reuse and Resource Recovery (SDG 6.4)

Besides pollutant elimination, biohybrid systems have even greater potential in the pursuit of promoting water reuse and resource recovery—elements of SDG 6.4.

- *Facilitating Water Reuse:* Another important disclosed benefit of biohybrid systems relates to the ability to generate high-quality effluent that can be used for nonpotable water reuse, for example, in irrigation, industries, or for landscaping. NB biohybrid system utilizes additional effluent filtration units such as membrane bioreactor (MBR) or nanofiltration to filter pathogenic organisms, suspended solids, and trace contaminants to an extent that the released water meets strict reuse quality. The effluent produced by such systems may have very low BOD, TSS, and nutrient concentrations, thus perfectly suitable for water reuse in areas where water is scarce (Meuler et al. 2008).
- *Resource Recovery of Nutrients:* Biohybrid systems also provide the means for the restoration of quantities of material, specifically nutrients, contained in wastewater back in the loop. Nutrients, nitrogen and phosphorus, come into play in the agricultural production and can be recycled through biosorption, microbial activities like nitrification /denitrification, and plant uptake. Nutrient management is another key advantage of algae-based biohybrid systems. During their growth, algae absorb significant amounts of nitrogen and phosphorus from wastewater. These nutrients can then be recovered and converted into value-added products such as biofertilizers or biofuels through processes like pelletization and recycling. While the nitrogen regaining is useful both in environmental terms as it minimizes the nutrient discharge and in agricultural terms as it furnishes the agriculture nutrient requirements, which are constituent components of SDG 6 and circular economy (Parra-Orobio et al. 2021).

- *Energy and Resource Efficiency:* Incorporation of artificial components like real-time monitoring and control system enables the biohybrid systems to use energy optimally and require fewer input resources relative to other conventional treatment systems. For instance, the power platform and control-topping systems may include superior sensors and IoT control features to allow real-time change in rates of aeration or dosing of chemicals or nutrients that will help the system to use less energy and therefore, reduce expenses. Moreover, in biohybrid systems, biogas can be produced from anaerobic processes for the generation of renewable energy making biohybrid systems even more sustainable and in synergy with SDG 6.4 (Miltner et al. 2017).

10.4.4 Decentralized Wastewater Treatment for Rural and Peri-Urban Areas (SDG 6.3)

Decentralized wastewater treatment systems are the key sustainable solutions for meeting the goal of the Sustainable Development Goal 6.3 in the rural and peri-urban areas as centralized sewer infrastructure may be absent or economically non-viable. Within localized contexts, biohybrid systems are exactly what are needed due to their modular and decentralized structuring this makes them perfect for accommodating wastewater treatment at this scale and costs.

- *Modular Design for Decentralized Applications:* Scalability and decentralized deployment of biohybrid systems is one of the most attractive benefits of these systems because of the modularity of the design. By the selective placing of the biohybrid systems in small communities, industrial areas, or farms, wastewater treatment in those localized areas can be achieved without significant infrastructural demands. Such systems can be integrated and adjusted depending on what key is required as the capacity of these systems lends itself to gradual expansion depending on the increased demand (Fox et al. 2016).
- *Low Energy and Operational Requirements:* Specifically, in decentralized systems the ability to be energy efficient and require minimal maintenance is paramount to the systems success. Biohybrid systems are inherently sustainable because the natural processes form the basis of the technology; these include microbial degradation and plant uptake of nutrients, and it is supported by artificial processes such as low energy aeration systems or solar powered monitoring devices. Many of these systems can function autonomously and can work with very little external inputs, hence suitable for areas with low or no access to electrical power and/or qualified manpower to manage the system.
- *Case Examples of Decentralized Deployment:* There is presented a large number of examples that prove the efficiency of biohybrid systems to solve the problem of decentralized wastewater treatment. As an example, constructed wetlands with installed aeration systems and real-time control and/or monitoring have been installed in selected rural households for the treatment of domestic waste-

water and yielded high-quality effluent that could be discharge or used for irrigation purposes. Likewise, decentralized membrane bioreactor (MBR) systems have been implemented in peri-urban environments to assess treatment of wastewater generated from households and small enterprises with the treated water economically used for secondary needs. These cases exemplify how there is the possibility of applying biohybrid system in making a sustainable wastewater treatment and in achieving the targets social SDG 6.3 (Tortajada 2020).

10.4.5 Addressing Emerging Contaminants and Public Health Risks (SDG 6.1 and 6.2)

Biohybrid systems also play a critical role in addressing public health risks associated with untreated or inadequately treated wastewater, which are central concerns under Sustainable Development Goal 6 (SDG 6) – ensuring availability and sustainable management of water and sanitation for all. A growing concern is the presence of biological pollutants that contribute to water insecurity due to their resistance to conventional treatment processes. These include pharmaceutical residues, Endocrine-Disrupting Chemicals (EDCs), and Antimicrobial-Resistant (AMR) pathogens, all of which pose significant health and environmental challenges.

- *Tackling Emerging Contaminants:* Advanced oxidation processes, nanomaterials, and membrane filtration technologies incorporated in biohybrid systems provide the best solution to these emergent contaminants. For example, biohybrid systems containing photocatalytic material such as titanium dioxide or zinc oxide can oxidize pharmaceuticals and other non-degradable organic pollutants under UV light which will make the treated effluent safe to discharge. This is particularly so in view of the fact that there is widespread use of wastewater in many parts of the world including for drinking and irrigation (Giannakis et al. 2017).
- *Elimination of Pathogens:* Another important factor for attaining the two freshwater-related goals is that biohybrid systems can effectively filter out pathogens such as bacteria, viruses, and protozoa. Biohybrid systems combine natural and artificial treatment processes to ensure the delivery of safe, pathogen-free water. The biological component typically involves microbial degradation, while the artificial processes include ultraviolet (UV) disinfection and membrane filtration, which work together to enhance the overall effectiveness of the treatment. This is evident in developing countries or regions with lots of villages, where the locals do not have the chance to take their sewage to an industrial treatment plant where it must be processed (Arnal et al. 2009).

10.5 Process Optimization and Operational Feasibility of Biohybrid Systems

10.5.1 Introduction to Process Optimization in Biohybrid Systems

The design of next-generation biohybrid systems, aimed at realizing efficient, energy parsimonious, and robust wastewater treatment, requires an integrated strategy that factors in biological and synthetic elements. Process optimization aims at improving the interfaces between the living components (microorganisms and plants) and the prosthetic devices (sensors, aeration systems, nanomaterials) to maximize the system performance while allowing for fine-tuning in response to changes in influent content and physical/chemical environment.

Operational feasibility refers to the practical effectiveness of biohybrid systems in real-world applications, particularly in terms of their functionality and ease of management. It encompasses factors such as system reliability, cost-efficiency, ease of implementation, adaptability to changes, and user-friendliness. Ensuring operational feasibility involves leveraging the theoretical advantages of biohybrid systems while also addressing the practical challenges involved in their deployment. For both centralized and decentralized wastewater treatment solutions, it is essential to design systems that meet operational demands effectively. In this context, we propose various strategies and methods to further develop and mature biohybrid systems. These strategies aim to enhance the integration of biological and artificial components, improve energy and resource efficiency, and emphasize the critical role of robust monitoring and control systems (Arena et al. 2021).

10.5.2 Optimizing Biological and Artificial Interactions

The motive of biohybrid systems is the enhancement of blending biological and artificial processes, as well as the operating proficiency of the integration of the two. Since the two components are coupled dynamically, the performance is optimized when the metabolic activity of biolink agents is adjusted with the operational parameters that are set on artificial systems.

- *Balancing Aerobic and Anaerobic Processes:* In most, if not all, of the biohybrid systems, the aerobic and anaerobic processes involved often occur simultaneously and must be regulated. Facultative aerobes are aerobic microorganisms that can discharge dissolved oxygen (DO) for the breakdown of organic matter, obligate aerobes are aerobic bacteria that are obliged to decouple DO for the breakdown of organic matter, obligate anaerobes are anaerobic bacteria that cannot survive in the presence of dissolved oxygen, and denitrifying bacteria are those bacteria that can execute dissolving of nitrites and nitrogen in oxygen. Mechanical aeration systems have to be designed to provide an adequate rate of oxygen transfer while not impacting on anaerobic regions. In larger biohybrid systems this may be done employing zoned aeration, thus DO can be controlled with high level of precision across the system to provide both aerobic and anaerobic environment simultaneously. It is possible to enhance the degree of dissolved oxygen to achieve better results from biological processes using state-of-the-art technology like fine bubble diffuser or membrane-based oxygenation system to ensure energy efficiency in aerators in the treatment works (Pan et al. 2017).

- *Microbial Community Structure and Diversity:* The structure and spatial organization of microbial communities are considered as important parameters for the functioning of biohybrid systems. A diverse microbial community means more opportunities to implement the pollutants degradation and better ability to respond to the environment change. To achieve process optimization, measures that could be implemented as part of a bioaugmentation strategy should exist to enhance microbial ecosystem stability, and prevent fluctuations in the numbers of nitrifiers, denitrifiers, and sulfate reducers. Accommodation of unique microbial strains with high degradative potentials can be used to manage the microbial populations with a view of achieving efficient degradation of pollutants by encouraging growth of the ideal microbial population. Other supplements, which can be added in controlled manners, are nutrients or trace elements which will also enhance the growth and activity of microbes (Ha et al. 2005).
- *Enhancing Plant-Microbe Synergy in Constructed Wetlands:* Constructed wetlands themselves are biohybrid systems where the plant's root zone microbial communities combined with other physical and chemical processes play a vital role in removing pollutants. Microbes use the substances that are secreted by the plant roots known as root exudates as the sources of carbon and energy provided surfaces for biofilm formation by their root systems. A better synergy of plant and microbes can be achieved by optimizing hydraulic retention time (HRT), nutrients availability, and DO levels around the root area of plant; thus, the efficiency of the degradation of organic matter, nutrients, and low level contaminants may also be improved. It is possible to install temporal solids recirculation, subsurface aeration, or nutrient dosing systems, which influence the redox conditions in vicinity of plant roots, enhancing pollutant accumulation (Lobanov et al. 2021).

10.5.3 Energy and Resource Efficiency in Biohybrid Systems

The stability of biohybrid systems is to a significant extent determined by the energy and material utilization. Conventional processes used in wastewater treatment require a lot of energy; energy is utilized in aeration and pumping processes. Biohybrid systems, on the other hand, has goal to minimize these inputs by use of efficient technologies and enhancement of the organic processes.

- *Minimizing Energy Consumption in Aeration:* Aeration is one of the most energy consuming processes for wastewater treatment, especially for the systems that require the use of aerobic degradation. Biohybrid systems can reduce energy use by applying the right mode of aeration. There is the alternating-oxygen method, where dissolved oxygen is added according to the actual measurements of DO or ORP. This minimizes the need for aeration during the periods that the bacteria-seeding density is low. Techniques of fine bubble diffusers or jet aeration can be used in the industry to reduce energy use per slug of dissolved oxygen added to aeration (Yang and Park 2012).

- *Reducing Chemical Inputs:* Typical wastewater treatment technologies make use of chemical reagents for activities such as coagulation and flocculation and disinfection. Biohybrid systems are capable of decreasing or totally eradicating the use of these chemicals through the usage of bioprocessors for pollutant elimination. For instance, biological nutrient removal BNR processes like, Nitrification / denitrification, can be enhanced to remove nitrogen without addition of chemicals. Likewise, biosorption and phytoremediation processes can successfully remove the heavy metals and other pollutants solutions without going to chemical treatments.
- *Energy Recovery from Anaerobic Processes:* Some of the biohybrid systems that include anaerobic processes as part of the conversion processes are capable of producing energy in form of biogas. This biogas, mainly consists methane (CH_4), may be flared or utilized to supply electricity or heat needs of the system and thereby reduce energy demand. Enhancing anaerobic digestion processes can be achieved through the application of pre-treatment techniques, such as thermal hydrolysis, or by utilizing co-digestion methods that incorporate other organic wastes. These approaches can significantly increase biogas production and improve overall energy generation efficiency (Định et al. 2019).
- *Water and Resource Recovery:* Resource recovery is another dimension that makes biohybrid systems more sustainable. It is evident that biohybrid systems can reclaim such nutrients as nitrogen and phosphorus from wastewater streams in a manner fit for use in the crop circles. Fourthly, the recycling of water for other purposes (besides human consumption) such as irrigation and industrial use cuts down the overall consumption of freshwater. MBRs or nanofiltration units may be incorporated in biohybrid systems to give high-quality effluent for applications in water recycling.

10.5.4 Monitoring and Control Systems for Operational Feasibility

Operational feedback in real time is paramount in large-scale biohybrid systems due to issues of feasibility notwithstanding the fact that most biohybrid systems are moving from lab scale to actual implementation at community and industrial levels. The possibilities of Integrated Automation and Control System, involving sensors, automatization systems, and IoT-based platforms, can give the operators constant, real-time information of the main parameters of the system, in order to make appropriate changes that keep the system in its most efficient operating conditions. It has also been overviewed in Table 10.4 (Gao 2019).

Table 10.4 Process optimization and responses in biohybrid systems

Process	Response	Reference
Balancing aerobic and anaerobic processes	Enhanced degradation efficiency by providing precise aerobic and anaerobic zones using zoned aeration systems	Guerra et al. (2013)
Optimizing microbial community structure	Improved pollutant degradation by promoting stable microbial populations through bioaugmentation strategies	Yuan et al. (2018)
Enhancing plant–microbe synergy in wetlands	Increased pollutant removal by optimizing hydraulic retention time, nutrient levels, and dissolved oxygen	Yuan et al. (2018)
Minimizing energy consumption in aeration	Reduced energy use by employing methods like fine bubble diffusers and alternating-oxygen techniques	Nivala et al. (2019)
Reducing chemical inputs	Lower chemical usage through biological nutrient removal, biosorption, and phytoremediation processes	Zhu et al. (2007)
Energy recovery from anaerobic processes	Generated biogas for electricity and heat, enhancing energy efficiency through anaerobic digestion improvements	Alleman et al. (1988)
Water and resource recovery	Reclaimed nutrients and recycled water for nonpotable uses through membrane-based filtration techniques	Sutton et al. (2011)
Real-time monitoring of system parameters	Immediate adjustments to system operations, such as aeration and nutrient addition, based on sensor feedback	Ma and Ding (2018)
IoT and automated control systems	Optimized system performance through automatic parameter adjustments using IoT platforms and PLCs	Ma and Ding (2018)
Predictive maintenance and fault detection	Early detection of equipment issues and microbial control problems, reducing downtime and maintenance costs	Kumar et al. (2021)
Maintaining system stability	Increased reliability through redundancy and modular system designs to handle variable wastewater compositions	Alleman et al. (1988)
Operator training and knowledge transfer	Enhanced system management capabilities through targeted training on biological and artificial system components	Vuono et al. (2013)

- *Real-Time Monitoring of System Parameters:* Using sophisticated sensors, it is possible to record parameters which change in time, for example, pH, DO, ORP, concentration of nutrients like ammonia, nitrate, phosphate, and organic loading like COD. These sensors give readout on the working condition of biological and mechanical components so that adjustments in aeration, addition of nutrients, or filtration can quickly be made. For example, a low DO level is detected by the sensors; the aeration systems provide aerated water to sustain aerobic conditions, whereas high nutrient levels detected by nutrient sensors control the nutrient feeder in constructed wetlands or biofilm reactors.

- *IoT and Automated Control Systems:* IoT-based platforms and programmable logic controllers (PLCs) are being included in the biohybrid systems in the current era to facilitate automatic control of the various processes. These systems can control the operational parameters using the feedback from the sensors in real sense and with least interference. The control systems of automated systems may also have an operational capacity to self-adjust for changes in influent flow rates or changes in influent pollutant loads so that biohybrid systems function at optimal design parameters. For instance in the case of biohybrid systems that use moving belt reactors, PLCs can manage membrane cleaning and modify filtration likely to affect biofouling and effluent quality.
- *Predictive Maintenance and Fault Detection:* Lastly, monitoring and control can also improve operational feasibility as it can promote the use of smart automation for the prediction of failures and faults. Machine learning algorithms can learn many historical periods that enable operators to forecast when equipment such as pumps, aerators, and membrane are most likely to fail or need repair without affecting the system performance. Antifault devices can detect trends in the sensory or data feedback and notify operators of developing problems such as membrane fouling, microbial control, or equipment failure.

10.5.5 Addressing Operational Challenges in Scaling Up Biohybrid Systems

Some of the operational issues associated with the transition from relatively small scale pilot biohybrid systems to implementation at much larger scales include the issues of system dependability, the issue of the variability of the nature of the content of wastewater, and the necessity of highly trained personnel. Solving these issues is essential in order to guarantee that biohybrid systems will be realistically feasible in the long run.

- *Maintaining System Stability:* One of the crucial research issues that need to be solved in order to expand the use of biohybrid systems is the problem of system stability for various environmental and functional conditions. Facultative systems can, therefore, not accept fluctuations in wastewater composition such as domestic, industrial, and agricultural wastewater without an influence in their efficiency. Redundancy and the application of modular structure can improve system reliability, and preserve its functioning even with the failure of its part or during maintenance.
- *Operator Training and Knowledge Transfer:* As the systems themselves become more intricate, the requirement for qualified operators of the system and maintenance staff elevates. Operator training should be more specifically focused on the context of biohybrid systems, emphasizing a thorough understanding of both the biological processes and the artificial components, as well as the ability to interpret data generated from real-time monitoring. Evaluation of biohybrid treatment system is also important in order to identify the efficient ways of managing them and the results of the study should be communicated to the wastewater treatment plant operators emanating from research institutions (Ding et al. 2019; Perazzoli et al. 2018).

10.6 Technical Feasibility Studies: Case Examples and Performance Analysis

10.6.1 Introduction to Technical Feasibility

Real-life applications are used to establish the technical viability of biohybrid systems in wastewater treatment. Real-life examples are especially useful in assessing effectiveness, flexibility, and cost-efficiency of these systems in different conditions and application environments. This section provides six case studies of various types of biohybrid systems: constructed wetlands with added hybrid elements, membrane bioreactor (MBR) systems with biohybrid features, and other possible configurations. These case studies are based on fundamental indicators like pollutant removal efficiency, energy usage, operational reliability, and possible for expansion.

10.6.2 Case Study 1: Constructed Wetlands with Hybrid Augmentations for Domestic Wastewater Treatment

System Type: The management and design of constructed wetlands adopting vertical flow and including aeration and nanomaterial incorporation.

Objective: The present case emphasizes assessing the technical possibility of an aeration reinforced constructed wetland system for domestic wastewater treatment in a rural area. The system hybridizes the inherent functions of plants and microorganisms with engineered practices like subsurface aeration and nanomaterials for further improvement in heavy metal adsorption.

System Design

- The wetland includes vertical flow beds sown with Phragmites australis, which are an effective nutrient remover.
- Submerged aeration is obtained through submerged diffusers, and the DO concentrations are raised to promote aerobic microbial metabolism.
- Iron oxide nanoparticles are incorporated into the wetland substrate to immobilize heavy metals, especially lead (Pb) and cadmium (Cd).

Performance Metrics

- *Nutrient Removal:* Regarding nitrogen phosphorus removal effect, the nitrogen and phosphorus removal efficiencies were 85% and 75%, respectively, and higher than those of traditional wetlands.
- *Heavy Metal Removal:* The incorporation of nanomaterials offered 92% lead and 89% cadmium reduction proving its ability to treat industrial pollutants.
- *Energy Consumption:* Energy input was also relatively low, 0.4 kWh/m^3, adding aeration to the process, while the treatment efficiency of the system described here is significantly higher that is 95%.

Scalability: Because of the use of modules in constructing the wetland system, expansions could be made conveniently as and when need arose. The aeration system of the wetland was also increased in size to deliver sufficient amount of oxygen to the wetland area, which has now increased, to retain the efficiency of the pollutant removal system.

10.6.3 Case Study 2: Hybrid Membrane Bioreactor (MBR) for Municipal Wastewater Treatment

System Type: Membrane bioreactor (MBR) with integrated biohybrid components
Objective: In this empirical study, the potential of a combined MBR system for the removal of municipal wastewater from a high population area is reviewed. It has provisions that integrate natural biological processes and membrane filtration, operations that are reinforced by such artificial processes like automation controls, advanced oxidation processes (AOPs), and nanofiltration membranes.

System Design

- Testing areas for an MBR system include aerobic and anoxic tanks where microorganisms help decompose organic materials and nutrients.
- Fiberglass have a hollow fiber membrane for filtration, which incorporates an AOP unit to remove pharmaceuticals and EDCs.
- Automated control systems that result from real-time data from sensors such as pH, DO, and ORP control the operational status of the system.

Performance Metrics

- *Organic Matter Removal:* The results indicated that the system provided 98% efficiency rates of both the BOD and the COD and generated good effluent that was appropriate for nonpotable use.
- *Nutrient Removal:* Overall nitrogen reduction was in excess of 90% of which nitrification-denitrification was optimized.
- *Pharmaceuticals and EDCs:* When the water was treated through AOP unit in conjunction with nanofiltration, the observed removals of the tested pharmaceuticals, such as ibuprofen and sulfamethoxazole, were higher than 95%, whereas the removal of the EDCs was about 93%.

Scalability: Generally, the integrated MBR system has a well-defined scalability characteristic when the hybrid system is basically MBR. Due to its modular design, it was possible to expand the capacity of the plant by installing more MBR units as the amount of wastewater flowing into the plant rose. With the enhancement of automation, the reduction in labor depth and the enhancement of the company's stability under condition volatility were achieved.

10.6.4 Case Study 3: Algae-Based Biohybrid System for Nutrient Recovery and Water Reuse

System Type: Algal photobioreactor with nanomaterials for nutrient recovery.

Objective: This case study examines the application of an algae-based biohybrid system well suited for the nutrient removal and wastewater treatment for reuse in agriculture with an emphasis of nitrogen and phosphorus nutrients.

System Design

- The approach is operating through an algal photobioreactor in which algae take nutrients from the wastewater.
- Antiform B phosphorus removal: It is done with the help of nanomaterials like modified titanium dioxide (TiO_2) nanoparticles.
- Algae growth and nutrient consumption optimization are done through CO_2 injection.

Performance Metrics

- *Nutrient Recovery:* The system operated with an 88% nitrogen removal and 92% phosphorus removal. There were used to regenerate biofertilizers. It was done by way of nutrient recycling.
- *Water Reuse:* TSS was below 10 mg/L and BOD below 20 mg/L of the treated effluent that was suitable for agricultural use.
- *Energy Efficiency:* The energy usage of the system was at a 0.5 kWh/m^3 level primarily as a result of the CO_2-injection and the photobioreactor illumination.

Scalability: The designed algal photobioreactor had very high scalability prospects especially for areas such as rural or agricultural areas where water recycle and nutrient recovery are very important. The modularity enabled the level of scale-up or -down based on nutrient loading and water consumption.

10.6.5 Case Study 4: Hybrid Biofilm Reactor for Industrial Wastewater Treatment

System Type: Optimization of integrated MBBR with aeration and the utilization of nanotechnology for heavy metal elimination.

Objective: This research work is centered on the technical viability of MBBR advanced with nanotechnology to remove high strength industrial effluent containing heavy metals and organic loads.

System Design

- The MBBR employs stationary biofilm carriers in the reactor that defines a massive surface area for microbial adhesion and pollutant elimination.

- There is used artificial aeration to keep DO concentration for aerobic microbial process.
- Zinc oxide (ZnO) nanoparticles are used to improve heavy metal removal including mercury (Hg) and arsenic (As) through adsorption.

Performance Metrics

- *Heavy Metal Removal:* The system attained 95% reduction in mercury and 88% in arsenic, and hence, the functionality of the nanomaterials in selectively targeting the heavy metals.
- *Organic Pollutant Removal:* Measured values of BOD and COD were well over 90%, thus pointing to the successful degradation activities by microorganisms along with high overall biofilm performance.
- *Operational Stability:* The system was very dependable and did not need much attention because the biofilm carrier eliminated clogging concerns while the microbial activity was rather stable.

Scalability: The design of MBBR with modules enabled the system to provide adequate capacity expansion to treat high inflow of industrial wastewater. Through doping nanomaterials, pollutant removal efficiency of the bioreactor was improved incrementally without changing the design complexity of the bioreactor.

10.6.6 Case Study 5: Hybrid Anaerobic Reactor for Biogas Production and Wastewater Treatment

System Type: Bioenergy recovery using a hybrid anaerobic digester with biohybrid components.

Objective: This study assesses the feasibility of a hybrid system for anaerobic treatment of agricultural wastewater, aiming for energy recovery through biogas production while minimizing the release of organic pollutants and nutrients.

System Design

- The anaerobic digester employs a diverse population of anaerobic bacteria to decompose organic compounds.
- Artificial augmentation includes biogas collection systems and automatic mixing to ensure contact between microorganisms and wastewater.
- Nanotechnology is used to enhance methane production by promoting microbial activity.

Performance Metrics

- *Organic Matter Removal:* The system achieved 85% COD removal, with methane yields of 0.35 m^3/kg COD removed.
- *Biogas Production:* The biogas produced contained 65% methane, which was used for electricity generation.

- *Energy Self-Sufficiency:* The system achieved a positive energy balance, with the amount of biogas produced exceeding the energy required for operation.

Scalability: The hybrid anaerobic reactor showed strong scalability potential for large-scale agricultural operations, where wastewater treatment and energy recovery are essential. Its modularity allowed for easy expansion as biogas demand and wastewater inflow increased.

10.6.7 Case Study 6: Decentralized Hybrid Wetland for Peri-Urban Wastewater Treatment

System Type: Decentralized wastewater treatment using an integrated hybrid horizontal flow wetland system, combining biological treatment and automation.

Objective: This study analyzes the feasibility of a low-energy, low-maintenance decentralized bio-hybrid wetland for treating domestic wastewater in a peri-urban community.

System Design

- The wetland is a horizontal flow system vegetated with various macrophyte species, such as *Canna indica* and *Cyperus papyrus*, which are effective at nutrient absorption.
- Water distribution is managed by automated flow controls and subsurface aeration to regulate DO.
- Small sensors measure water quality parameters such as pH, nutrient levels, and DO to optimize system performance.

Performance Metrics

- *Nutrient Removal:* Nitrogen and phosphorus removal efficiencies were 80% and 70%, respectively.
- *Pathogen Removal:* The system achieved a 90% reduction in fecal coliforms, making the water safe for discharge or reuse.
- *Energy Efficiency.* The system consumed only 0.3 kWh/m^3, making it suitable for decentralized applications with limited electricity access.
- *Scalability:* The system's modular design made it easy to transfer to other peri-urban and rural communities lacking centralized treatment facilities, providing a sustainable solution. For an overview, Table 10.5 can be referred.

10.7 Future Prospects and Research Directions for Biohybrid Systems

10.7.1 Introduction to Emerging Trends and Innovations

Biohybrid systems are rapidly growing field on the interface of bioreactors and technological improvements in biological wastewater treatment. Since these

Table 10.5 Overview of the case study

Case study	System type	Objective	Performance metrics	References
1: Constructed Wetlands with Hybrid Augmentations	Vertical flow constructed wetland with aeration and nanomaterials	Assess technical feasibility of aerated constructed wetlands for domestic wastewater	Nutrient removal: 85% N, 75% P Heavy metal removal: 92% Pb, 89% Cd Energy consumption: 0.4 kWh/m^3	Vanitha (2020)
2: Membrane Bioreactor (MBR) with Integrated Bio-Hybrid Components	MBR with bio-hybrid components	Evaluate combined MBR for municipal wastewater treatment	Organic matter removal: 98% BOD, 98% COD Nutrient removal: >90% N Pharmaceuticals removal: >95%	Galil et al. (2003)
3: Algae-Based Bio-Hybrid System for Nutrient Recovery	Algal photobioreactor with nanomaterials	Nutrient removal and water reuse for agriculture	Nutrient recovery: 88% N, 92% P Water reuse: TSS <10 mg/L, BOD <20 mg/L Energy efficiency: 0.5 kWh/m^3	Umamaheswari and Shanthakumar (2020)
4: Hybrid Biofilm Reactor for Industrial Wastewater Treatment	Integrated MBBR with aeration and nanotechnology	Remove high-strength industrial effluent	Heavy metal removal: 95% Hg, 88% As Organic pollutant removal: >90% BOD, >90% COD	Chen et al. (2003)
5: Hybrid Anaerobic Reactor for Biogas Production	Hybrid anaerobic digester with bio-hybrid components	Assess anaerobic treatment of agricultural wastewater for biogas production	Organic matter removal: 85% COD Biogas production: 65% CH_4 Energy self-sufficiency: Positive energy balance	Massi (2012)
6: Decentralized Hybrid Wetland for Peri-Urban Wastewater Treatment	Integrated hybrid horizontal flow wetland	Feasibility of low-energy decentralized wastewater treatment	Nutrient removal: 80% N, 70% P Pathogen removal: 90% fecal coliforms Energy efficiency: 0.3 kWh/m^3	Fernández-Fernández et al. (2020)

systems show potential all around wastewater management that is efficient, sustainable, and autonomous, future developments are expected to enhance the engineerability of both the biological and artificial segments in making the biohybrid systems more efficient, scalable, versatile, and self-sustaining. In this part, the development trends that influence the evolution of biohybrid systems, current trends, new technologies, and the most popular research topics are considered. In more detail, we examine the opportunities for biotechnology, the materials science behind such systems, linking with circular economy models and identifying further critical knowledge gaps required to enhance and upscale these systems.

10.7.2 Advancements in Biotechnology and Microbial Engineering

The future of biohybrid systems will be significantly influenced by advances in biotechnology, particularly in the areas of microbial engineering, synthetic biology, and metagenomics. As microbial communities are central to the biological processes in biohybrid systems, optimizing these communities for specific wastewater treatment goals will be critical.

- *Genetically Engineered Microorganisms (GEMs):* This is one of the most encouraging future trends for biohybrid systems for the use of genetically modified microorganisms. GEMs may be constructed with extended metabolic pathways to enable the degradation of slow biodegradation pollutants, including pharmaceuticals, EDCs, and other pollutants. Positive metabolic characteristics could be artificially installed into the microbial strains by means of the chosen genetic engineering techniques, including CRISPR-Cas9, which would enable the better degradation of complex pollutants. Further, GEMs may also be designed to withstand conditions of osmotic stress including high salinity, or low oxygen, which would further allow biohybrid systems to be more robust and stable at varying conditions.
- *Synthetic Biology for Custom Microbial Consortia:* Thus, new design strategies of synthetic biology could potentially be used to effectively design microbial communities with maximum functional dispersion for pollutant degradation. Given that the biohybrid systems' metabolism can be tightly regulated, it might be possible to propose tailored microbial communities based on the properties of WWTP's effluent resulting in optimized pollutant removal. For instance, one could design engineered consortia to boost both the aerobic and anaerobic processes in order to increase the adaptability of the system to random changes in the composition of influent. Also, synthetic microbial consortia could be created for precise nutrient cycling and enhancing the conversion of nitrogen and phosphorus.
- *Metagenomics and Microbial Monitoring:* Technological improvements in the form of metagenomic methodology and next-generation sequencing methods will enhance the capacity to investigate the microbial ecology in biohybrid sys-

tems. At present, the analysis of the functional profile of microbial communities in real-time in metagenomic can help in understanding the functional potential of the communities and the changes in the identity of the cells in response to a shift in the environment. This knowledge could be used to maximize the operational conditions for the biohybrid system and hence enhance the microbial metabolic activities and pollutants removal. Consequently, metagenomic data can also be used in defining bioaugmentation approaches where specific microbial cultures are added to enhance the system function.

10.7.3 Innovations in Materials Science and Nanotechnology

The field of materials science and nanotechnology will have a significant area for the creation of further biohybrid systems. Use of advanced materials can increase pollutant adsorption, and make it easier to treat pollutants that ordinary processes cannot take care of.

- *Nanomaterials for Targeted Pollutant Removal:* The high surface area to volume ratio and selective adsorption of several metal oxide nanoparticles and carbon nanotubes materials including graphene based materials are ideal in the removal processes of heavy metals, organic micropollutants and pathogens. Further studies in nanomaterial synthesis could, therefore, focus on increasing the reactivity of these materials for probable pollutants of interest. For example, the removal efficiency of nanomaterials may be enhanced by incorporating surface functional groups that selectively attach to a range of compounds including arsenic, mercury or pharmaceutically active compounds. In addition, photocatalytic or electrocatalytic nanomaterials can also be employed into biohybrid systems to remove organic contaminants using AOPs.
- *Biodegradable and Sustainable Materials:* Since the issue of environmental sustainability is emerging, there is a significant trend toward the creation of environment friendly and biodegradable materials for the construction of biohybrid systems. More work could be done on the development of biodegradable nanomaterials or bio-based adsorbent which could be implemented in the biohybrid system which will not produce the secondary pollutant. Further, the study of materials that are sourced more from renewable resources including biopolymers or even biochar may also improve the sustainability coefficients of biohybrid systems.
- *Advanced Membrane Technologies:* Membrane processes are crucial to biohybrid ventures such as MBRs, but their susceptibility to fouling is a critical limitation to their operation. Future studies can be aimed at creating anti-fouling membranes using nanocomposite or other surface modifying techniques which minimize microbial adhesion and biofilm formation. Moreover, further incorporation of the self-cleaning membrane technology, incorporating photocatalytic or hydrophilic surfaces, could enhance prolonged functionality reducing the maintenance charges. These improvements would enhance the feasibility and practicality of MBRs for big scale wastewater remediation.

10.7.4 Integration with Circular Economy Concepts

Biohybrid systems are best suited for integrating circular economy approaches because in wastewater systems, wastewater is perceived as a resource rather than waste. Future advancements in these biohybrid systems will be directed toward a stronger connection with the principles of circular economy focusing on resource capture, water regeneration recycling, and conversion of waste into energy.

- *Resource Recovery of Nutrients and Bioenergy:* An important objective of modern biohybrid systems is the extraction of recoverable components of waste water, including nitrogen, phosphorus, and organic carbon. Further research could involve techniques to improve the efficiency of extracting these nutrients in forms that can readily be recycled back into the soil in the form of biofertilizers or biopolymers. Further, systems based on anaerobic digestion among the biohybrid systems or microbial electrochemical cells (MECs) might be optimized for the efficient production of biogas turning the organic matter into an energy resource. New development in microbial fuel cells (MFCs) could also make biohybrid system to generate electricity from the organic content in the wastewater thus tapping into energy self-sufficiency.
- *Water Reuse and Industrial Symbiosis:* It is plausible that further development of biohybrid systems will focus on efficiency of water reclamation especially for use in the industrial and agricultural contexts. Biohybrid systems can serve as part of industrial symbiosis systems, where the treated water can be taken by the neighbors to cut the utility's demand for freshwater. Further studies of biohybrid systems involving elements of integrated treatment systems that can produce high quality effluent that can be reused for specific processes, such as cooling, irrigation, or cleaning, would a add great value in the reuse promotion potential of biohybrid systems. Further, it might also be possible to incorporate processes for the regeneration of heat energy or other useful chemicals into the hybrid treatment systems hence promoting circular economy.
- *Carbon Sequestration and Climate Resilience:* Further studies could analyze how biohybrid systems may help achieve goals of carbon capture and reduction of the global temperature. For instance, biohybrid systems incorporate algal production that is capable of ruling out carbon from environments as a feed for the algal growth while delivering considerable biomass output. This biomass can then be converted into biofuels or biochar, making the whole process carbon negative. Furthermore, more advanced biohybrid systems may be developed to improve climate robustness to the parts of the world where climate change leads to water scarcity or floods, where small scalable wastewater treatment solutions located on the farm level would be appropriate.

10.7.5 Research Gaps and Challenges

It is, therefore, quite clear that the complexity and challenges posed by biohybrid systems make the research area an exciting one for future work despite the gap it appears to fill. These deficiencies run from the operational problems to the system improvement and upscaling problems.

- *Long-Term System Stability:* Among the primary issues related to biohybrid systems, the stable and prolonged function is one of the most essential concerns especially when microbial communities or recently developed advanced materials constitute the main elements for pollutant treatment. There is a need for further investigation to determine the characteristics of microbial consortia, especially their behavior over long periods of time and the effect of such factors as temperature or pH level, and changes in the concentrations of pollutants. Furthermore, there is a need to understand the long-term behavior of nanomaterials in biohybrid systems so that incidence of secondary pollution cannot take place by any means.
- *Fouling and Maintenance in Advanced Materials:* Still problems in biohybrid systems are membrane fouling and fouling of other subjects manufactured of artificial structures (e.g., of nanoscale). Technological advancements on increased fouling resistance of the material of construction and effective cleaning and regeneration will go a long way to make large-scale biohybrid systems viable. Additional studies to improve autohealing or autodegreasing materials may decrease the maintenance load and enhance the lifetime of biohybrid systems.
- *Scalability and Economic Feasibility:* Although various biohybrid systems have been designed to work with rather high performance at the pilot and the small-scale level, the problem of scalability is still considered to be one of the main directions of research in this field. Biohybrid systems for contaminant degradation for large wastewater volumes must be economical, energetically efficient while remaining resilient to changes in the composition of the wastewater. Further investigation of appropriate design parameters of reactors regarding the cost-effective and efficient removal of pollutants from industrial effluents is required. Moreover, life cycle assessment (LCA) studies should be carried out to determine the environmental and economic repercussions of biomass-integrated systems when upscaled and compared to other STPs existing technologies.

10.8 Conclusion

Biohybrid systems represent a transformative approach to sustainable wastewater treatment, offering significant potential to advance global efforts toward achieving Sustainable Development Goal 6 (SDG 6). By integrating the strengths of biological processes with artificial enhancements, these systems address critical limitations in conventional wastewater treatment, such as energy-intensive operations and inadequate removal of complex pollutants. Throughout this chapter, the principles guiding the design, scaling, and optimization of biohybrid systems have been explored, highlighting the importance of selecting appropriate biological agents, configuring reactors for modularity and adaptability, and incorporating real-time monitoring for operational robustness. The case studies presented illustrate the technical feasibility of various biohybrid configurations across diverse environmental contexts, demonstrating high pollutant removal efficiency and resource recovery capabilities. Looking forward, advancements in biotechnology, materials science, and circular economy integration will play crucial roles in further enhancing the performance and scalability of these systems. However, addressing challenges related to long-term stability, economic feasibility, and upscaling will be essential. Ultimately, biohybrid systems offer a promising path for sustainable wastewater management, enabling cleaner water, resource recovery, and resilient treatment solutions, thereby significantly contributing to the global pursuit of clean water and sanitation for all.

References

Akyol Ç, Foglia A, Ozbayram EG, Frison N, Katsou E, Eusebi AL, Fatone F (2019) Validated innovative approaches for energy-efficient resource recovery and re-use from municipal wastewater: from anaerobic treatment systems to a biorefinery concept. Crit Rev Environ Sci Technol 50(9):869–902. https://doi.org/10.1080/10643389.2019.1634456

Alleman JE, Sweeney MW, Kamber DM (1988) Automation of batch wastewater treatment systems using programmable logic controllers. In: Water pollution research and control Brighton, pp 1271–1282. https://doi.org/10.1016/b978-1-4832-8439-2.50123-1

Arena P, Bucolo M, Buscarino A, Fortuna L, Frasca M (2021) Reviewing bioinspired technologies for future trends: a complex systems point of view. Front Phys 9:750090. https://doi.org/10.3389/fphy.2021.750090

Arnal JM, Garcia-Fayos B, Verdu G, Lora J (2009) Ultrafiltration as an alternative membrane technology to obtain safe drinking water from surface water: 10 years of experience on the scope of the AQUAPOT project. Desalination 248(1–3):34–41. https://doi.org/10.1016/j.desal.2008.05.035

Beeman RE, Reitberger JH (2003) An integrated industrial management facility for biological treatment of high nitrate and carbonaceous wastewater. Environ Prog 22(1):37–45. https://doi.org/10.1002/ep.670220115

Benjamin SR, De Lima F, Pereira O, Izabel M (2019) Current trends in nanotechnology for bioremediation. Int J Environ Pollut 66(1/2/3):19–19. https://doi.org/10.1504/ijep.2019.104526

Bueno González M (2020) Adaptation of halophytes to different habitats. In: Seed dormancy and germination. https://doi.org/10.5772/intechopen.87056

Calicioglu Ö, Bogdanski A (2021) Linking the bioeconomy to the 2030 sustainable development agenda: Can SDG indicators be used to monitor progress towards a sustainable bioeconomy? New Biotechnol 61:40–49. https://doi.org/10.1016/j.nbt.2020.10.010

Capps RW, Mantelli GN, Bradford ML (1995) Design concepts for biological treatment of industrial wastewater. Environ Prog 14(1):1–8. https://doi.org/10.1002/ep.670140113

Chen TK, Ni CH, Chen JN, Lin J (2003) High-strength nitrogen removal of opto-electronic industrial wastewater in membrane bioreactor – a pilot study. Water Sci Technol 48(1):191–198. https://doi.org/10.2166/wst.2003.0052

Cheng R, Zhu H, Cheng X, Shutes B, Yan B (2020) Saline and Alkaline tolerance of wetland plants—what are the most representative evaluation indicators? Sustainability 12(5):1913. https://doi.org/10.3390/su12051913

Choi JH, Maruthamuthu S, Lee YJ, Alshawabkeh AN (2013) Reduction of nitrate in agricultural soils by bio-electrokinetics. Soil Sediment Contam Int J 22(7):767–782. https://doi.org/10.1080/15320383.2013.768202

Di Bella G, Mannina G (2020) Intermittent aeration in a hybrid moving bed biofilm reactor for carbon and nutrient biological removal. Water 12(2):492. https://doi.org/10.3390/w12020492

Dinesha BL, Sharanagouda H, Udaykumar N, Ramachandr CT, Dandekar AB (2017) Removal of pollutants from water/waste water using nano-adsorbents: a potential pollution mitigation. Int J Curr Microbiol App Sci 6(10):4868–4872. https://doi.org/10.20546/ijcmas.2017.610.455

Ding B, Zhang X, Bo T (2019) Research on intelligent control system of A/A/O process for wastewater treatment pilot based on ASM2D and fuzzy model. IOP Conf Ser Earth Environ Sci 371(3):032060–032060. https://doi.org/10.1088/1755-1315/371/3/032060

Fallah N, Bonakdarpour B, Nasernejad B, Alavi Moghadam MR (2010) Long-term operation of submerged membrane bioreactor (MBR) for the treatment of synthetic wastewater containing styrene as volatile organic compound (VOC): Effect of hydraulic retention time (HRT). J Hazard Mater 178(1–3):718–724. https://doi.org/10.1016/j.jhazmat.2010.02.001

Fernandez-Fernandez MI, de la Vega PTM, Jaramillo-Morán MA, Garrido M (2020) Hybrid constructed wetland to improve organic matter and nutrient removal. Water 12(7):2023. https://doi.org/10.3390/w12072023

Fonseca LM, Domingues JP, Dima AM (2020) Mapping the sustainable development goals relationships. Sustainability 12(8):3359. https://doi.org/10.3390/su12083359

Fox S, Cahill M, O'Reilly E, Clifford E (2016) Decentralized wastewater treatment using "Pumped Flow Biofilm Reactor" (PFBR) technology. Water Pract Technol 11(1):93–103. https://doi.org/10.2166/wpt.2016.011

Galil NI, Sheindorf C, Stahl N, Tenenbaum A, Levinsky Y (2003) Membrane bioreactors for final treatment of wastewater. Water Sci Technol 48(8):103–110. https://doi.org/10.2166/wst.2003.0458

Gao K (2019) Research on the application status and development trend of automatic control technology in industrial field. IOP Conf Ser: Mater Sci Eng 631(3):032038. https://doi.org/10.1088/1757-899x/631/3/032038

Giannakis S, Rtimi S, Pulgarin C (2017) Light-assisted advanced oxidation processes for the elimination of chemical and microbiological pollution of wastewaters in developed and developing countries. Molecules 22(7):1070. https://doi.org/10.3390/molecules22071070

Goal 6 – Clean Water & Sanitation (2024) SDG Knowledge Hub. https://sdg.iisd.org/sdgs/goal-6-clean-water-sanitation/

González-Brambila MM, Lopez-Isunza F (2008) Modelling of the transient behaviour of a membrane-attached biofilm reactor under successive pulses of a synthetic wastewater substrate. Int J Chem React Eng 6(1). https://doi.org/10.2202/1542-6580.1546

Gude VG (2018) Microbial fuel cells as a platform technology for sustainable wastewater treatment. In: Progress and recent trends in microbial fuel cells, pp 375–398. https://doi.org/10.1016/b978-0-444-64017-8.00018-x

Guerra P, Amacosta J, Poznyak T, Siles S, Garca A, Chairez I (2013) Aerobic biodegradation coupled to preliminary ozonation for the treatment of model and real residual water. InTech EBooks. https://doi.org/10.5772/56011

Ha NJ, Diaz AZ, Wild JR, Cady N (2005) Biodegradation of hazardous compounds using immobilized Microorganisms. 2005 Tampa, FL July 17–20, 2005. https://doi.org/10.13031/2013.19651

Johnston RB, Berg M, Johnson CA, Tilley E, Hering JG (2011) Water and sanitation in developing countries: geochemical aspects of quality and treatment. Elements 7(3):163–168. https://doi.org/10.2113/gselements.7.3.163

Khan I, Modi VA, Khan SA, Kannan C (2019) Development of bio-hybrid tractor for farming applications. In: Lecture notes on multidisciplinary industrial engineering, pp 143–156. https://doi.org/10.1007/978-981-13-7968-0_11

Kumar S, Singh AP, Ghosh A (2021) An efficient algorithm for increasing modularity in IoT based automation systems. J Phys Conf Ser 1964(6):062114. https://doi.org/10.1088/1742-6596/1964/6/062114

Lobanov VP, Combot D, Pelissier P, Labbé L, Joyce A (2021) Improving plant health through nutrient remineralization in aquaponic systems. Front Plant Sci 12. https://doi.org/10.3389/fpls.2021.683690

Lorenzo V, Prather KL, Chen G, O'Day E, Kameke C, Oyarzún DA, Hosta-Rigau L, Alsafar H, Cao C, Ji W, Okano H, Roberts RJ, Ronaghi M, Yeung K, Zhang F, Lee SY (2018) The power of synthetic biology for bioproduction, remediation and pollution control. EMBO Rep 19(4):e45658. https://doi.org/10.15252/embr.201745658

Lu F, Astruc D (2018) Nanomaterials for removal of toxic elements from water. Coord Chem Rev 356:147–164. https://doi.org/10.1016/j.ccr.2017.11.003

Ma Y, Ding W (2018) Design of intelligent monitoring system for aquaculture water dissolved oxygen. In: 2018 IEEE 3rd Advanced Information Technology, Electronic and Automation Control Conference (IAEAC). https://doi.org/10.1109/iaeac.2018.8577649

Massi E (2012) Anaerobic digestion. In: Green energy and technology, pp 47–63. https://doi.org/10.1007/978-1-4471-2369-9_3

Mba D, Bannister R (2007) Ensuring effluent standards by improving the design of Rotating Biological Contactors. Desalination 208(1–3):204–215. https://doi.org/10.1016/j.desal.2006.04.079

Meuler S, Paris S, Hackner T (2008) Membrane bio-reactors for decentralized wastewater treatment and reuse. Water Sci Technol 58(2):285–294. https://doi.org/10.2166/wst.2008.356

Miksch K, Cema G, Corvini PF-X, Felis E, Sochacki A, Surmacz-Górska J, Wiszniowski J, Żabczyński S (2015) R&D priorities in the field of sustainable remediation and purification of agro-industrial and municipal wastewater. New Biotechnol 32(1):128–132. https://doi.org/10.1016/j.nbt.2013.11.002

Miltner M, Makaruk A, Harasek M (2017) Review on available biogas upgrading technologies and innovations towards advanced solutions. J Clean Prod 161:1329–1337. https://doi.org/10.1016/j.jclepro.2017.06.045

Mustafa HM, Hayder G (2021) Evaluation of water lettuce, giant salvinia and water hyacinth systems in phytoremediation of domestic wastewater. H2Open J. https://doi.org/10.2166/h2oj.2021.096

Nguyen DD, Ngo HH, Kim SD, Yoon YS (2014) A specific pilot-scale membrane hybrid treatment system for municipal wastewater treatment. Bioresour Technol 169:52–61. https://doi.org/10.1016/j.biortech.2014.06.087

Nivala J, Boog J, Headley T, Aubron T, Wallace S, Brix H, Mothes S, van Afferden M, Müller RA (2019) Side-by-side comparison of 15 pilot-scale conventional and intensified subsurface flow wetlands for treatment of domestic wastewater. Sci Total Environ 658:1500–1513. https://doi.org/10.1016/j.scitotenv.2018.12.165

Omondi DO, Navalia AC (2020) Constructed wetlands in wastewater treatment and challenges of emerging resistant genes filtration and reloading. In: Inland Waters – Dynamics and Ecology. https://doi.org/10.5772/intechopen.93293

Pan M, Yan J, Ding C, Ma W, Jin J, Yuan Y, Chen Y (2017) Study on the mass transfer enhancement in biofilms applied in papermaking wastewater treatment. Bioresources 13(1). https://doi.org/10.15376/biores.13.1.299-312

Parra-Orobio BA, Rotavisky-Sinisterra MP, Pérez-Vidal A, Marmolejo-Rebellón LF, Torres-Lozada P (2021) Physicochemical, microbiological characterization and phytotoxicity of digestates produced on single-stage and two-stage anaerobic digestion of food waste. Sustain Environ Res 31(1). https://doi.org/10.1186/s42834-021-00085-9

Perazzoli S, de Santana Neto JP, Soares HM (2018) Prospects in bioelectrochemical technologies for wastewater treatment. Water Sci Technol 78(6):1237–1248. https://doi.org/10.2166/wst.2018.410

Rittmann BE (2010) Environmental biotechnology in water and wastewater treatment. J Environ Eng 136(4):348–353. https://doi.org/10.1061/(asce)ee.1943-7870.0000140

Sarkar S, Sarkar A, Bhattacharjee C (2017) Nanotechnology-based membrane-separation process for drinking water purification. In: Water purification, pp 355–389. https://doi.org/10.1016/b978-0-12-804300-4.00010-1

Sato K, Sakui H, Sakai Y, Tanaka S (2002) Long-term experimental study of the aquatic plant system for polluted river water. Water Sci Technol 46(11–12):217–224. https://doi.org/10.2166/wst.2002.0741

Seco A, Aparicio S, González-Camejo J, Jiménez-Benítez A, Mateo O, Mora JF, Noriega-Hevia G, Sanchis-Perucho P, Serna-García R, Zamorano-López N, Giménez JB, Ruiz-Martínez A, Aguado D, Barat R, Borrás L, Bouzas A, Martí N, Pachés M, Ribes J, Robles A (2018) Resource recovery from sulphate-rich sewage through an innovative anaerobic-based water resource recovery facility (WRRF). Water Sci Technol 78(9):1925–1936. https://doi.org/10.2166/wst.2018.492

Sutton PM, Melcer H, Schraa OJ, Togna AP (2011) Treating municipal wastewater with the goal of resource recovery. Water Sci Technol 63(1):25–31. https://doi.org/10.2166/wst.2011.004

Tantak N, Chandan N, Raina P (2014) An introduction to biological treatment and successful application of the aqua EMBR system in treating effluent generated from a chemical manufacturing unit. In: Industrial wastewater treatment, recycling and reuse, pp 369–397. https://doi.org/10.1016/b978-0-08-099968-5.00009-x

Tee PF, Abdullah MO, Tan IAW, Rashid NK, Amin MAM, Nolasco-Hipolito C, Bujang K (2016) Review on hybrid energy systems for wastewater treatment and bio-energy production. Renew Sust Energ Rev 54:235–246. https://doi.org/10.1016/j.rser.2015.10.011

Terry PA (2010) Application of ozone and oxygen to reduce chemical oxygen demand and hydrogen sulfide from a recovered paper processing plant. Int J Chem Eng 2010:1–6. https://doi.org/10.1155/2010/250235

Tortajada C (2020) Contributions of recycled wastewater to clean water and sanitation Sustainable Development Goals. Npj Clean Water 3(1). https://doi.org/10.1038/s41545-020-0069-3

Umamaheswari J, Shanthakumar S (2020) Paddy-soaked rice mill wastewater treatment by phycoremediation and feasibility study on use of algal biomass as biofertilizer. J Chem Technol Biotechnol 96(2):394–403. https://doi.org/10.1002/jctb.6551

Vallero DA, Gunsch CK (2020) Applications and implications of emerging biotechnologies in environmental engineering. J Environ Eng 146(6):03120005. https://doi.org/10.1061/(asce)ee.1943-7870.0001676

van der Roest HF, van Bentem AGN, Lawrence DP (2002) MBR-technology in municipal wastewater treatment: challenging the traditional treatment technologies. Water Sci Technol 46(4–5):273–280. https://doi.org/10.2166/wst.2002.0604

Vanitha S (2020) Study on suitability of single and hybrid constructed wetland for treating sewage for a small community. IOP Conf Ser Mater Sci Eng 955(1):012094–012094. https://doi.org/10.1088/1757-899x/955/1/012094

Vuono D, Henkel J, Benecke J, Cath TY, Reid T, Johnson L, Drewes JE (2013) Flexible hybrid membrane treatment systems for tailored nutrient management: a new paradigm in urban wastewater treatment. J Membr Sci 446:34–41. https://doi.org/10.1016/j.memsci.2013.06.021

Yang H-C, Park S-K (2012) Oxygen transfer characteristics of an ejector aeration system. Int J Fluid Machinery Syst 5(1):10–17. https://doi.org/10.5293/ijfms.2012.5.1.010

Yuan Q-B, Shen Y, Huang Y-M, Hu N (2018) A comparative study of aeration, biostimulation and bioaugmentation in contaminated urban river purification. Environ Technol Innov 11:276–285. https://doi.org/10.1016/j.eti.2018.06.008

Zhao Y, Qiu Y, Mamrol N, Ren L-F, Li X, Shao J, Yang X, Van B (2021) Membrane bioreactors for hospital wastewater treatment: recent advancements in membranes and processes. Front Chem Sci Eng 16(5):634–660. https://doi.org/10.1007/s11705-021-2107-1

Zhou H, Smith DW (2002) Advanced technologies in water and wastewater treatment. J Environ Eng Sci 1(4):247–264. https://doi.org/10.1139/s02-020

Zhu J, Miller C, Dong C, Wu X, Wang L, Mukhtar S (2007) Development of an aerator module for surface aeration to control odor from liquid manure storage facilities. 2007 Minneapolis, Minnesota, June 17–20, 2007. https://doi.org/10.13031/2013.23246

Microbial Fuel Cell (MFC) Technology in Bioenergy Generation from Wastewater Treatment

11

Moksh Rajpal, Prajna Ghosh, Anoushka Singh, and Naveen Dwivedi

Abstract

Microbial fuel cell (MFC) is a cutting-edge multifunctional technology residing at the interface of power generation and wastewater treatment. By leveraging the bio-electrochemical capabilities of electroactive microorganisms, MFCs offer a dual advantage: addressing environmental concerns with regard to wastewater and participating in generation of renewable energy. This chapter discusses the general and subsystems of MFCs, as well as how the bidirectional syntropy process enables the MFC's ability to biodegrade organic matter and generate bioelectricity. The talk continues with the general overview of the MFCs usage depending on the wastewater type and their comparability with traditional treatment technologies. MFCs are clearly related to the Sustainable Development Goal 6 (SDG 6) and SDG 7 focusing on clean water and sustainable energy, but the common barriers that technology-based business models experience are specific and general constraints to scale. Recent developments in microbial engineering, material selection and design, and system design, however, suggest viable avenues through which they can be incorporated into efficient smart wastewater management systems and smart circular economy systems. The real-life application of energy recovery pathways and future prospects discussed in this chapter demonstrate that MFCs have the ability to revolutionize conventional thinking and practice in bioenergy generation and wastewater treatment. Finally, it postulates that only with the policy support, innovation, and industrial adoption the potential of MFC technology can be harnessed to its real goal of providing solutions for sustainable development.

M. Rajpal (✉) · P. Ghosh · A. Singh · N. Dwivedi
Department of Biotechnology Engineering and Food Technology, University Institute of Engineering, Chandigarh University, Chandigarh, Punjab, India

N. Dwivedi et al. (eds.), *Wastewater Treatment Through Nature-Based Solutions*,
https://doi.org/10.1007/978-981-96-6699-7_11

Keywords

Microbial fuel cell (MFC) · Bioenergy generation · Wastewater treatment · Energy recovery pathways · Renewable energy

11.1 Introduction to MFC Technology and Its Relevance

Microbial fuel cell (MFC) technology is an innovative bio-electrochemical system that combines wastewater treatment with bioenergy generation. By utilizing the metabolic activities of electroactive microorganisms, MFCs can simultaneously degrade organic pollutants in wastewater and recover energy in the form of electricity, hydrogen, or other biofuels. This dual functionality positions MFCs as a sustainable solution to two of the most pressing global challenges: environmental pollution and energy scarcity. Various microorganisms are responsible to produce biofuel. Single-celled microorganisms not only act as catalysts but also synthesize biodiesel by using alcohol and fatty acid as a substrate (Kim et al. 2013).

11.1.1 Addressing Global Wastewater Challenges

Overview of Wastewater as a Critical Environmental Issue

Wastewater has, therefore, become a major global environmental and social issue in the twenty-first century. Globally, around 80% of wastewater is released without treatment into water streams, becoming a major source of water pollution with giant impacts on aquatic life. This has become especially true in areas where there is limited development, more so in the developing world where raw sewage affects the community's health.

Their presence is very common in areas where the treatment of wastewater to acceptable standards is lacking such as cholera, dysentery, and typhoid. Further, courtesy of industries and agriculture wielding calcium hypochlorite, heavy metals, organic compounds, and nitrates, wastewater does not have a friendly stay in the environment but can take up to three decades.

Also, current types of wastewater treatment that use the activated sludge process and anaerobic digestion are energy demanding; and this leads to high operation costs as well as high emission of carbon. To achieve these objectives, effective technologies that can solve the conflict between pollutant removal and energy sustainability have been developed to solve this problem as brought about by the high demand for technologies such as microbial fuel cells (MFCs) (Aquino Neto et al. 2018; Capodaglio et al. 2016).

The Dual Role of MFCs in Wastewater Treatment and Energy Recovery

They are actually eco-friendly in nature and among several technologies known as microbial fuel cells (MFCs); wastewater can be turned from being a problem into an asset. MFCs do that through bio-electrochemical processes in which microorganisms feed on the organic load in the water and emit electrons as a by-product of their metabolism. These electrons are then collected by electrodes for use in the generation of electrical energy.

- *Pollutant Removal*: The efficiency of MFCs in the degradation of organic matters was evidenced by the marked decrease in the chemical oxygen demand (COD) of water. It has been established that COD reductions in some environments were up to 80% possible and this places MFCs at par with the normal treatments.
- *Energy Recovery*: Unlike conventional systems which in most cases draw heavily on energy, MFCs provide energy. It can meet the power needs of wastewater treatment plants or even be used to supply the power grid helping fulfil other renewable power objectives.

Case Study

In the pilot study carried out in urban India, a laboratory scale MFC removed 85% of COD from the municipal wastewater and simultaneously produced sufficient electrical energy to light LED bulbs of small wattage. Basing with such examples, there is a sign of hope that MFCs could considerably achieve or solve wastewater issues in the developing world (He et al. 2017).

11.1.2 SDG 7: Alignment with SDG 6

Importance of MFCs in Achieving Clean Water and Sanitation Goals (SDG 6)

SDG 6 focuses on water and sanitation including providing sustainable management of water and sanitation for all. This quest entails the use of advanced wastewater treatment technologies that can effectively remove different types of pollutants, and/or eliminate them in a manner most suited for recycling the water for use. The latter is directly supported by MFCs since they offer an economic and ecological solution to these goals.

- *Water Security*: MFCs can easily be integrated into the wastewater treatment facilities to enhance pollutant removal efficiency hence increasing water quality and decreasing the public health risks.
- *Energy Efficiency*: Modern treatment plants require 2–3% electrical power of the total global annual electrical power supply. On the other hand, MFCs have the possibilities to produce electricity that will lower operational energy requirements and develop the process of self-powered treatment systems.

Alignment with Affordable and Clean Energy Goals (SDG 7)

Achievement of the goals set in the Affordable and Clean Energy SDG 7. The emphasis on affordable, reliable, and modern energy service is made by the SDG 7. By recovering energy from wastewater, MFCs contribute to this goal in multiple ways (Li et al. 2008; Narayan et al. 2018):

1. *Renewable Energy Source*: MFCs generate electricity with no use of the traditional fuels hence incorporates with the international trend of developing the systems of nonconventional power.

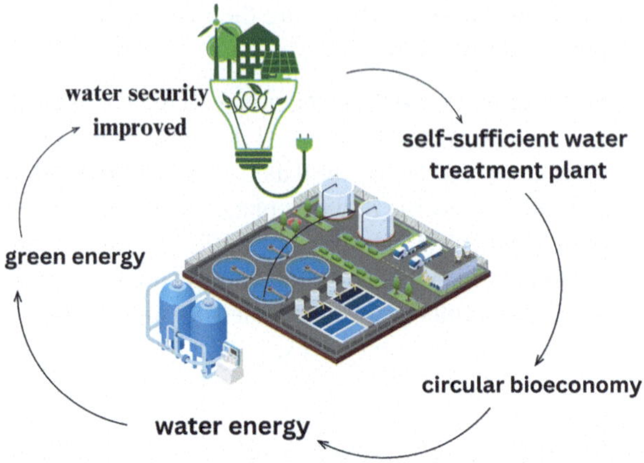

Fig. 11.1 Synergistic impact of SDG 6 and SDG 7

2. *Decentralized Energy Production*: For off-grid or rural areas, MFCs can complement the energy delivery system in a localized way and solve energy accessibility problems of underdeveloped areas.
3. *Hybrid Systems*: MFCs increase the effectiveness of hybrid renewable energy systems, such as solar panel or wind turbine in generating energy when incorporated into the system to guarantee a stable energy flow.

Synergistic Impact of SDG 6 and SDG 7

In that way, the interaction between MFCs and the water and energy infrastructure sufficiently facilitate the attainment of multiple SGDs. For instance, deep in urban areas, MFCs can turn wastewater treatment facilities into net energy producers, and supply purified water in a sustainable manner as referred to Fig. 11.1.

Case Study

A wastewater treatment plant in Denmark has adopted MFC in parallel with the other plant applications and the plant became energy self-sufficient. The plant processed 100,000 cubic meters of wastewaters per day while transmitting 50% of its energy needs through MFCs and other renewable energy sources (Mathuriya and Yakhmi 2016).

11.2 Fundamentals of Microbial Fuel Cells (MFCs)

Microbial fuel cells (MFCs) are a type of bio-electrochemical system, whereby microorganisms directly participate in the conversion of chemical energy from organic compounds in to electricity as depicted in Fig. 11.2. It is necessary to explain the main structural and functional aspects of MFCs to reveal their potential in the potential application of wastewater treatment and bioenergy production (AlSayed et al. 2020).

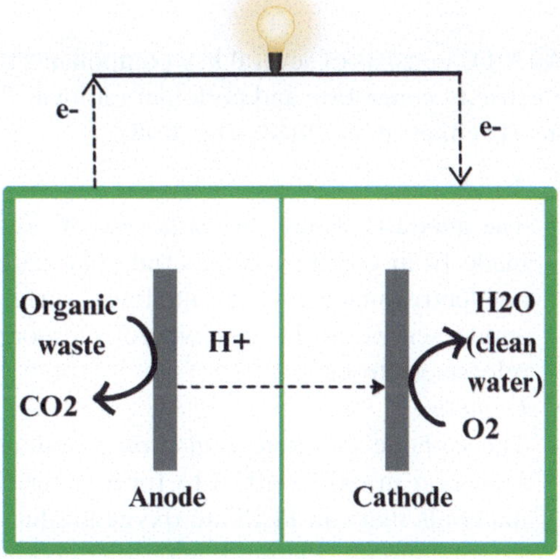

Fig. 11.2 Working mechanism of MFCs

11.2.1 Principles of MFC Operation

MFCs depend on bio-electrochemical activities, where microorganisms catalyze the *oxidation organic compounds*. The main functionality of the process includes the use of *electroactive bacteria* to transmute chemical energy from the organic contaminants in the wastewater into electrical energy.

- *Electron Transfer Mechanism*: In an MFC, *electrons* are liberated during microbial metabolism (the oxidation of organic matter). These electrons move from the *anode* (where oxidation occurs) through the external circuit to the *cathode* (where reduction occurs), creating electrical current.
- *Proton Movement:* In addition to the transfer of electrons, protons (H^+) migrate through the *membrane* separating the anode and cathode chambers. This establishes an electrochemical gradient and a proton motive force, essential for completing the circuit and maintaining microbial activity.
- *Microbial Metabolism:* Microorganisms, such as *Geobacter sulfurreducens* and *Shewanella oneidensis*, are often used in MFCs for their ability to transfer electrons directly to the anode. This electron transfer is facilitated by specialized proteins, such as *cytochromes*, found on the bacterial surface.

The overall chemical reaction in the anode chamber can be simplified as (Fan and Xue 2016):

$$C_6H_{12}O_6 + 6O_2 \rightarrow 6CO_2 + 6H_2O + \text{Energy}$$

This reaction represents the oxidation of glucose (a common organic substrate) by microorganisms, producing CO_2, H_2O, and releasing energy in the form of electrons.

11.2.2 Components of MFC Systems

An MFC consists of several key components that work together to facilitate efficient electricity generation and pollutant removal. These components include the following (Elviliana et al. 2018; Liu 2008):

- *Anode:*
  The anode is where the *oxidation* of organic substrates occurs. It is typically made from conductive materials like carbon paper, graphite, or other porous, conductive materials that maximize surface area for microbial attachment. The anode serves as the site where microorganisms degrade organic matter and release electrons.
- *Cathode:*
  The cathode is where *reduction* reactions take place, typically involving the reduction in oxygen (O_2) to form water (H_2O). The cathode must be made of materials that can facilitate oxygen reduction and allow electron flow from the external circuit. Platinum and other conductive catalysts are often used for this purpose.
- *Proton Exchange Membrane (PEM):*
  The *membrane* separates the anode and cathode chambers, allowing protons (H^+) to pass through while preventing the mixing of wastewater with oxygen at the cathode. A good membrane must have high proton conductivity, low resistance, and be durable in the presence of wastewater contaminants.
- *Substrates:*
  The substrates are the organic materials present in wastewater that serve as the fuel for the microbial metabolic processes. Common substrates include sugars, fatty acids, and proteins, depending on the type of wastewater being treated. The choice of substrate significantly affects the performance of the MFC (Table 11.1).

11.2.3 Role of Microbial Communities

The function of *microbial communities* is crucial for the overall effectiveness of MFC technology. *Exoelectrogens* are electroactive bacteria that can directly transfer these electrons in the cellular metabolic processes to the anode. These bacteria attach to the anode surface and form what is known as the biofilm and with the organic content of the raw wastewater they carry out the transfer of electrons (Palanisamy et al. 2024).

- *Types of Electroactive Bacteria:*
  - *Geobacter sulfurreducens* is among the most researched electroactive bacteria, capable of donating electrons to metals.
  - *Shewanella oneidensis*, another species on the list, can lower various electron acceptors such as metals and oxygen.
  - Other more common bacteria that share similar functions in the MFC are *Pseudomonas* and *Enterobacter* in the conductivity of the transfer of electron.

Table 11.1 Comparison of MFC components and their functions

Component	Material examples	Function	Reference
Anode	Carbon paper, graphite, carbon cloth	Site of organic matter oxidation; releases electrons and protons	Logan et al. (2006) and Zhang et al. (2012)
Cathode	Platinum, activated carbon	Facilitates reduction reactions (e.g., O_2 reduction to H_2O)	Zhou et al. (2014)
Proton exchange membrane (PEM)	Nafion, cation-exchange membranes	Separates anode and cathode chambers; allows proton transport	Thornton and Armstrong (2006)
External circuit	Copper wires, conductive pathways	Transfers electrons from the anode to the cathode, generating electricity	Liu et al. (2005)
Substrates	Glucose, acetate, wastewater	Fuel for microbial metabolism; drives bioelectricity generation	Logan (2013) and Xie et al. (2013)

- *Microbial Community Dynamics:*
  A conventional MFC features the presence of many sorts of microorganisms such that a synergistic microbial consortium that accelerates the biodegradation of the organic pollutants is manifested. This is because the various microbial species can have the ability to pull apart the individual compounds in the wastewater differently from the other species.
- *Biofilm Formation:*
  The microbial biofilms are extremely useful in enhancing the dimensions along the surface for microbial adhesion and increasing the conductivity of electron transfer. By forming a biofilm, the microorganisms can concentrate and function effectively under different conditions; the environment may have low pH or high temperature.

Case Study

Logan et al. (2006) employed *Geobacter sulfurreducens* in an MFC used in wastewater treatment and electricity generation. This experiment was an inference that MFCs could obtain COD degradation rates of over 70% from wastewater and an eco-friendly electric power density of about 0.5 W/m^2. This particular case will also show how individual microbial species can improve MFC's performance once it is being applied to wastewater treatment and energy generation (Amirdehi et al. 2020) (Table 11.2).

11.3 MFCs in Wastewater Treatment

Microbial fuel cell (MFCs) is a novel technology that has the potential of achieving pollutant removal and power production from wastewater. The Bio-EET process involves degrading organic compounds through a technique associated with MFCs, therefore enhancing the sustainability of using MFCs in wastewater management.

Table 11.2 Comparison of electroactive microorganisms in MFCs

Microorganism	Electron transfer mechanism	Power density achieved	Advantages	Reference
Geobacter sulfurreducens	Direct electron transfer via pili (nanowires)	1.8 W/m^2	High efficiency, adaptable to various substrates	Logan et al. (2006)
Shewanella oneidensis	Uses cytochromes for electron transfer to the anode	0.8 W/m^2	Versatile in anaerobic environments	Zhang et al. (2012)
Pseudomonas species	Produces mediators for indirect electron transfer	0.5–1.0 W/m^2	High biofilm formation capability	Pant et al. (2010)
Enterobacter species	Oxidizes diverse substrates for enhanced bioenergy	0.6 W/m^2	Efficient in treating complex wastewater	Liu et al. (2005)

This section focuses on the operation and performance of pollutant removal, the general performance of the MFC with different types of wastewaters, and the comparison of MFC with conventional wastewater treatment (Hu et al. 2018).

11.3.1 Mechanisms of Pollutant Removal

Another advantage of operation of MFCs in wastewater treatment is that they possess the efficiency of removing organic pollutants. MFCs serve as bio-electrochemical systems as microbial anodic metabolism breaks down organic compounds thus decreasing contaminants such as *chemical oxygen demand (COD)*, *biochemical oxygen demand (BOD)*, and *nutrient levels* (e.g., nitrogen and phosphorus) (Yap et al. 2021).

- *Organic Matter Degradation:*
  Some of the common degradation products of the organic pollutants that include sugars, fatty acids, and proteins are CO_2, H_2O, and electrons notable in the microbial communities of the anode chamber. The organic compound can in this sense be regarded as the "fuel" for the microbial activity, and electrons are released during different biomass transformation processes. These electrons are transferred to the anode, hence increasing bioelectricity production while at the same time reducing the organic content.
  - Example: Logan (2013) made a study on the ability of *Geobacter sulfurreducens* in degrading organic matter in wastewater to attainable COD decrease and the sustainable electricity generation.
- *COD and BOD Reduction:*
  MFCs show a highly efficient trend with regard to COD and BOD, which are parametric indicators of the total organic strength of the wastewater. This chemical attack lowers the concentration of large organic soluble substances provided by the various microorganisms thus enhancing water quality.

  - For instance, in industrial wastewater, MFCs can remove 80% COD level enhancing the capability to eradicate organic pollutants.
- *Nutrient Removal (Nitrogen and Phosphorus):*
  Some MFC configurations have also been developed to operate with efficiency in the removal of nitrogen and phosphorus containing compounds in wastewater. This is usually carried by *nitrification–denitrification* processes, as the ammonium (NH_4^+) is oxidized to nitrites (NO_2^-) and nitrates (NO_3^-), and reduced back to nitrogen gas (N_2) under anoxic conditions at the cathode. Likewise, *phosphorus* may be removed through microbial action in some of the systems as well (Thornton and Armstrong 2006).
  - Case Study: Pant et al. (2010) employed an MFC in treating synthetic wastewater containing high concentration of nitrogen and phosphorus, thereby achieving high removal efficiency of both the nutrients.

11.3.2 Performance in Treating Different Wastewater Types

MFCs are also very adaptable systems applicable for wastewater treatment ranging from domestic, industrial, and agricultural effluents. The nature of the wastewater has a strong impact in the performance of MFCs for both the pollutant degradation and electricity production (Tharali et al. 2016; Zhou et al. 2014).

- *Domestic Wastewater:*
  MFCs show high efficiency when treating domestic wastewater, which contain miscellaneous organic contaminants such as sewage, washing, and food remnants. The high organic content means that domestic wastewater could be the perfect medium for microbial growth.
  - Example: MFCs treating domestic sewage in urban areas have been reported to remove more than 70% of organic pollutants (COD) while generating power for small-scale application.
- *Industrial Wastewater:*
  Industrial effluents are frequently comprised of a diverse group of organic compounds dissolved or suspended solids and heavy metal/inorganic chemicals. There are MFCs that can be used to treat effluents from food processing, textiles, and paper mill industries in addition to exploiting energy. However, effluent released through industries could be noticed to require some modifications such as shifting from the general microbial consortium to specific one and/or some form of pretreatment for enhancement of their functionality (Zhang et al. 2012).
  - Case Study: Using MFCs, Zhang et al. (2014) established that food processing industrial effluent could be treated effectively to remove 60–80% COD and produce electrical energy that can be utilized in far-off locations.
- *Agricultural Wastewater:*
  Another application for MFCs is in agriculture, especially in managing runoff water, containing such nutrients as nitrates and phosphates, or organic droppings from livestock. MFCs can contribute to the diminution of nutrient levels as well as the production of bioelectricity.

– Example: MFCs have been applied to treat wastewater generated from dairy farms where high concentrations of organic load and nitrogen containing compounds exist, and the achievement has been the reduction in the pollutant load as well as power that can be harnessed for use by the farm.

11.3.3 Comparative Analysis: MFCs Versus Traditional Wastewater Treatment Technologies

The utilization of MFCs for wastewater treatment has several benefits over more conventional wastewater treatment processes most of which are inclined toward the dual functionality of the MFCs in *pollutants removal* and *energy recovery* as explained in Fig. 11.3. Nonetheless, there are issues related to feasibility and the costs of the overall process (Kalathil et al. 2018; Schröder and Harnisch 2013).

- *Traditional Wastewater Treatment Technologies:*
  Conventional systems, for example, the *activated sludge* and *constructed wetlands* entail physical and biological approach for removal of pollutants. These systems do well in treating organic matter, but lack the ability to produce bioelectricity. However, these systems usually need a lot of energy to supply for aeration, and to pump water, so is not as energy efficient.
  – Example: Activated sludge systems, for instance, are expensive to operate this, because of the intensive energy hauled in aeration. They are useful in reducing COD as well as BOD but are not useful in energy recovery.

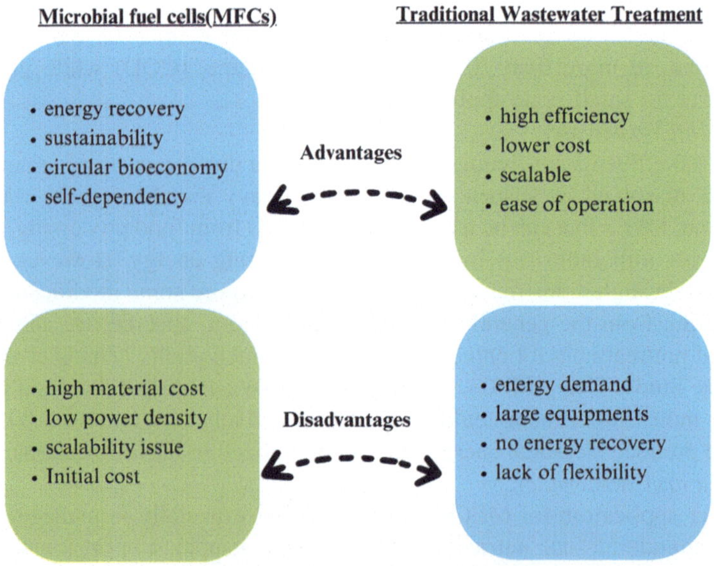

Fig. 11.3 Comparison between MFCs and traditional wastewater treatment technologies

- *MFCs in Comparison:*
  Besides, MFCs do not only remove the wastewater but also recover the energy from the process, which makes them viable choices in wastewater treatment. Though they produce less energy compared with a conventional energy recovery technique such as biogas production, MFCs have the benefits of being able to provide power in remote areas inaccessible by supply chain means of conventional suppliers. Moreover, to boost energy production, MFCs are flexible and can be connected to *hybrid renewable energy system* like solar or wind systems.
  - Case Study: MFCs treating domestic sewage had efficiencies up to 50% lower when compared with activated sludge systems further augmented with a consistent though small level of electricity production.
- *Advantages of MFCs Over Traditional Methods:*
  - *Energy Recovery:* MFCs produce bioelectricity for powering little gadgets or for integration into the power system; this aspect is energy-independent unlike traditional techniques.
  - *Sustainability:* MFCs are relatively energy efficient and are good candidates for distributed, stand-alone systems.
  - *Small Footprint:* MFCs are usually slightly smaller than traditional systems; they can be used effectively in places where space is limited or costs a lot of money.
- *Limitations:*
  Despite many benefits of MFCs, their main deficiencies are those related to *scaling*, *high costs of materials*, and *low power density* that must be solved before MFCs can become the continuation of traditional systems. Second, the general performance of MFCs may significantly depend on a type of wastewater and the microbial consortium.

11.4 Bioenergy Generation Through MFC Technology

Microbial fuel cells (MFCs) are unique technology for bioenergy production that facilitates operation of electrical devices using organic matter or wastewater as the main source of power. This technology harnesses the electrochemical processes of microbial system to produce electricity and other forms of biophoton bioenergy. Since MFCs operate to provide wastewater treatment and energy generation service, they can be regarded as promising technology to address environmental issues as well as contribute to efficient energy generation. Here, it expands the details of the pathways to the energy recovery process, the relationship between the energy yield and energy parameters, and the other applications and possible integration of MFCs into bioenergy systems (Ali et al. 2018; Nandy and Kundu 2018).

11.4.1 Energy Recovery Pathways

The major purpose of MFCs is to produce electricity through the biotransformation of chemical energy from organic materials. Nevertheless, MFCs are more flexible and further bioenergy outputs in a single system are possible by varying the configuration of the system, the type of microbial selection, and substrates to be employed.

- *Electricity Generation:* MFCs work based on the conversion of electrical energy by microbial decomposition of organic materials. In an MFC, organic substrates in the anode chamber are oxidatively reduced by microorganisms and, thus, released electrons and protons. The electrons movement takes place through an external circuit and finally reaches the cathode to form an electric current. This electricity can be collected and utilized by powering small appliances, sensors, or help feed local standalone electricity systems. This process is, however, governed by the type of microbial species, reactor configuration, and the type of organic matter under consideration.
  - *Example:* Liu et al. (2005) have also conducted a series of laboratory studies showing that the MFCs could generate a stable voltage of around 0.6 V, sufficient to operate sensors and other low power applications. This electricity generation can, therefore, be directly attributed to microbial oxidation of carbonaceous materials soluble in the wastewater.
- *Hydrogen Production:* MFCs can also produce *hydrogen* as well through *biohydrogen production* with some of the MFCs. This happens at the cathode where protons (H^+) are reduced to hydrogen gas and also eliminate the counter ions. This hydrogen may be then subsequently collected for use in light hydrocarbons like hydrogen fuel cells or in any other energy utilization. Hydrogen production keeps pace with the production of traditional fossil fuels but advances the clean energy initiative.
  - *Example:* Zhang et al. (2012) explored MFC-based hydrogen production, achieving a hydrogen production rate of *3.2 mL H_2/L/day* from wastewater. The combination of bioelectrochemical reactions and microbial metabolic pathways that lead to hydrogen generation in MFCs holds considerable potential for future renewable hydrogen production.
- *Biogas Production:* MFCs can also be coupled with anaerobic digestion processes to produce *biogas* (primarily methane) in addition to electricity. In hybrid MFC-biogas systems, the MFC can help degrade organic matter that is difficult to break down anaerobically, while also contributing to electricity generation. The synergy between microbial fuel cell processes and anaerobic digestion enhances the overall bioenergy recovery.
  - *Case Study*: Xie et al. (2013) found that integration of MFCs with anaerobic digestion improved the yield of biogas. The electricity produced by the MFC enhanced the microbial action during the process of anaerobic digestion and as a result enhanced methane production and consequently wastewater treatment (Table 11.3).

11.4.2 Factors Influencing Energy Yield

Based on the work done above, it can be concluded that the energy yield of MFCs depends on the activity of the microbial community imposing on it, quality of

Table 11.3 Energy recovery pathways in MFCs

Pathway	Process description	Energy output	Applications	Reference
Electricity generation	Electrons released during microbial metabolism are transferred to an external circuit	0.2–1.8 W/m^2	Powering sensors, small devices, off-grid systems	Logan et al. (2006)
Hydrogen production	Protons reduced at the cathode to generate hydrogen gas	3.2 mL H_2/L/day	Hydrogen fuel cells, clean energy applications	Zhang et al. (2012)
Biogas production	Coupling with anaerobic digestion to enhance methane yield	Enhanced methane production	Bioenergy generation, industrial applications	Xie et al. (2013)
Combined energy recovery	Simultaneous production of electricity, hydrogen, and methane through system integration	Multienergy recovery	Hybrid energy systems, circular economy setups	Zhang et al. (2014)

substrata used, and type of reactor constructed for MFC. Knowledge and mastery of these elements are critical in the enhancement of MFCs' bioenergy performance (Liu 2008; Yap et al. 2021).

- *Microbial Activity:* It is important to note that the two aspects which describe the *microbial community* play a central role in the function of MFCs. The bioactive organelles which can transfer electrons directly to the surface of the electrodes are the electroactive bacteria. Such species as *Geobacter sulfurreducens* and *Shewanella oneidensis* reveal high rate of electron transferring during organic matter oxidation. Sequence activity of microbes can also affect the energy capture as some microbes can generate high power densities than others.
  - *Example*: A model electroactive bacterium *Geobacter sulfurreducens* was found to shuttle electrons toward the anode surface through its pilus structure to enhance power output. Its use in MFCs has led to achieving power densities of approximately *1.8 W/m²* (Logan et al. 2006), showing how important it is toward maximizing energy production.
- *Substrate Quality:* One of the most important factors that have been identified to affect power density in MFCs is the nature of the substrate. MFCs generate electricity through the microbial degradation of organic matter. When substrates with higher carbon content—such as glucose, acetate, and other short-chain fatty acids—are used, they tend to produce higher power densities. Different wastewaters sourced from domestic, industrial, and agriculture sources are characterized with dissimilar organic strength and composition, a factor that influences the MFC.
  - *Example:* Wastewater with high concentration of easily biodegradable substrate such as glucose will generate high power output as compared to that containing difficult substrate such as lignocellulose. The latter in most cases leads to development of lower power densities.

- *Reactor Design:* Therefore, the design of the MFC reactor itself has a very important influence on the energy yield improvement. This includes the size and shape of *electrodes*, *type of membrane*, or the *distance between the anode and cathode* and *flow conditions* all in one way or the other have an effect on the efficiency of electrochemical reactions occurring in the MFC. For instance, larger electrode surface area gives additional microbial attachment hence greater microbial activity and higher power production.
  - *Case Study:* Logan et al. (2006) carried out tests that proved that power density could be enhanced considerably by increasing the surface area of the MFC electrodes. *Larger anode surface area* and *shorter anode-cathode distances* gave MFCs a power density of up to ~1.3 W/m^2 which was higher than in systems with smaller size of electrodes and greater distance.
- *Environmental Conditions:* The efficiency of the microbial communities in MFCs may be influenced by the physical and chemical characteristics of the environment, for instance, *temperature*, *pH,* and *salinity*. Mesophilic MFCs are proven to have optimal strategies for efficiency because microbes are more active at moderate temperatures of around 30 °C. Either microbial metabolism is affected negatively or else the formation of undesirable products is caused at extreme temperatures. Likewise, pH values have to be regulated for sustaining the activity of the microbial consortium because the extremes in the pH level are destructive to the electroactive bacteria.
  - *Example:* In conditions which are above 40 °C for instance, the use of thermophilic bacteria that can thrive with high metabolic rate which improves power generation.

11.4.3 Applications and Integration

The low-cost design of MFCs means they can be used flexibly in several forms of wastewater treatment and for bioenergy generation where centralized utilities may not be available, off-grid systems are required, or where they can be part of a hybrid energy system. That they are capable of both treating wastewater and generating electricity makes them ideal for use in the sustainable development (AlSayed et al. 2020; Mathuriya and Yakhmi 2016).

- *Off-Grid Energy Systems:* MFCs are ideal for *off-grid systems* whereby conventional sources of energy are incontrollable or inaccessible. Besides, due to electricity produced using wastewater, MFCs can act as the efficient source of energy in developing, rural or isolated regions. This can serve low power consumer devices, sensors, or possibly small power system communities.
  - *Case Study:* In a study undertaken in *India*, MFCs were used in the rural setting to supply power for lighting and water pumping and at the same time treating domestic sewage. This dual-functionality is best suited for off-grid solutions by enhancing the energy and sanitation situation in rural areas.

- *Hybrid Renewable Energy Systems:* Like any other renewable energy systems, MFCs can be interconnectable with other forms of renewable energy systems including the *solar* and *wind* systems for the purpose of forming a combination of systems. In such systems, MFCs can complement energy generated by solar or wind by providing a secondary energy source that can sustain power delivery in remote sites.
  - *Example:* In *Japan*, an experimental project connected MFCs to *solar panels* in an area with no centralized power supply. The MFCs gave out power in the evening, or whenever the sun was out and the solar power was not sufficient, to provide power for vital needs for the community.
- *Large-Scale Wastewater Treatment Plants:* In large-scale municipal or industrial *wastewater treatment plants*, MFCs can be incorporated as a way of using the energy from the MFCs while promoting the efficiency of the treatment process. Compared to the conventional aeration and chemical treatment where MFCs both produce electricity and at the same time decompose organic pollutants, the energy demands and therefore the operational costs decrease.
  - *Example:* A case study, based on *Germany*, the implementation of MFCs in a local wastewater treatment plant has *reduced energy bills by 30%*. The MFCs did not only manage to treat the wastewater but also to generate additional energy for the facility thereby making the system more efficient.
- *Resource Recovery and Circular Economy:* MFCs can also provide benefits in a circular economy scenario by providing additional products at the wastewater treatment step, such as *nutrients (nitrogen, phosphorus)* or *biopolymers*, in addition to energy. Energy production and resources recovery compatible concepts are the modern advanced approaches to wastewater treatment that allow minimizing emissions and wastes.
  - *Example:* Integrated technologies comprised MFCs and nutrient recovery systems in a *resource recovery project* in *Netherlands* where nitrogen and phosphorus from wastewater were recovered and used as fertilizers. The combination of MFCs with nutrient recovery systems as done by the researchers who developed the butterfly shaped anode closely completes the cycle on wastewater treatment.

11.5 Challenges and Opportunities in Scaling MFCs

Microbial fuel cells (MFCs) offer an ideal wastewater treatment and bioenergy producer technology with several drawbacks hindering their large-scale implementation. Such disturbances must be resolved to extend MFC technology for vast usage in wastewater treatment plants and energy systems. Nevertheless, there are substantial opportunities for the enhancement of MFCs with elimination of barriers in their practical application for wastewater treatment, electricity generation, and resources recycling (Wrighton et al. 2008).

11.5.1 Technical and Economic Barriers

Even today, many technological and economical constraints hamper the usage MFCs in large-scale applications. Addressing these challenges is critical for the future of MFCs as a profitable technology for wastewater treatment as well as bioenergy production.

- *Low Power Output:* A second technical challenge limiting MFCs performance is a *relatively low output*, particularly when utilized in natural conditions. Studies of MFCs have shown that the device is capable of producing electricity in a laboratory situation, although the power density can be lower in larger applications. For instance, MFCs generate energy densities of power ranging from *0.1 to 1 W/m*^{2}, which is considerably lower than in conventional sources of energy. This hampers their capacity to produce adequate power to meet high utilization applications without supplementary systems or energy storage devices.
  - *Example*: Logan et al. (2006) reported that MFCs achieved a maximum power density of only around 0.68 W/m^2. However, practical applications of electricity generation require significantly higher power densities to be cost-effective. This highlights the need for enhanced production of microbial strains, improved electrode materials, and optimized microbial fuel cell designs.
- *High Material Costs:* The *materials* used in MFCs—particularly the electrodes and membranes—can be *expensive* and are a significant contributor to the overall system costs. Most MFCs use *precious metals* like platinum or gold for the cathode, and the proton exchange membranes (PEMs) used for ion conduction are often costly. This makes MFCs less economically feasible for large-scale wastewater treatment and energy production systems.
- The *components* used in MFCs especially the electrodes and membranes are costly and contribute to *high costs* of entire systems. A majority of MFCs employ the cathode of *precious metals* such as platinum or gold, and PEMs through which the ions flow are expensive. This makes the MFCs less economically viable for the large-scale wastewater treatment and energy production systems.
  - *Example: Platinum electrodes*, which offer excellent conductivity, are commonly used in MFCs but are costly and can account for a significant portion of the total operational expenses. Alternative materials, such as *carbon-based electrodes* and *biomaterial-based membranes*, are being researched to reduce costs and improve system efficiency.
  - *Platinum electrodes* have higher conductance and are widely used in MFCs; they are, however, expensive and may contribute to most of cost of operating the MFC. *Carbon-based electrodes* and *biomaterial-based membranes* have also been studied to bring down the cost and make the system more effective.
- *Scalability and System Design:* Challenges in scaling MFCs primarily stem from the fact that MFCs are developed as small-scale laboratory models that are scaled up to systems. Typically, MFC systems provide a high level of *environmental control* such as temperature, pH, and microbial culture which may be hard to

emulate in large-scale system. Furthermore, building up multicentimeter-scale reactors, providing adequate biosurface and efficient conductivity for microbial immobilization and electron shuttling without compromising cost-effectiveness, is a technological challenge.

 - *Example:* Full-scale development of MFCs in the context of wastewater treatment plants entails great advancements in *reactors configuration*, surface area, and the electrode location to enhance the contact of microbes with other fluids and improve power output. These engineering concerns are unaddressed today and current systems actually fail to provide higher efficiency and better performance as the capacity increases.

11.5.2 Strategies for Optimization

Nonetheless, there has been a lot of advancement in the formulation of efficient strategies to enhance the MFCs to accomplish the desired results at optimum cost as well as on a large-scale production basis. Several strategies are currently being sought in order to overcome the technical and the economic challenges regarding MFCs (Hu et al. 2018).

- *Advances in Electrode Materials:* Research today is also aimed at finding *new cheap electrode* type instead of precious platinum. Graphite, carbon cloth, and carbon nanotubes are some examples of *carbon-based materials* that are presently under consideration because they can transmit electrons and are cheaper in the long run. Also, other *biomaterials* like the conductive polymers and biocomposites have been considered from the aspect of environmental friendliness.
  - *Example:* For the flexible transparent electrode, *graphene-based electrode* material is currently under research due to high conductivity, large surface area, and cheap. Zhou et al. (2014) specifically showed that the use of graphene-based electrodes increases the power production of MFCs and decreases the cost of the electrodes as compared to massive production, meaning the technology has a potential application in larger scale.
- *Microbial Engineering:* Another area of optimization is the *engineering of microbial communities;* a number of linked tasks can be mentioned here. In *electroactive bacteria* cases, researchers try to select for, or genetically engineer, bacteria that will improve their electron transferring capabilities and convert more organic compounds. Applying *synthetic biology* tools to strengthen the genetically tailored metabolic functionalities of microorganisms also helps reach better power density values and minimize startup times of MFCs (Kumar et al. 2018).
  - *Example:* Electroactive bacterium *Geobacter sulfurreducens* that has been widely researched has been genetically enhanced to increase electron transfer rates and hence higher power densities in MFC. Other works have also investigated manipulation of *Shewanella oneidensis using genetic engineering* to improve conductivity of bioelectricity.

- *Reactor Design Improvements:* Several areas require improvement in order to increase the efficiency of MFCs. *Reactor design* is a vital area because improvements can be made to the MFC design. The *multichamber reactors*, *biofilm reactors,* and *stacked MFC systems* have also been seen to enhance the surface area available for the microbial adhesion with a view of enhancing the electron transfer. Moreover, advances in *ion-conducting membranes* and *separator materials* will also enhance the efficiency of MFCs through decreasing the internal resistance and enhancing the ionic conductivity (Mohammed and Ismail 2018).
  - *Example:* From the study of the characteristics of MFCs with *hollow-fiber membranes*, researchers demonstrated that utilizing this structure of membranes was able to overcome the main problem of high internal resistance, improve the ion transport, and enhance the overall efficiency of MFCs. This design modification is particularly helpful for large volume, fully closed cycle systems typical of wastewater treatment facilities.
- *Hybrid Systems and Integration:* Hybrid systems that combine MFCs to other forms of renewable energy technologies like the solar or wind energy systems may be regarded as another effective way of minimizing low power densities and ensuring a reliable energy source. The combination of MFCs with energy storage solutions such as batteries or supercapacitors will allow electricity storage for usage where MFC power output is insufficient. Smart capabilities can also be incorporated in these hybrid systems to enable it function in *smart grids* and support the energy transition through decentralized clean electricity generation.
  - *Example*: An experimental model of MFC was established in conjunction with solar panels and *energy storage systems* in *California* for powering far-flung communities with renewable energy. This system provided wastewater treatment as well as electricity for the necessities of a community; this showed the applicability of MFCs as part of required global sustainable energy.

11.5.3 Future Opportunities

In the future, there are several principal opportunities to reach further development and commercialization of MFC technology in wastewater treatment and bioenergy applications. With reference to the existing challenges and possible improvements in the system structure, MFCs can contribute greatly to sustainable energy and environmental conservation across the world (Elviliana et al. 2018; Tharali et al. 2016).

- *Smart Wastewater Management Systems:* MFCs can be incorporated into *intelligent wastewater treatment systems* where in addition to addressing water treatment needs, they pull double duty as a real-time monitoring and data acquiring device. As *Internet of things (IoT)* and *sensor technologies* progressively improve in their capabilities, the MFCs can be integrated in a network of self-organized systems for wastewater treatment and environmental control. They could be established in urban environments, factories, and farms to build *circular economy model* in which wastes to energy and other resources are recycles.

  - *Example:* As implemented in the *IoT sensors* for MFC systems, microbial performance, energy generation, and wastewater characteristics could be monitored. This data could be used to fine-tune operation of the systems, increase energy recovery, and participate in the development of intelligent water systems.
- *Integration into the Circular Economy:* MFCs operate under the *circular economy* paradigm because they offer an approach through which waste could be converted to useful commodities like electricity and hydrogen, and nutrients for use in the farming sector. When integrated with MFCs, the technology can provide closed-loop solutions that include wastewater treatment, energy generation, and recovery valuable byproducts thus lowering impact that industrial and municipal wastewater treatment have on the environment.
  - *Example: Nutrient recovery systems* coupled with MFCs allows recycling of nutrients like nitrogen and phosphorus from the wastewater and using them as a useful fertilizer that can replace synthetic fertilizer.
- *Commercialization and Policy Support:* The growth of MFCs as a commercial product still depends on access to *policy incentives and support* from governments and other bodies of regulation. In this context, governments bear a crucial responsibility, both through direct financing of scientific investigation and through the support of subsidies for the implementation of MFCs in sewage treatment plants and other measures aimed at the promotion of sustainable technologies. The integration of *green technologies* in wastewater treatment especially through the use of biogas technology and renewable energy use for power generation contributes to achievement of the *sustainable development goals (SDGs)* or improving on water and sanitation hurdle (SDG 6) and affordable and clean energy (SDG 7).
  - *Example:* Countries like *Germany* and *South Korea* have provided funding and incentives for renewable energy projects, which include MFC technology. With supportive policies, these regions have seen greater investments in sustainable wastewater treatment and bioenergy solutions (Palanisamy et al. 2024).

11.6 Conclusion and Future Scope

Microbial fuel cells (MFCs) still remain a fairly innovative technology that deals with the *management of wastewater* and *energy creation* at the same time and is a landmark solution for some of the most challenging planetary problems. Yahmed et al. sustain that MFCs have shown in studies conducted over the years to eliminate different types of wastewaters and at the same time recoup energy in electricity energy, hydrogen energy, or other sorts of bioenergy. But for enhanced applications of the MFCs in the real-life situations, there are some technical and economic barriers, which need to be addressed. Through further overcoming these barriers through material selection, microbial engineering, system design, and technology integration, MFCs can substantially contribute to the sustainable water and power scenario of the world (Aquino Neto et al. 2018; Capodaglio et al. 2016; Firdous et al. 2018).

Key Takeaways

- *Dual Functionality*: Such facility is capable of handling and treating wastewaters while at the same time generating energy as well which make MFCs the ideal solution to answering the two major problems of increasing scarcity of water and inadequate energy. It has been believed that they afford both the options to save *ecosystems* and to *produce renewable power*.
- *Technological Advancements*: There exists substantial progress in improving MFCs efficiency and that of its related components the *electrode materials*, *engineered microbes*, and the *design of circuits*. That has increased efficiency of MFCs and reduced costs by bringing the technology closer to the commercial market.
- *Potential Applications*: In simple terms, the applicability of MFCs is not only limited to treatment of wastewater. Their application in *off-grid electricity* and *water management*, as well as *renewable hybrid systems* indicates future potential of decentralized energy and water systems.
- *Challenges*: Nevertheless, the development of MFCs has its drawbacks based on the *low power generation*, *high material cost*, and *scalability* issues. All these factors present a hurdle to MFCs to be economical at large scales, and hence, must be tackled. To overcome these barriers, *efficient* and *low-cost materials* together with *cost-effective reactor designs* must be investigated.

Pathways to Sustainable Development

The future of MFC technology is intertwined with *sustainable development goals (SDGs)*, particularly *SDG 6 (clean water and sanitation)* and *SDG 7 (affordable and clean energy)*. The global demand for clean water and renewable energy is rising, and MFCs offer a viable path toward meeting these needs in a sustainable and integrated manner (He et al. 2017; Mohan et al. 2009).

- *Policy and Regulatory Support*: It was, therefore, shown that policy will play a significant part in promoting the development of MFC technology. Governments can support financing of related research, give individuals and organizations a push toward using MFCs in wastewater treatment plants, and encourage the use of sustainable technologies. Neural public–private partnerships could extend commercialization of MFCs, which would guarantee that these systems are both technologically and economically sustainable.
- *Innovation and Industrial Adoption*: Continued *innovation* will be vital for overcoming the existing limitations of MFCs. *Collaborative research* involving academia, industry, and government will facilitate breakthroughs in materials, microbial cultures, and reactor systems. Industrial adoption will drive mass production, reducing costs and improving the overall efficiency of MFCs. Furthermore, integrating MFCs into *circular economy* models, where waste is treated and energy is recovered, will encourage broader adoption across sectors like agriculture, industrial manufacturing, and urban wastewater management.
- The current limitations in the design of MFCs will also be the main focus for further improvements in the future. *Joint academic–industrial–government partnership* in research can lead to advancements in the material science, microbial

consortia, and the reactor systems. This means adoption of MFCs in industries because when produced at large, costs will be minimized and efficiency will be enhanced. Moreover, the use of MFCs in *circular economy* systems, where waste is exploited, and energy is generated, will promote its applications in different industries such as agriculture, industrial production, and wastewater systems in dense urban areas.

- *Global Scaling*: The possibility of scaling MFCs to accommodate the demands of *developing regions* and the *emerging markets* is inevitable. In these areas, MFCs can develop as a *cost-effective* and *custom-designed systems* providing the needed amounts of freshwater at a variety of locations as well as renewable energy to support further development of decentralized infrastructure. MFCs inclusion in *off-grid energy system* could offer power solutions in the developing countries as well as *treatment of wastewater* in areas where clean water is hard to come by (Narayan et al. 2018).

References

Ali J, Sohail A, Wang L, Rizwan Haider M, Mulk S, Pan G (2018) Electro-microbiology as a promising approach towards renewable energy and environmental sustainability. Energies 11(7):1822. https://doi.org/10.3390/en11071822

AlSayed A, Soliman M, Eldyasti A (2020) Microbial fuel cells for municipal wastewater treatment: from technology fundamentals to full-scale development. Renew Sust Energ Rev 134:110367. https://doi.org/10.1016/j.rser.2020.110367

Amirdehi MA, Khodaparastasgarabad N, Landari H, Zarabadi MP, Miled A, Greener J (2020) A high-performance membraneless microfluidic microbial fuel cell for stable, long-term bench-top operation under strong flow. ChemElectroChem 7(10):2227–2235. https://doi.org/10.1002/celc.202000040

Aquino Neto S, Reginatto V, De Andrade AR (2018) Microbial fuel cells and wastewater treatment. In: Electrochemical water and wastewater treatment. Elsevier, pp 305–331. https://doi.org/10.1016/B978-0-12-813160-2.00012-2

Capodaglio AG, Molognoni D, Pons AV (2016) A multi-perspective review of microbial fuel-cells for wastewater treatment: Bio-electro-chemical, microbiologic and modeling aspects. 030032. https://doi.org/10.1063/1.4959428

Elviliana, Toding OSL, Virginia C, Suhartini S (2018) Conversion banana and orange peel waste into electricity using microbial fuel cell. IOP Conf Ser Earth Environ Sci 209:012049. https://doi.org/10.1088/1755-1315/209/1/012049

Fan L, Xue S (2016) Overview on electricigens for microbial fuel cell. Open Biotechnol J 10(1):398–406. https://doi.org/10.2174/1874070701610010398

Firdous S, Jin W, Shahid N, Bhatti ZA, Iqbal A, Abbasi U, Mahmood Q, Ali A (2018) The performance of microbial fuel cells treating vegetable oil industrial wastewater. Environ Technol Innov 10:143–151. https://doi.org/10.1016/j.eti.2018.02.006

He L, Du P, Chen Y, Lu H, Cheng X, Chang B, Wang Z (2017) Advances in microbial fuel cells for wastewater treatment. Renew Sust Energ Rev 71:388–403. https://doi.org/10.1016/j.rser.2016.12.069

Hu J, Zhang Q, Lee D-J, Ngo HH (2018) Feasible use of microbial fuel cells for pollution treatment. Renew Energy 129:824–829. https://doi.org/10.1016/j.renene.2017.02.001

Kalathil S, Patil SA, Pant D (2018) Microbial fuel cells: electrode materials. In: Encyclopedia of interfacial chemistry. Elsevier, pp 309–318. https://doi.org/10.1016/B978-0-12-409547-2.13459-6

Kim K-Y, Chae K-J, Choi M-J, Yang E-T, Hwang MH, Kim IS (2013) High-quality effluent and electricity production from non-CEM based flow-through type microbial fuel cell. Chem Eng J 218:19–23. https://doi.org/10.1016/j.cej.2012.12.018

Kumar R, Singh L, Zularisam AW, Hai FI (2018) Microbial fuel cell is emerging as a versatile technology: a review on its possible applications, challenges and strategies to improve the performances. Int J Energy Res 42(2):369–394. https://doi.org/10.1002/er.3780

Li Z, Zhang X, Lei L (2008) Electricity production during the treatment of real electroplating wastewater containing Cr6+ using microbial fuel cell. Process Biochem 43(12):1352–1358. https://doi.org/10.1016/j.procbio.2008.08.005

Liu H, Ramnarayanan R, Logan BE (2005) Production of electricity during wastewater treatment using a single chamber microbial fuel cell. Environ Sci Technol 38(7):2281–2285. https://doi.org/10.1021/es034923g

Liu H (2008) Microbial fuel cell: novel anaerobic biotechnology for energy generation from wastewater. In: Anaerobic biotechnology for bioenergy production. Wiley, pp 221–246. https://doi.org/10.1002/9780813804545.ch10

Logan BE, Hamelers B, Rozendal R, Schröder U, Keller J, Freguia S, Aelterman P, Verstraete W, Rabaey K (2006) Microbial fuel cells: methodology and technology. Environ Sci Technol 40(17):5181–5192. https://doi.org/10.1021/es0605016

Logan BE (2013) Microbial fuel cells. John Wiley & Sons, Hoboken

Mathuriya AS, Yakhmi JV (2016) Microbial fuel cells – applications for generation of electrical power and beyond. Crit Rev Microbiol 42(1):127–143. https://doi.org/10.3109/1040841X.2014.905513

Mohammed AJ, Ismail ZZ (2018) Slaughterhouse wastewater biotreatment associated with bioelectricity generation and nitrogen recovery in hybrid system of microbial fuel cell with aerobic and anoxic bioreactors. Ecol Eng 125:119–130. https://doi.org/10.1016/j.ecoleng.2018.10.010

Mohan SV, Raghavulu SV, Peri D, Sarma PN (2009) Integrated function of microbial fuel cell (MFC) as bio-electrochemical treatment system associated with bioelectricity generation under higher substrate load. Biosens Bioelectron 24(7):2021–2027. https://doi.org/10.1016/j.bios.2008.10.011

Nandy A, Kundu PP (2018) Configurations of microbial fuel cells. In: Progress and recent trends in microbial fuel cells. Elsevier, pp 25–45. https://doi.org/10.1016/B978-0-444-64017-8.00003-8

Narayan M, Solanki P, Srivastava RK (2018) Treatment of sewage (domestic wastewater or municipal wastewater) and electricity production by integrating constructed wetland with microbial fuel cell. In: Sewage. InTech. https://doi.org/10.5772/intechopen.75658

Palanisamy D, Chockalingam LR, Murugan D (2024) Microbial fuel cell for effluent treatment and sustainable power generation. Energy Sour Part A Recovery Utilization Environ Eff 46(1):11952–11964. https://doi.org/10.1080/15567036.2020.1796844

Schröder U, Harnisch F (2013) From microbial bioelectrocatalysis to microbial bioelectrochemical systems. pp 137–162. https://doi.org/10.1002/9783527680436.ch4

Tharali AD, Sain N, Osborne WJ (2016) Microbial fuel cells in bioelectricity production. Front Life Sci 9(4):252–266. https://doi.org/10.1080/21553769.2016.1230787

Thornton RJ, Armstrong RD (2006) An evaluation of proton exchange membranes for use in microbial fuel cells. J Power Sources 160(2):896–903. https://doi.org/10.1016/j.jpowsour.2006.02.077

Wrighton KC, Agbo P, Warnecke F, Weber KA, Brodie EL, DeSantis TZ, Hugenholtz P, Andersen GL, Coates JD (2008) A novel ecological role of the Firmicutes identified in thermophilic microbial fuel cells. ISME J 2(11):1146–1156. https://doi.org/10.1038/ismej.2008.48

Xie Y, Feng Y, Wang X, Logan BE (2013) Design of stainless steel mesh cathodes for improving power generation in microbial fuel cells. Environ Sci Technol 45(2):930–935. https://doi.org/10.1021/es102413z

Yap K-L, Ho L-N, Ong S-A, Guo K (2021) Effect of carbon materials as cathode on wastewater treatment and bioelectricity generation in a double chambered microbial fuel cell. IOP Conf Ser Earth Environ Sci 646(1):012001. https://doi.org/10.1088/1755-1315/646/1/012001

Zhang J, Zhang E, Scott K, Burgess JG (2012) Enhanced electricity production by use of reconstituted artificial consortia of estuarine bacteria grown as biofilms. Environ Sci Technol 46(5):2984–2992. https://doi.org/10.1021/es2020007

Zhou M, Yang J, Wang H, Jin T, Hassett DJ, Gu T (2014) Bioelectrochemistry of microbial fuel cells and their potential applications in bioenergy. In: Bioenergy research: advances and applications. Elsevier, pp 131–152. https://doi.org/10.1016/B978-0-444-59561-4.00009-7

Integrative Omics Approaches for Improving the Effectiveness of Plants for Phytoremediation

12

Gbotemi Jerry Oni, Samuel Adeniyi Oyegbade, Zuwa Yoowa Duyann, Emmanuel Ojochegbe Mameh, Austine Atokolo, Queen Elizabeth Sule, and Ahmed Wopa Wurie

Abstract

As industrialization evolves, it is accompanied by vast utilization of natural resources and synthetic compounds which often results in nature distortion and consequently poses heavy environmental concerns. The rigorous search for sustainable solutions to these issues paved the way for the emergence of an affordable, safe, and eco-friendly plant-based approach, widely known as phytoremediation, which has proven to be potent in getting rid of toxic environmental contaminants and pollutants such as radioactive compounds, pharmaceuticals, xenobiotics, pesticides, oil spills, heavy metals, and medical wastes. Further insights into phytoremediation at the genetic level have been gained and advanced via omics technologies. This chapter explores diverse omics technologies including metabolomics, genomics, metaproteomics, phenomics transcriptomics, signalomics, proteomics, metagenomics, fluxomics, meta-transcriptomics, and metabolomics highlighting their integration, recent developments, limitations, and future perspectives toward improving the effectiveness of plants for higher phytoremediation efficiency and sustainable wastewater treatment endeavors. Recent technological efforts and advances toward the discovery of plant biosynthetic pathways via omics technologies

G. J. Oni · Z. Y. Duyann · A. W. Wurie
Department of Molecular Biology and Bioinformatics, Africa Centre of Excellence for Mycotoxins and Food Safety (ACEMFS), Federal University of Technology, Minna, Nigeria
e-mail: pg4412905.oni@st.futminna.edu.ng; duyann.ac2117305@st.futminna.edu.ng; ahmed.2117636@st.futminna.edu.ng

S. A. Oyegbade (✉) · E. O. Mameh · A. Atokolo · Q. E. Sule
Department of Biological Sciences, College of Science and Technology, Covenant University, Ota, Nigeria
e-mail: samuel.oyegbadepgs@stu.cu.edu.ng; emmanuel.mamehpgs@stu.cu.edu.ng; austine.atokolopgs@stu.cu.edu.ng; queen.sulepgs@stu.cu.edu.ng

N. Dwivedi et al. (eds.), *Wastewater Treatment Through Nature-Based Solutions*,
https://doi.org/10.1007/978-981-96-6699-7_12

and integration avail new paths and opportunities. Proteome-based methods have also been more vastly explored in plant pathology studies, and tracking of phosphorylation alterations in enhancing signal transduction pathways identification and protein function, while metabolomics and transcriptomics have helped understand mechanisms underlying phytoremediation, identify emerging plant biosynthetic pathway members, and also target specific ones.

Keywords

Integrative omics approaches · Phytoremediation · Plants · Environment · Wastewater

12.1 Background, Definition, and Relevance of Phytoremediation

Over several decades, the substantial utilization of plants for the elimination and neutralization of environmental pollutants has undergone extensive change. The idea came about due to the sixteenth-century discovery of plants that could withstand metals and the 1940s observation of plants metabolizing organic pollutants (Stephenson and Black 2014). One effective, well-recognized, affordable, and environmentally benign method for facilitation of ecosystems' restoration is phytoremediation. According to Kuppusamy et al. (2016), the field took off in the late 1980 when hyperaccumulator plants extract large heavy metal concentrations of from the soil. As a result of basic restrictions, phytoremediation confronts difficulties in commercial implementation despite its promise and potential (Nwogwu et al. 2022). It is interesting to know that phytoremediation was first used by Chaney (1983) (Vasavi et al. 2010). Phytoremediation has become a popular, low-cost substitute for traditional clean-up methods. A few examples of the potential outcomes of various phytoremediation forms include heavy metal buildup in particular plant organs, leaf surfaces' volatilization, changes to the form or availability of organic chemicals in the soil or within the plant, and the active removal of chemicals from plant tissues and their removal from the food chain. Phytoremediation has benefits, but it also has drawbacks. One of these is the requirement for handling contaminated plant material safely after harvest (Naz et al. 2022). Although the utilization of phytoremediation advances, ecological traits of plants employed have not received much attention (Li et al. 2021a). The terms "phytoremediation" are derived from "remedium" and "phyto." Remedium, implies "to correct," while phyto means "plant." (Islam et al. 2022). Industrialization advances spanning several decades have brought about a substantial rise in water and soil pollution levels. Addressing this age-long issue demands efficient, dynamic, cost-effective, scalable, and eco-friendly tools and strategies (Grobelak et al. 2019). Plants' utilization in extracting pollution and lowering water and pollution levels, often referred to as phytoremediation, has proven to be a far less expensive, easily controlled, nutrient-providing, and safer clean-up method in diverse aquatic and soil environments. They can also facilitate the provision of wildlife habitat even after completing their activities on degraded sites. Phytoremediation has been previously utilized in the treatment of toxic contaminants which could include petroleum and heavy metals solvents, etc. by participating via diverse techniques. Amongst several other scenarios, metals undergo translocation above-ground and rid of via phytoextraction. This is consequent to the fact that their inability to get involved in metabolic processes poses a substantial challenge to their phytoremediation treatment processes (Ren et al. 2018). Despite advancements in phytoremediation studies, implementation on a large scale is still elusive due to its rather extremely slow process, particularly with regulatory bodies demanding substantial impacts of decontamination on pollution sites after some years. This constraint has instigated interest by researchers toward improving phytoremediation efficiency. Previously, the development of particular strategies via evolving technologies for the utilization of plant species as bioindicators and remediation purposes has revolved around standard toxicity tests' usage for the assessment of pollutant persistence levels and phytotoxicity in plant tissues.

Enhancement of phytoremediation processes and approaches in the future will necessitate accurately identifying mechanisms of metabolisms that take part in detoxification, uptake, and tolerance of pollutants (Chaâbene et al. 2018). This mechanism of metabolism comprises sophisticated cellular paths modulated by vast gene families. This is where omics technology applications come into play to facilitate the accurate identification of peculiar characteristics for selected plant species as well as accompanying symbiotic microorganisms. Next-generation sequencing (NGS) technologies showed up in 2005 and were geared toward addressing these issues, particularly as regards researching plants as viable tools for attaining robust and upgraded understanding of metabolic processes and interactions between symbiotic microbes and plants. Initially, the translation of vast omics data into technologies directly applicable to the medical field was limited due to their high cost. As the years roll by, there has been a drastic drop in next-generation sequencing machineries' costs, thereby extending the frontiers of applications into myriads of fields, with emphasis on tackling water and water pollution issues. Rapid technological advancements yielded the evolution of omics concepts advances in advances in methodical cross-examination of the cellular system (Dai and Shen 2022).

This chapter investigates advances in omics applications in the remediation processes of polluted soils. Proteomics and transcriptomic studies and applications pave the way for gene identification linked to specific detoxification paths when they are exposed to organic contaminants and heavy metals. Recent advances of omics research utilization in synthesizing plants that are engineered genetically to serve phytoremediative uses. Next-generation sequence technologies can be adopted in advancing plant research, particularly for developing phytoremediation approaches. Investigation of high-throughput techniques is vastly carried out for several environmental functions. Costs linked to chemical and physical methods in treating polluted sites are potentially high, resulting in the abandonment of polluted land sites by private enterprises (Ren et al. 2018). Implementing cost-efficient remediation techniques such as phytodegradation and phytoextraction has immense potential to decontaminate vast polluted areas.

12.2 Phytoremediation Effectiveness

Phytoremediation employs plants to purify contaminated air and aid in the implementation of healthy habits. In this twenty-first century, there are several approaches to heavy metal cleanup; nevertheless, they are all inadequate because of the changing characteristics of the soil, high outlay, interruption of the microflora in the soil, and high labor requirements. Interestingly, phytoremediation is reasonably more effective than any other remediation method in resolving this thoughtful matter. Using plants and associated soil bacteria, phytoremediation lowers the heavy metal concentration and their hazardous level (Devi and Kumar 2020). However, several factors affect how well a phytoremediation works, including the degree and type of contamination (Wang et al. 2020) (Fig. 12.1).

12.2.1 Type of Contaminants

It is an established fact that air, soil, and water pollution have aggravated over the years largely due to anthropogenic activities and population influx. Nevertheless,

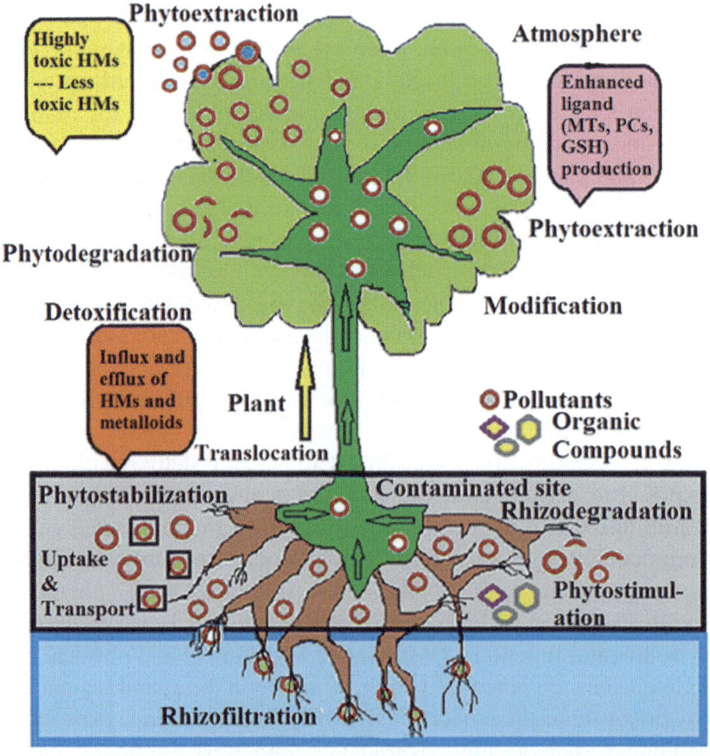

Fig. 12.1 Representation of different phytoremediation approaches (Sikandar et al. 2019)

Table 12.1 Contaminants examples

Contaminant	Example	Reference
Nutrients	(e.g., nitrogen, phosphorus)	De Groot et al. (2013)
Heavy metals	Cadmium and lead	Rahman and Singh (2019)
Organic Pollutants	(e.g., petroleum hydrocarbons)	Varjani et al. (2017)

the necessity for remediation tools needs to be seriously considered (Karami and Shamsuddin 2010).

12.2.1.1 Heavy Metals

Ali et al., 2013, stressed that heavy metals' inability to biodegrade buildup in the environment and pollute the nutrition chain, and cause damage to human health and entire ecosystem. While some heavy metals cause behavioral and neurological anomalies, particularly in young adults, some others cause endocrine disruption, mutagenicity, carcinogenicity, and teratogenicity.

12.2.1.2 Organic Pollutants

The impact of bio-accumulative and long-lasting characteristics, persistent organic pollutants, or POPs, is mentioned as soundless destroyers (Alharbi et al. 2018). It is ear sensitive and essential to understand that explosives, petroleum hydrocarbons, and chlorinated diluters are the three types of substances that have traditionally been the focus of phytoremediation of organic pollutants (Kaur et al. 2020) (Table 12.1).

12.3 Plant Species

It is significant to realize that more than 400 plant species have been found; hyperaccumulators such as Brassica and Thlaspi have been vastly investigated. According to pot studies, ryegrass, tall fescue, green gram, sunflower, and oat plants can all thrive in mixed polluted soils; sunflower has the best rate of contamination elimination (Chirakkara et al. 2015). In highly polluted soils, hyperaccumulation behavior for cadmium and zinc in extremely contaminated soils has been observed in wild plant species such as *Stellaria holostea* and *Artemisia vulgaris* (Antoniadis et al. 2020).

12.3.1 Environmental Conditions

One may argue that anthropogenic factors contribute to environmental conditions only. However, several environmental aspects such as stomatal conductance, temperature, light intensity, and species of microorganisms affect the efficacy of phytoremediation. Concurrently, soil pH is influenced by plants bioavailability, climate reaction toward temperature, sunlight and rainfall, by influencing the growth of plants and not forgetting the soil texture which affects water retention, root penetration, etc. with regard to other environmental conditions (Madhav et al. 2024).

12.3.2 Time Frame

Plants may become phytotoxic when exposed to varying concentrations of heavy metals, which can adversely affect their ability to carry out phytoremediation (Bortoloti and Baron 2022). Besides, it potency appears like a very difficult process to undertake phytoremediation of heavy metals since it is influenced by the environment, bioavailability of the heavy metals, plant development and selection, and environmental conditions (Khan et al. 2021).

12.3.3 Limitations

Some plant species are not as good at eliminating particular contaminants as others are which limit the circumstances in which phytoremediation may be used.

Limitations on Complexity

Since phytoremediation is more successful at shallow soil pits, it is not as suitable for deep contamination (Jaskulak et al. 2020). Phytoremediation has its technical restrictions. The long-lasting development times required for plant growth or acclimation are one strategic issue to take into account. The need for the pollutants to be near the plant roots raises another tactical issue. Furthermore, the attentiveness of the toxins affects how well the roots absorb or break down the contaminants (Shen et al. 2022).

12.4 Principles of Phytoremediation

The basic principle of phytoremediation is that plant roots absorb pollutants from the soil or that pollutants are broken down in the soil through plant-associated processes. The contaminants are then stored in the plant's shoots and leaves. In contrast, traditional soil remediation methods are often more invasive, costly, and environmentally disruptive.

12.4.1 Traditional Soil Remediation Methods

Conventional methods of cleaning up polluted soil typically involve isolating the contamination or simply disposing of the affected soil. This is unsustainable. An age-old issue, soil contamination, or land pollution has grown increasingly complicated as cities and industries have grown.

Soil contamination is often understood to be the presence of hazardous compounds in soil at concentrations that threaten the environment and health of humans. Certain contaminants, such as sulfates or arsenic, are naturally occurring toxins. Nonetheless, the majority of cleanup initiatives concentrate on man-made pollutants originating from diverse sources. Water pollution refers to the presence of harmful substances in various water bodies. Anthropogenic water pollution has been a long-standing issue, significantly worsened by global industrialization through chemical and plastic production, as well as urban development. However, this intensification occurred primarily in modern history—not during the Paleolithic period (Chakraborty and Chakraborty 2021).

12.4.2 Traditional Groundwater Remediation

It is possible to carry out either in situ or ex situ conventional groundwater cleanup methods. Ex situ remediation is prevalently carried out by pumping out polluted water available in the earth via a well and collecting it in vessels for disposal or treatment purposes. The "inexpensive and simple to operate" features of pumping systems makes them suitable for poor countries (Li 2021). Unfortunately, there isn't a single, effective method to extract water from soil. Site factors, such as the kind of soil and where water is located in the soil profile, substantially affect well designs as well as pumping systems (Singhal et al. 2010).

12.4.3 Traditional Surface Waters Remediation

Regeneration of surface water under various circumstances would increase in the current body of knowledge about surface water cleanup. Purification is the most fundamental traditional method for rehabilitating river zones and it entails bringing out pure water into aquatic platform to facilitate downstream removal of contaminants. Purification method simply dilutes the contaminant concentration levels to minimal and less poisonous range to individuals and ecosystem; it encourages the processes of natural biological, chemical, and physical deterioration (Simon and Joshi 2021).

12.4.4 Conventional Phytoremediation Techniques

Techniques for phytoremediation are divided into five parts: (1) rhizofiltration technique which entails treatment of water via absorption of impurities by plant roots; (2) phytoextraction techniques entail intake from the soil; (3) phytotransformation is a water and soil technique which entails the annihilation of contaminants through plant metabolism; (4) plant-assisted bioremediation, also known as phytostimulation, which is applied to both soil and water. It entails utilizing plants in the root zone to encourage microbial biodegradation; and (5) phytostabilization, which uses plants to lessen the potential for contaminants to migrate and move through the soil (Qadir et al. 2021).

12.4.5 Limitations and Gaps in Current Phytoremediation Techniques

As indicated by Oladoye et al. (2022), determining the best remediation strategies and lowering soil pollution by utilizing cheaper strategies to facilitate decontamination caused by heavy metals, the current evaluation filled a research void. Additionally, heavy metal pollution exceeding the baseline or regulated limits has affected almost 20 million hectares of terrestrial areas. Nickel, chromium, zinc, mercury, selenium, arsenic, cobalt, cadmium, and lead are some of these metals. Reclaiming land is good for the economy and the environment.

12.4.6 Essence of Plant Selection and Genetic Modification

A viable approach to enhancing phytoremediation of organic xenobiotics and heavy metals is genetic modification. In particular, for elements like Cd, Pb, Cu, As, Se, and Hg, transgenic plants have been developed to increase metal absorption, accumulation, and volatilization. Although there are ethical and technological issues with releasing genetically engineered microbes into the environment, they can also help break down dangerous chemicals (Pant et al. 2021). Again, the phytoremediation efficacy can be further enhanced by plant–microbe partnerships, such as endophytic and rhizospheric interactions.

12.5 Omics Research and Technologies

Omics research advances have yielded an enhanced understanding of metabolic processes especially on exposure to plant–microbial communities' interactions and poisonous substances' exposure. Omics technology encompasses diverse disciplines targeted at elucidating biochemical principles of an organism's function in response to exposure to chosen factors by utilizing molecular biology methods. The names and terminology used in various omics research areas are typically linked to the specific class of biological molecules being studied. Transcriptomics has to do with alterations manifesting in an organism's transcriptome' level. Proteomics entails finding out robust and holistic details on the constituents of the proteome of a particular sample. Incorporation of omics technologies with novel bioinformatic techniques fosters understanding and specific patterns' determination in crop plants responding to diverse stress factors (Jaskulak et al. 2018a).

12.5.1 The Dawn of Omics Methods

Omics technologies have evolved to surmount the constraint of studying an insignificant proportion of overall taxa through approaches that rely on culture. It paves the way for compositional analysis of the entire microbial community (Fagorzi 2021). Omics techniques beam robust understanding into unravelling molecular mechanisms of Polycyclic Aromatic Hydrocarbons (PAH) breakdown at varying population levels (El Amrani et al. 2015). Next-generation sequencing technologies emerged almost two decades ago, resulting in vast dataset creation, thereby stirring the need for data translation into utilizable technologies. Understanding signaling networks and cellular processes about environmental stress response has been fostered by bringing in high-throughput technologies with mass spectrometry incorporation (Fagorzi 2021).

Severe environmental imbalances are widespread consequences of indiscriminate pesticide applications in soil ecosystems. Severe environmental imbalances are common consequences of indiscriminate pesticide application in soil ecosystems. Despite its immense potential, phytoremediation remains underutilized as a sustainable solution for restoring pesticide-contaminated environments. This makes the deep investigation of the omics tools to foster a robust understanding of mechanistic actions in improving the remediation of contaminants imperative (Fagorzi 2021).

In analyzing omics datasets, researching integrative system biology is also expedient, as it can enhance bacterial-aided phytoremediation technologies' interpretation of intricacies in interactions between soil and plant microbiomes. Diverse efforts have been geared in recent decades by researchers toward enhancing phytoremediation via the utilization of recombinant strains of bacteria, and transgenic plants to eliminate an array of contaminants and areas polluted with crude oil also (Fagorzi 2021). System biology omics tackles complexity issues with the conventional approach by bringing on essential tools to analyze robustly. Sequencing potentials in recent years have transformed studies of genomics, transcriptional profiling, and genetics. Complex biochemical processes relating myriads of molecules are present at varying degrees levels in biological systems and understudied via mass spectrometry-based analysis, thereby facilitating metabolomics, lipidomics, glycomics, glycoproteomics, and high-throughput proteomics analysis. Widespread deforestation and urbanization have resulted in enormous release of toxic substances, thereby heightening soil and air pollution rates. Traditional soil remediation approaches are expensive, tedious, and time-consuming. The consequence is that they cannot support sustainable agriculture in the economic sense. The way out is to explore phytoremediation by leveraging plants and microorganisms. Gene manipulation and editing approaches can be utilized in plant efficacy enhancement for phytoremediation purposes (Gutierrez et al. 2024).

12.5.2 Single Organism Omics

Screening of bacteria that are efficient in remediation processes and also rely on culture as seen in rhizospheric bacteria in the course of its isolation and characterization paves the way for the identification of pollutants degraded strains by fostering buildup via the chromium hyperaccumulator. Evaluation of the analysis of transcriptome and transcriptome can be carried out for detection of differentially depicted genes in contaminants (Uchiyama and Miyazaki 2009).

12.5.3 Multiple Organism Omics

DNA, RNA, and proteins meta-omics studies are often utilized in getting a robust perception of structure of diverse microbial populations in situ, thereby eliminating the need to grow microbes. Constructing plasmids alongside huge genomic portions to be moved in *Escherichia coli* has proven to be a likely approach to the application and testing of novel findings and discoveries in the study of meta-omics. Multi-layered data can be created on plant stress response induced by integrating diverse meta-omics methods (Uchiyama and Miyazaki 2009).

12.5.4 Metagenomics

For the identification and analysis of microbial composition within an environment at varying degrees, metagenomics can be utilized. Additionally, it is deployed in studying the community's mobilome allot functional characteristics to microbes and explore changes in the structure of microbes and functional paths involved when soil pollutants are present (Feng et al. 2017).

12.5.5 Metatranscriptomics

Beyond providing details on microbial community's functional conformation as demonstrated via metagenomic analysis, transcriptomics fosters an understanding of how single genes play crucial roles in the degradation of pollutants and also provides insights into changes found in microbial genes' expression in contaminated environments. In a bioremediation background, gaining a robust understanding of varying gene expression across diverse weather settings is made possible via transcriptomic shifts (Klatt et al. 2013).

12.5.6 Metaproteomics

Metaproteomics is utilized in the definition of plant–microorganisms and microorganisms–microorganisms interactions in various environmental settings.

Metaproteomics offers details on the specific microbes and synthesized proteins' presence in an environment with little diversity (Delmotte et al. 2009). The metaproteomics method integrated with stable probing can facilitate which specific microbes act in pollutants' metabolism.

12.5.7 Metabolomics

The metabolome analysis of a target microbe acting in the process of bioremediation and investigation of the microbial community' meta-metabolome linked with various plant organs can foster mechanisms unraveling toward improving resistance to heavy metals in crop plants. Metabolomic analysis facilitates bioremediation applications linked with molecular connectivity analysis (Feng et al. 2017).

12.6 Case Studies of Successful Integrative Omics Application

12.6.1 Standard Approaches for Phytoremediation of Polluted Sites

Due to high costs incurred by private enterprises, contaminated sites are frequently abandoned (Ren et al. 2018). Therefore, application of other economical soil remediation techniques may proffer enhanced results particularly in vast lands that have previously been contaminated by widespread anthropogenic sources (Marchand et al. 2016). Identifying plant species that are most suitable for each phytoremediation technique is essential to achieving favorable outcomes (Azab et al. 2016). Jaskulak et al. (2018a, b) affirmed that metal hyperaccumulators are well suited for phytoextractive processes due to their capacity to concentrate vast metals concentrations. *Petris vittata* was deployed recently for arsenic elimination by hyperaccumulation. *P. vittata* can extract significant amounts of arsenic from soil at a concentration of 97 ppm. Arsenic is a poison that is both acute and fatal. Furthermore, according to Liu et al. (2017), a large chunk of arsenic was moved to plant branches that were simple to harvest. *Thlaspi caerulescens* another hyperaccumulation facilitated cadmium in its shoots to a concentration (Bayçu et al. 2017). According to Azab et al. (2016), phytodegradation is a suitable option. Currently, chlorinated hydrocarbons and polycyclic aromatic hydrocarbons are the main substances broken down by this technique (Cevher-Keskin et al. 2018). Hybrid poplars, for instance, have been employed to break down trichloroethylene, a carcinogenic, and hepatotoxic pollutant (BenIsrael et al. 2019). Phytodegradation was successfully deployed by demonstrating several plant species' use for the decomposition of perchloroethylene and carbon tetrachloride (CT) (Ma et al. 2015). Trinitrotoluene and other explosives, such as polycyclic aromatic hydrocarbons, have also been demonstrated to be successfully degraded by phytodegradation (BenIsrael et al. 2019; Kumar et al. 2018). The concentration of PAHs in the environment has been rising steadily over the past few decades. According to

Marchand et al. (2016), Populus nigra is one plant that is effective at degrading a variety of PAHs, including pyrene, benzopyrene, anthracene, phenanthrene, and chrysene. Augmented bioremediation using particular bacterial species and strains capable of transforming or degrading specific pollutants are an additional option for contaminant clean-up (Grobelak et al. 2019). However, a significant disadvantage of this method is the great sensitivity of microbial communities; as a result, to achieve successful restoration, rigorous conditions must be maintained at an ideal level. The main factors influencing the effectiveness of these efforts are the species' persistence and bioavailability, the accessibility and bioavailability of particular microorganisms, their capacity to thrive in a particular environment, the existence of carbon, the presence of other dominant species, and the availability of all necessary micro- and macronutrient sources (Bramhachari et al. 2017).

12.6.2 Metagenomic Approach to Enhance Phytoremediation Efficiency

Plants can increase the number and quality of microbial populations in bulk soil by being introduced to severely degraded and contaminated soils (Balcom et al. 2016). Plants can favor particular microbes that either suppress the activity of diseases or encourage their growth, even though plant–microbe interactions are extremely complex and sensitive to many environmental variables. Furthermore, microbes can influence the possible uptake of contaminants, such as the uptake and translocation of heavy metals from plant roots to shoots (Kumar et al. 2018). In general, microbial communities that may form relationships with plants can affect those plants' growth and development in both good and negative ways, which will change the plants' capacity to improve the soil (Ma et al. 2015). Heavy metals and other contaminants can be taken up and moved by plants or altered by certain microbes in the rhizosphere. By releasing root exudate, plants can enhance the stimulation of microbial activity (Bramhachari et al. 2017). Tests of such intricate connections can greatly benefit from metagenomic research. Metagenomics, to put it briefly, is the study of whole genetic material extracted directly from environmental samples. Thus, the term "metagenome" refers to the entirety of the genetic material found in the chosen environmental sample, which includes the genomes of numerous distinct creatures (Kumar et al. 2018). Targeted amplicon sequencing was used in a few recent phytoremediation studies, including field trials, to demonstrate how the phylogeny of the plants and the concentration of the contaminant in the soil shape the bacterial and fungal communities (BenIsrael et al. 2019; Phillips et al. 2012). Additional research has also shown that the concentration of metal influences the composition of arbuscular mycorrhizal (Chen et al. 2017). Accordingly, research indicates that abiotic stress in plants can determine whether or not microbial bioremediation is stimulated. According to Phillips et al. (2012), an experiment conducted in 2012 revealed that the exudate produced by wild rye exposed to hydrocarbons was not as repressive as the exudate from wild rye grown on an uncontaminated medium. This finding suggests that plants can modify the degree to which they promote total rhizosphere biodegradation. It is imperative to acknowledge that the extent to which plants can regulate bacterial and fungal populations is contingent upon the concentration of a particular pollutant and the overall severity of its harmful effects (Hao et al. 2018). It is well known that soil microbes are essential to the mineralization of organic materials in soils and can change certain pollutants into less hazardous and more stable chemical forms (Hou et al. 2017). Simultaneously, over 99% of the microbial taxa identified are frequently not yet cultivated, which is one of the most common issues with research on soil microbial communities. Thus, while maintaining the environmental elements (both biotic and abiotic) that created these microbial communities in situ, the use of NGS technology can enable a substantially more extensive investigation of microbial composition and activity (Song et al. 2016). Additionally, shotgun sequencing and amplicon targeting of genes can be utilized as a potential approach to sufficiently probe the plant-associated microbial communities and examine plant-dependent habitats, including plants rhizosphere, phyllosphere, and endosphere (Kumar and Pannu 2018). The isolated DNA is randomly divided into tiny segments during the shotgun sequencing procedure, and these segments are then sequenced using the chain termination method (Ren et al. 2018). The overlapping ends of various readings are then combined into a single continuous sequence using bioinformatics methods. Shotgun sequencing may now be applied to metagenomic samples thanks to the innovative developments in bioinformatics, which have significantly facilitated the study of DNA sequences extracted from complicated environmental samples. This method can be applied to the sequencing of several cultivated microorganisms. All of the genes found in environmental samples are revealed overall by shotgun sequencing (Ma et al. 2015). Shotgun high throughput metagenomic sequencing was employed in the Hao et al. investigation to identify the characteristics of the Taxus rhizosphere microbiome that were being negatively impacted by PAHs. Numerous functional genes linked to xenobiotic metabolism and biodegradation were found in the study, along with other genes engaged in various defense mechanisms against organic contaminants (Hao et al. 2018). Concurrently, as polluted soils and plant rhizospheres are known to be hotspots for rapid plasmid exchange, plasmid DNA has garnered a lot of attention in phytoremediation research in recent years (Chen et al. 2017). Plasmids can serve as a storehouse for genetic innovation during abiotic stress, allowing symbiotic bacteria and fungi to adjust to changing environmental conditions. In one study, genes implicated in heavy metal resistance are disproportionately numerous in plasmid DNA compared to the overall genome, as shown by metagenomic sequencing of plasmid DNA from municipal wastewater treatment plants. The same study demonstrated that plasmids contain the majority of the genes involved in the transformation and breakdown of organic contaminants (Balcom et al. 2016). Plasmids containing the toluene degradation genes were successfully transmitted to aerial plant endophytes, increasing the overall degradation of the pollutant that would have volatilized. This knowledge has already been put to use (Hao et al. 2018). Furthermore, metagenomic studies conducted under varied environmental circumstances revealed a connection between the genetic and taxonomic makeup of microbial communities and their capacity to metabolize carbon. Furthermore, particular taxa implicated in the inhibition of plant diseases in soils were revealed by metagenomic DNA sequencing (BenIsrael et al. 2019). Studies conducted by Balcom et al. have shown how microorganisms can metabolize xenobiotics and other micropollutants throughout the wastewater treatment process (Balcom et al. 2016). Microorganisms and particular genes involved in the breakdown of benzoate were identified through metagenomic research. The research as a whole demonstrated the high concentration of xenobiotic metabolism genes found in the biofilm of the constructed wastewater treatment facility. In a different study, the metagenomic method was used to look into how rice rhizosphere bacteria responded to PAH exposure. The overall distance from the surface of the roots and the concentration of PAHs had an impact on the microbial communities in the rice rhizosphere, according to the results. Furthermore, the likelihood of PAH breakdown in the rice rhizosphere was reflected in the number of genes linked to PAH degradation, such as dioxygenase genes (Ma et al. 2015). A 2019 study employed compound-specific isotope analysis and molecular metagenomic techniques to examine the mechanisms and effectiveness of toluene biodegradation in the unsaturated zone where poplars were deployed as a soil remediation tool. In addition to prior evidence for toluene biodegradation in the unsaturated zone, experiments showed that hybrid poplar roots contained spatially varying populations of toluene degraders (BenIsrael et al. 2019). Table 12.2 shows how metagenomic research's environmental applications varied from 2014 to

Table 12.2 Primary metagenomic studies deployed in remediating environmental pollution between 2014 and 2018

Goals	Pollutants	Year	Year
Wastewater treatment plant microbial communities' analysis and their potential for pharmaceuticals' metabolism	Pharmaceuticals	2016	Balcom et al. (2016)
Understanding the mechanism of plant growth-promoting rhizobacteria	None	2017	Bramhachari et al. (2017).
Analysis of the response of microbial metagenome to polycyclic aromatic hydrocarbons breakdown in the rice rhizosphere	PAHs	2015	Ma et al. (2015)
Perspectives of lindane (γ-hexachlorocyclohexane) breakdown by biological processes	Lindane	2018	Kumar et al. (2018)
Identification of the potential for biodegradation of toluene in the vadose zone of a poplar	Toluene	2019	BenIsrael et al. (2019)
Analysis of the influence of cadmium contamination on microbial composition and bioenergy cropping of cadmium-polluted soil	Cadmium	2017	Chen et al. (2017)
Analysis of the possibilities of Taxus rhizosphere microbiome for phytoremediation	None	2018	Hao et al. (2018)
Rhizosphere microbes' analysis of PAHs polluted soil planted with alfalfa and barley	PAHs	2018	Kumar et al. (2018)
Identification of the effect of microbial community on metal phytoremediation efficiency by *Sedum plumbizincicola*	Zinc and cadmium	2017	Hou et al. (2017)

2018. In summary, metagenomic analyses have multiple applications in environmental research, such as (i) identifying functional characteristics of particular microbial assemblages; (ii) identifying and elucidating the role of microbial genes linked to a wide range of functions, such as plant growth promotion or hydrocarbon biodegradation; and (iii) identifying specific environmental conditions that will support our target symbiotic microorganisms (BenIsrael et al. 2019; Kumar et al. 2018). Furthermore, single-cell isolation and sequencing methods make it possible to ascertain the potential of uncultured microbes in degrading pollutants. These methods could find new targets to be exploited in accurately determining the entire genome of certain uncultured strains (Ma et al. 2015).

12.6.3 Metal Hyperaccumulators

Metal hyperaccumulators are plant species utilized for phytoextraction, which fosters metal concentration in their biomass (Jaskulak et al. 2018a, b). Removal of arsenic, a lethal poison, was made possible via its hyperaccumulation through the utilization of Petris vittata extract. Harvesting of a large proportion of arsenic was carried out with ease via the translocation to plant shoots. Cadmium absorption mechanisms in the family, Brassicae were investigated and found to involve metal transporters which when exposed demonstrate vast expression (Kumar and Panda 2018). Via this method, plant metabolism degrades organic substances via secreted or internal enzymes. This approach also degrades polycyclic aromatic hydrocarbons (Cevher-Keskin et al. 2018). For instance, hybrid poplars are utilized in the degradation and uptake of trichloroethylene, a prevalent carcinogenic, and hepatoxic groundwater contaminant (Benisrael et al. 2019).

12.6.4 Plant Transcriptomic and Genomic Studies in Relation to Environmental Contamination

12.6.4.1 Transcriptomics

At present, most plant omics research and analyses conducted in environmental pollution zone have been made on plants' exposure to different heavy metals. The study by Chen et al. (2018) determined myriads of genes linked to tolerance of cadmium, and also determined buildup, uptake, and transport (Chen et al. 2018). The results of another researchers (Cevher-Keskin et al. 2018) also revealed that inducement of regulatory mechanisms after *Zea mays* was exposed to petroleum hydrocarbon stress.

12.6.4.2 Genomics

Integrating genomic approaches into environmental sciences has resulted into identifying several genes that participate in phytoremediative processes and abiotic stress tolerance to plant. It is a known fact that high-throughput approaches are particularly applicable in the identification of plant traits for the identification of plant characteristic that are dependent on specific blend of diverse genes. It is well established that high-throughput approaches are particularly useful for identifying plant traits that depend on complex interactions among diverse genes. These methods have also been applied in the identification of quantitative trait loci (QTLs), such as those associated with arsenic accumulation in maize stems and leaves. Besides, this study depicted that one quantitative trait locus was found in the tissues. Accumulation of arsenic was attributed to diverse genes in various plant tissues (Cevher-Keskin et al. 2018).

12.7 Conclusion

Social acceptability, cost-efficiency, and reliability are factors that make phytoremediation stand out in the treatment of polluted areas. Enhancing phytoremediation efficiency via omics technologies has attracted researchers' interests within the present decade. This chapter captured that identification of distinct environmental conditions has been made possible via vast omics datasets collection. Beyond their direct role in phytoremediation, the introduction of certain plant species can alter the soil microbial community, which may also contribute to the control of invasive species during crop production. A major benefit of omics technology is utilization of growing rise in datasets and incessant drop in costs. Consequently, translating robust knowledge and data obtained into utilizable technologies is significantly dependent on the kind of research that omics analyses are put into use accurately.

12.8 Future Perspectives

About 2.5 billion residing in rural areas across diverse developing countries depend largely on agricultural activities for their survival and sustenance. Agriculture contributes substantially to gross domestic product, yielding about 30% of economic growth (FAO 2012). By mid-twenty-first century, about ten billion humans will be on the planet, and there will likely be severe food insecurity (Smith et al. 2010; Naika et al. 2013), and might be aggravated due to rise in human actions that has led to unfavorable environmental changes. Because population growth and pollution are placing an ever-growing strain on crop productivity worldwide, there is a need for crop varieties that are stress-tolerant and adaptive. The molecular characterization of the proteins and genes participating in tolerating stress due to heavy metal has been aided by developments in a variety of functional methods, resources, and "omics." Furthermore, unexpected findings have been obtained from the heavy metal-responsive genes, metabolites, and proteins' genetic engineering; however, its full potential has not yet been realized. Future research must be planned with a concerted, interdisciplinary, and collaborative research in view and properly incorporated alongside such as proteomics, metabolomics, and transcriptomics in order to maximally surge the up abiotic stress and heavy metal tolerance of economically significant crop plants.

References

Alharbi OM, Khattab RA, Ali I (2018) Health and environmental effects of persistent organic pollutants. J Mol Liq 263:442–453

Ali, H., Khan, E. and Sajad, M.A. (2013) Phytoremediation of heavy metals—concepts and applications. Chemosphere, 91(7), pp. 869–881.

Antoniadis J, Chanlaridis S, Gräfener G, Langer N (2020) Type Ia supernovae from non-accreting progenitors. Astronomy Astrophys 635:A72

Azab E, Hegazy AK, El-Sharnouby ME, Abd Elsalam HE (2016) Phytoremediation of the organic xenobiotic simazine by p450-1a2 transgenic Arabidopsis thaliana plants. Int J Phytoremediation 18(7):738–746

Balcom IN, Driscoll H, Vincent J, Leduc M (2016) Metagenomic analysis of an ecological wastewater treatment plant's microbial communities and their potential to metabolize pharmaceuticals. F1000Research 5:1881

Bayçu G, Gevrek-Kürüm N, Moustaka J, Csatári I, Rognes SE, Moustakas M (2017) Cadmium-zinc accumulation and photosystem II responses of Noccaea caerulescens to Cd and Zn exposure. Environ Sci Pollut Res 24:2840–2850

BenIsrael M, Wanner P, Aravena R, Parker BL, Haack EA, Tsao DT, Dunfield KE (2019) Toluene biodegradation in the vadose zone of a poplar phytoremediation system was identified using metagenomics and toluene-specific stable carbon isotope analysis. Int J Phytoremediation 21(1):60–69

Bortoloti GA, Baron D (2022) Phytoremediation of toxic heavy metals by Brassica plants: a biochemical and physiological approach. Environ Adv 8:100204

Bramhachari PV, Nagaraju GP, Kariali E (2017) Metagenomic approaches in understanding the mechanism and function of PGPRs: perspectives for sustainable agriculture. In: Agriculturally important microbes for sustainable agriculture: volume I: plant-soil-microbe nexus. Springer Nature Singapore, pp 163–182

Cevher-Keskin B, Selçukcan-Erol Ç, Yüksel B, Ertekin Ö, Yıldızhan Y, Onarıcı S, Kulen O, Memon AR (2018) Comparative transcriptome analysis of Zea mays in response to petroleum hydrocarbon stress. Environ Sci Pollut Res 25:32660–32674

Chaâbene Z, Rorat A, Hakim IR, Bernard F, Douglas GC, Elleuch A, Vandenbulcke F, Mejdoub H (2018) Insight into the expression variation of metal-responsive genes in the seedling of date palm (Phoenix dactylifera). Chemosphere 2018(197):123–134

Chakraborty SK, Chakraborty SK (2021) River pollution and perturbation: perspectives and processes. In: Riverine ecology volume 2: biodiversity conservation, conflicts and resolution. Springer, Cham, pp 443–530

Chaney (1983) should be replaced with Singh, J., Hembram, P. and Basak, J. (2014) Potential of Vigna unguiculata as a phytoremediation plant in the remediation of Zn from contaminated soil. American Journal of Plant Sciences, 2014.

Chen Y, Zhi J, Zhang H, Li J, Zhao Q, Xu J (2017) Transcriptome analysis of Phytolacca americana L. in response to cadmium stress. PLoS One 12(9):e0184681

Chirakkara RA, Reddy KR, Cameselle C (2015) Electrokinetic amendment in phytoremediation of mixed contaminated soil. Electrochim Acta 181:179–191

Dai X, Shen L (2022) Advances and trends in omics technology development. Front Med (Lausanne). https://doi.org/10.3389/fmed.2022.911861

De Groot CC, Marcelis LF, van den Boogaard R, Kaiser WM, Lambers H (2013) Interaction of nitrogen and phosphorus nutrition in determining growth. Plant Soil 248:257–268

Delmotte N et al (2009) Community proteogenomics reveals insights into the physiology of phyllosphere bacteria. Proc Natl Acad Sci USA 106:16428–16433

Devi P, Kumar P (2020) Concept and application of phytoremediation in the fight of heavy metal toxicity. J Pharm Sci Res 12(6):795–804

El Amrani A, Dumas A, Wick LY, Yergeau E, Berthomé R (2015) "Omics" Insights into PAH degradation toward improved green remediation biotechnologies. Environ Sci Technol 49(19):11281–11291. https://doi.org/10.1021/acs.est.5b01740

Fagorzi, C. (2021) Challenges of Multi-omics in Improving Microbial-assisted Phytoremediation. In Rhizomicrobiome Dynamics in Bioremediation (pp. 69-79). CRC Press.

FAO (2012) Yearbook world food and agriculture. Food and Agriculture Organization of the United Nations, Rome, p 15

Feng J, Jia W, Lv S, Bao H, Miao F, Zhang X, Li Y (2017) Comparative transcriptome combined with morpho-physiological analyses revealed key factors for differential cadmium accumulation in two contrasting sweet sorghum genotypes. Plant Biotechnol J 16(2):558–571

Grobelak A, Świątek J, Murtaś A, Jaskulak M (2019) Cadmium-induced oxidative stress in plants, cadmium toxicity, and tolerance in plants: from physiology to remediation. In: Cadmium toxicity and tolerance in plants. Academic Press, pp 213–231

Gutierrez RCD, Alejo-Jacuinde G, Perez Sanchez B, Chavez Reyes J, Onigbinde S, Mogut D, Hernández-Jasso I, Calderón-Vallejo D, Quintanar JL, Mechref Y (2024) Multi omics applications in biological systems. Curr Issues Mol Biol 2024(46):5777–5793. https://doi.org/10.3390/cimb46060345

Hao DC, Zhang CR, Xiao PG (2018) The first Taxus rhizosphere microbiome was revealed by shotgun metagenomic sequencing. J Basic Microbiol 58(6):501–512

Hou J, Liu W, Wu L, Hu P, Ma T, Luo Y, Christie P (2017) Modulation of the efficiency of trace metal phytoremediation by sedum plumbizincicola by microbial community structure and function. Plant Soil 421:285–299

Islam MS, Akter R, Rahman MM, Kurasaki M (2022) Phytoremediation: background, principle, and application, plant species used for phytoremediation. In: Design of materials and technologies for environmental remediation. Springer Nature Singapore, Singapore, pp 199–224

Jaskulak M, Inżynieria C, Środowiska O (2018a) Implementation of omics research to enhance phytoremediation efficiency – a review. 21(4):361–373 p-ISSN 1505–3695. Engineering and Protection of Environment e-ISSN 2391–7253 https://ios.is.pcz.pl/ https://doi.org/10.17512/ios.2018.4.4. Open Access (CC BY-NC 4.0) University of Technology

Jaskulak M, Rorat A, Grobelak A, Kacprzak M (2018b) Antioxidative enzymes and expression of rbcL gene as tools to monitor heavy metal-related stress in plants. J Environ Manag 218:71–78

Jaskulak M, Grobelak A, Vandenbulcke F (2020) Modelling assisted phytoremediation of soils contaminated with heavy metals–main opportunities, limitations, decision making, and prospects. Chemosphere 249:126196

Karami A, Shamsuddin ZH (2010) Phytoremediation of heavy metals with several efficiency enhancer methods. Afr J Biotechnol 9(25):3689–3698

Kaur K, Badru R, Singh PP, Kaushal S (2020) Photodegradation of organic pollutants using heterojunctions: a review. J Environ Chem Eng 8(2):103666

Khan AHA, Kiyani A, Mirza CR, Butt TA, Barros R, Ali B, Iqbal M, Yousaf S (2021) Ornamental plants for the phytoremediation of heavy metals: present knowledge and future perspectives. Environ Res 195:110780

Klatt CG et al (2013) Temporal metatranscriptomic patterning in phototrophic Chloroflexi inhabiting a microbial mat in a geothermal spring. ISME J 7:1775–1789

Kumar D, Pannu R (2018) Perspectives of lindane (γ-hexachlorocyclohexane) biodegradation from the environment: a review. Bioresour Bioprocess 5(1):1–18

Kumar V, AlMomin S, Al-Aqeel H, Al-Salameen F, Nair S, Shajan A (2018) Metagenomic analysis of rhizosphere microflora of oil-contaminated soil planted with barley and alfalfa. PLoS One 13(8):e0202127

Kuppusamy S, Palanisami T, Megharaj M, Venkateswarlu K, Naidu R (2016) In-situ remediation approaches for the management of contaminated sites: a comprehensive overview. Rev Environ Contam Toxicol 236:1–115

Li, X., Li, Y., Li, Y. and Wu, J. (2021) The phytoremediation of water with high concentrations of nitrogen and phosphorus contamination by three selected wetland plants. Journal of Water Process Engineering, 40, p.101828.

Li P, Karunanidhi D, Subramani T, Srinivasamoorthy K (2021a) Sources and consequences of groundwater contamination. Arch Environ Contam Toxicol 80:1–10

Li Z, Xu X, Sheng X, Lin P, Tang J, Pan L, Kaneti YV, Yang T, Yamauchi Y (2021b) Solar-powered sustainable water production: state-of-the-art technologies for sunlight–energy-water nexus. ACS Nano 15(8):12535–12566

Liu X, Fu JW, Tang N, Da Silva EB, Cao Y, Turner BL, Chen Y, Ma LQ (2017) Phytate-induced arsenic uptake and plant growth in arsenic-hyperaccumulator Pteris vittata. Environ Pollut 226:212–218

Ma B, Lyu XF, Zha T, Gong J, He Y, Xu JM (2015) Reconstructed metagenomes reveal changes in microbial functional profiling during PAH degradation along a rice (Oryza sativa) rhizosphere gradient. J Appl Microbiol 118(4):890–900

Madhav S, Mishra R, Kumari A, Srivastav AL, Ahamad A, Singh P, Ahmed S, Mishra PK, Sillanpää M (2024) A review of sources identification of heavy metals in soil and remediation measures by phytoremediation-induced methods. Int J Environ Sci Technol 21(1):1099–1120

Marchand L, Sabaris CQ, Desjardins D, Oustrière N, Pesme E, Butin D, Wicart G, Mench M (2016) Plant responses to a phytomanaged urban technosol contaminated by trace elements and polycyclic aromatic hydrocarbons. Environ Sci Pollut Res 23:3120–3135

Naika M, Shameer K, Mathew OK, Gowda R, Sowdhamini R (2013) STIFDB2: an updated version of plant stress-responsive transcription factor database with additional stress signals, stress-responsive transcription factor binding sites, and stress-responsive genes in Arabidopsis and rice. Plant Cell Physiol 54(2):e8–e8

Naz M, Raza MA, Tariq M, Zaib S, Rajper SA, Jaskani MJ, Ahsan M, Dai Z, Du D (2022) Sustainable management of environmental contaminants: factors, control, and phytoremediation. In: Sustainable management of environmental contaminants: eco-friendly remediation approaches. Springer International Publishing, Cham, pp 1–16

Nwogwu NA, Ajala OA, Ajibade FO, Ajibade TF, Adelodun B, Lasisi KH, Ugya AY, Kumar P, Omotade IF, Babalola TE, Adewumi JR (2022) Phytoremediation mechanisms of heavy metal removal: a step towards a green and sustainable environment. In: Innovative bio-based technologies for environmental remediation. CRC Press, pp 207–236

Oladoye PO, Olowe OM, Asemoloye MD (2022) Phytoremediation technology and food security impacts of heavy metal contaminated soils: a review of the literature. Chemosphere 288:132555

Panda, D., Mandal, L. and Barik, J. (2020) Phytoremediation potential of naturally growing weed plants grown on fly ash-amended soil for restoration of fly ash deposit. International Journal of Phytoremediation, 22(11), pp.1195–1203.

Pant G, Garlapati D, Agrawal U, Prasuna RG, Mathimani T, Pugazhendhi A (2021) Biological approaches practised using genetically engineered microbes for a sustainable environment: a review. J Hazard Mater 405:124631

Phillips LA, Greer CW, Farrell RE, Germida JJ (2012) Plant root exudates impact the hydrocarbon degradation potential of a weathered-hydrocarbon-contaminated soil. Appl Soil Ecol 52:56–61

Qadir A, Hameed MA, Zafar MSB, Farooqi ZUR, Younas F, Hussain MM, Mohy-Ud-Din W (2021) Phytoremediation of inorganic pollutants: an eco-friendly approach, its types and mechanisms. Plant Environ 2(01):110–130

Rahman Z, Singh VP (2019) The relative impact of toxic heavy metals (THMs)(arsenic (As), cadmium (Cd), chromium (Cr)(VI), mercury (Hg), and lead (Pb)) on the total environment: an overview. Environ Monit Assess 191:1–21

Ren H, Wan Y, Zhao Y (2018) Phytoremediation of polychlorinated biphenyl-contaminated soil by transgenic alfalfa-associated bioemulsifier AlnA. In: Twenty years of research and development on soil pollution and remediation in China. Springer, Singapore, pp 645–653

Saxena G, Purchase D, Mulla SI, Bharagava RN (2019) Phytoremediation of heavy metal-contaminated sites: eco-environmental concerns, field studies, sustainability issues and future prospects. Rev Environ Contamin Toxicol

Shen X, Dai M, Yang J, Sun L, Tan X, Peng C, Ali I, Naz I (2022) A critical review on the phytoremediation of heavy metals from the environment: performance and challenges. Chemosphere 291:132979

Simon M, Joshi H (2021) A review on green technologies for the rejuvenation of polluted surface water bodies: field-scale feasibility, challenges, and future perspectives. J Environ Chem Eng 9(4):105763

Singhal BBS, Gupta RP, Singhal BBS, Gupta RP (2010) Groundwater contamination. In: Applied hydrogeology of fractured rocks: second edition. Springer Science & Business Media, pp 221–236

Smith P, Gregory PJ, Van Vuuren D, Obersteiner M, Havlík P, Rounsevell M, Woods J, Stehfest E, Bellarby J (2010) Competition for land. Philos Trans R Soc B: Biol Sci 365(1554):2941–2957

Song F, Li J, Fan X, Zhang Q, Chang W, Yang F, Geng G (2016) Transcriptome analysis of Glomus mosseae/Medicago sativa mycorrhiza on atrazine stress. Sci Rep 6(1):20245

Stephenson C, Black CR (2014) One step forward, two steps back: the evolution of phytoremediation into commercial technologies. Biosci Horiz 7:hzu009

Uchiyama T, Miyazaki K (2009) Functional metagenomics for enzyme discovery: challenges to efficient screening. Curr Opin Biotechnol 20:616–622

Varjani SJ, Gnansounou E, Pandey A (2017) A comprehensive review on toxicity of persistent organic pollutants from petroleum refinery waste and their degradation by microorganisms. Chemosphere 188:280–291

Vasavi A, Usha R, Swamy PM (2010) Phytoremediation–an overview review. J Ind Pollut Control 26(1):83–88

Wang L, Hou D, Shen Z, Zhu J, Jia X, Ok YS, Tack FM, Rinklebe J (2020) Field trials of phytomining and phytoremediation: a critical review of influencing factors and effects of additives. Crit Rev Environ Sci Technol 50(24):2724–2774

Life Cycle Assessment of Microalgae-Based Systems for Wastewater Treatment and Resource Recovery

13

Riti Kumari, Prashant Saxena, Aline Viancelli, and Naveen Dwivedi

Abstract

Life cycle assessment (LCA) of microalgae-based systems for wastewater treatment and resource recovery examines the potential of microalgae-based systems for treating wastewater and extracting resources. Microalgae, photosynthetic microorganisms found in diverse aquatic environments, offer significant ecological and technological benefits. Their metabolisms are such that they can remove nutrients and contaminants from wastewater thus providing a sustainable and less expensive alternative to conventional treatment methods. Through this assimilation process, nitrogen and phosphorus levels can be controlled thereby reducing eutrophication, while heavy metal levels are reduced lowering the discharge of hazardous substances into the rivers, lakes, or seas. Life cycle assessment (LCA) methodologies are employed to evaluate the environmental sustainability of these systems, revealing trade-offs and opportunities for improvement across different stages of the product life cycle. This chapter addresses the biological characteristics, traditional uses, as well as modern day applications of microalgae with an emphasis on its significance to environmental sustainability. It explores recent developments in technology, integrated system approaches, practical applications, and offers guidance for future urbanization based on these trends.

R. Kumari · P. Saxena (✉) · N. Dwivedi
Department of Biotechnology Engineering, University Institute of Engineering, Chandigarh University, Mohali, Punjab, India

A. Viancelli
Universidade Do Contestado, Concordia, SC, Brazil

N. Dwivedi et al. (eds.), *Wastewater Treatment Through Nature-Based Solutions*, https://doi.org/10.1007/978-981-96-6699-7_13

Keywords

Microalgae · Wastewater treatment · Biomass conversion · Resource recovery · Life cycle assessment

13.1 Overview of Wastewater Treatment

While clean water is a must for survival, the situation is gradually worsening as the availability of water is becoming more and more limited. The population growth along with urbanization and the industrial revolution has put a strain on freshwater sources while increasing the amount of wastewater. This wastewater that is not disposed of or treated in any way contaminates rivers, lakes, and the oceans and destroys ecosystems, threatening the health of millions of people in the process. A variety of chemical processes and energy demanding machines among other techniques have been deployed to deal with wastewater (Purba et al. 2022). However, there are major problems with these systems. They are costly, inefficient, and can't eliminate all pollutants, such as pharmaceutical drugs and heavy metals. Additionally, they usually disregard the possibility of recovering useful materials, and instead only concentrate on water purification. In a world that is becoming more and more focused on conservation, these drawbacks demonstrate the importance of discovering smarter and more effective methods.

Recycling water that is extracted from waste is important for the quality of water and the health of the environment. Most forms of recycling that are used for untreated water include biological, chemical, and mechanical measures to get rid of impurities and pollutants (Purba et al. 2022). The Aztecs, on the other hand, harvested spirulina from Lake Texcoco as their main source of protein (Manning et al. 2020). Among other things, they are used in producing biofertilizers and animal feeds. However, today's studies and technological breakthroughs have shown that microalgal applications are far ahead of these traditional ones. They have proved crucial and novel answers to contemporary ecological questions like the role of microalgae in wastewater treatment and resource recovery processes.

Treating wastewater is essential for preserving both water quality and environmental well-being (Fig. 13.1). Conventional methods of wastewater treatment typically involve mechanical, chemical, and biological processes to remove contaminants. However, these approaches often demand significant energy consumption, incur high costs, and may struggle to effectively address emerging pollutants such as pharmaceuticals and personal care products (Sun et al. 2018). These systems use the natural metabolic activities of microalgae to extract nutrients and pollutants from wastewater. Microalgae incorporate nitrogen and phosphorus—important nutrients that lead to eutrophication—into their biomass, restricting their discharge into water bodies and minimizing the need for chemical treatments (Wang et al. 2020). Furthermore, microalgae may absorb heavy metals and break down organic contaminants, including medicines, via multiple metabolic processes (Purba et al. 2022).

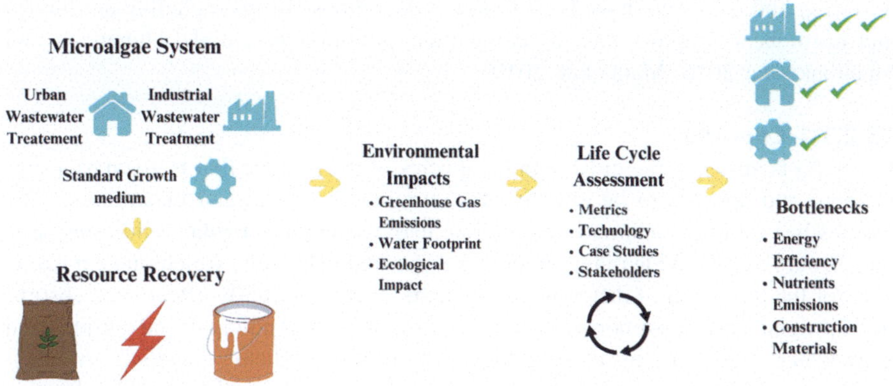

Fig. 13.1 Microalgae-based system

13.2 Microalgae-Based Systems for Wastewater Treatment

Systems that utilize microalgae are inexpensive and environmentally friendly means of treating sewage wastes. Microalgae utilize nitrogen and phosphorus nutrients present in sewage and hence, the pollutants are removed. Water is treated and simultaneously, useful biomass is produced for biofuels, fertilizers, and other products.

Open ponds and photobioreactors (PBR) are two of the more popular designs for these systems. Open ponds are low-tech, inexpensive approaches in which microalgae are allowed to bloom in broad shallow ponds that receive sunlight. However, there are disadvantages such as contamination and loss of water through evaporation. In contrast, photobioreactors are sealed containers and this makes them more costly; however, they are easier to manage, enhancing the growth of biomass as well as the effectiveness of treatment. These units do not allow contamination of the culture, but the cost of construction and maintenance of these units is higher among other disadvantages.

13.2.1 Types of Microalgae

Microalgae are single-celled bottom-up plant organisms that have a great potential in the treatment of wastewater due to their unique biological and ecological characteristics. The use of microalgae in the treatment of wastewater treats organic wastes since they are autotrophic and can use sunlight and carbon dioxide to produce organic biomass and at the same time releasing oxygen. The oxygenation activity assists aerobic bacteria that are essential in the biodegradation of organic matter from wastewater (Cai et al. 2017; Arashiro et al. 2018). In addition, microalgae have the ability to absorb heavy metals and degrade organic pollutants. They can also absorb and store heavy metals like cadmium, lead, and arsenic in their biomass which makes it hard for these metals to get into the water supply. For example,

Spirulina and *Chlorella* have been found to decompose drugs including antibiotics and hormone disruptors thus reducing their pollution potential (Abinandan and Shanthakumar 2015; Mata et al. 2010).

13.2.1.1 *Chorella*

Chlorella is amongst the most popular species of microalgae that is given as a supplement and appears to be effective for stimulating wastewater cleanliness. This species has rapid growth and can tolerate extreme environmental conditions. It is able to survive in a wide range of temperatures and pH levels, which make it favorable for the treatment of different wastewaters. It is particularly effective at absorbing nitrogen and phosphorus, common pollutants in municipal and agricultural wastewater. These organisms have the ability to remove heavy metals from industrial effluents (Khan et al. 2019). Their robustness also makes them highly effective for treating municipal wastewater, industrial effluents, and agricultural runoff. Also, the biomass produced by Chlorella can be harvested for biofuel production, fertilizers, or animal feed.

13.2.1.2 *Spirulina*

Apart from its noteworthy nutritional benefits, *Spirulina* also presents itself as an important contributor in wastewater treatment. Moreover, it is widely known for its tolerance to high pH and can be efficiently used for sewage water treatment that results from certain industrial activities (Arun et al. 2020). Under these circumstances, it, moreover, protects nutrients and organic substances quite efficiently while generating protein enrich biomass and hence making the treatment process even more worthwhile. It is particularly effective in industrial wastewater treatment because of its extreme environmental tolerance (Fernandez et al. 2021). In addition to this, its biomass can also be converted into other useful high valuable items such as biofuels, animal food, and supplements (Lee et al. 2021).

13.2.1.3 *Scenedesmus*

Scenedesmus is famous for wastewater treatment because it is efficient in removing nutrients and is robust. This species has an excellent tolerance for variations in water and has the ability to remove nitrogen and phosphorus (Chinnasamy et al. 2015). *Scenedesmus* is the most promising tolerant species for outdoor usage because it can withstand most of the environmental stresses, such as light or temperature, which are non-suitable for most of the temperate species that are used for wastewater treatment (Yadav et al. 2018). It is perfect for purifying agricultural runoff that is nutrient-dense and municipal wastewater. Its biomass can be harvested for biofuels or fertilizers, thus aiding in resource recovery (Rawat et al. 2016).

13.2.1.4 *Nannochloropsis*

Nannochloropsis is a microalgae species that is considered valuable for biofuel production due to its high lipid content. This species is energetic, but most importantly, it's found to survive in hypersaline habitats, thus making it useful for bioprospecting (Huo et al. 2018). It effectively cleans out the nutrients while stockpiling lipids, which are later converted into biofuels. It is used prominently in the treatment of aquaculture wastewater and bioenergy production. It is adaptable and can withstand salinity conditions, which add to its versatility (Wang et al. 2019).

13.2.1.5 *Dunaliella*

Dunaliella is a microalgae species distinctly adapted to extreme salinity and is thus particularly useful for selective treatment of niche wastewaters. This particular species is able to flourish in extremely saline conditions, whether it is salt lakes or seawater. It is known to be very high in carotenoids, which are the primary ingredients in the pharmaceutical and cosmetic industries (Goh et al. 2022). It is perfect for treating saline or brine water waste, such as from the desalination and food industries. This biomass is not useless, as it can be used for the extraction of pigments or for the making of biofuels, making the treatment process more economically feasible (Goh et al. 2022) (Table 13.1).

13.2.2 Open Pond System

Among the techniques to cultivate microalgae for use in the treatment of wastewater, the open pond system is the most widely used (Smith et al. 2021). These systems are structured as shallow basins or raceways wherein wastewaters are levered to multicultural conditions, and microalgae grow naturally while nutrient and other pollutants are being depleted.

The simplicity and low-cost of open pond systems are some of their key benefits. Construction and startup of these systems perform better off and need less expense than advanced systems such as photobioreactors. Open ponds need plenty of sunshine, as it is the primary energy source needed for photosynthesis, thereby

Table 13.1 Types of microalgae (Gupta et al. 2024)

Microalgae species	Characteristics	Applications in wastewater treatment
Chlorella	Rapid growth, nutrient uptake, tolerant to diverse conditions	Removes nitrogen, phosphorus, and heavy metals; treats municipal and industrial wastewater
Spirulina	Thrives in alkaline environments, produces high-value biomass	Treats specific industrial effluents; biofuel and animal feed production
Scenedesmus	Adaptable to various water conditions, resilient to stress	Removes nitrogen and phosphorus; treats agricultural runoff and municipal wastewater
Nannochloropsis	High lipid content, saline-tolerant	Treats saline wastewater, supports aquaculture, biofuel production
Dunaliella	High-salinity tolerance, carotenoid production	Treats saline and brine wastewater; pigment production and industrial wastewater treatment

lowering energy requirements (Kumar and Sharma 2020). These systems can also be used to manage large volumes of wastewater, especially in areas with abundant land.

13.2.2.1 Raceway Ponds

One of the best methods for growing colonies of microalgae is raceway ponds which are oval, shallow, and have paddlewheels fitted in them. When it comes to the mass production of algal biomass, the output in these ponds is lower than that of closed systems, but this method is extremely useful when there is a need to treat large amounts of wastewater. It does geopolitics and removes heavy metals, nutrients, and even organic pollutants, and at the same time, it is able to yield massive quantities of algal biomass that can be used as biofuel or any other purpose (Mehariya et al. 2023). High-rate algal ponds (HRAPs) are enhanced and evolved form of raceway ponds that help facilitate the growth of microalgae while increasing its productivity as well as efficiency of wastewater treatment. They are typically taller with partitions or baffles to enhance hydraulic retention time and improve mixing, leading to higher biomass productivity and better nutrient removal. They are also effective in removing heavy metals and organic contaminants from wastewater and continuous research and development efforts aim to enhance HRAP technology to broaden its suitability for various environmental scenarios through collaboration between government, business, and academia (Mehariya et al. 2023).

13.2.2.2 Circular Ponds

Circular ponds, or circular tank ponds, integrating the technique of open pond systems, are structures that are used for microalgae production in wastewater treatment and resource recovery purposes. These ponds have a centrally located rotating arm or agitator for consistent mixing of the microalgal culture, so that nutrients and light are available to phytoplankton evenly, so photosynthesis can be more effective while reducing sedimentation. This also achieves oxygen saturation in dissolved form and ensures effective transfer of gases which are important for microalgae cultures. Circular ponds are advantageous for urban settlements as they consume less space. Mechanically, construction can be complex and hence expensive, incorporating durable materials and proper engineering to ensure durability, but operation and maintenance are simpler compared to sophisticated systems such as HRAPs (Mehariya et al. 2023).

13.2.2.3 Unstirred Ponds

Unstirred ponds, or static ponds, are the open pond systems that are the least advanced and the cheapest for microalgae production in the process of wastewater treatment and resource recovery. Wind, convection currents, and diffusion are systems that are used for mixing the ponds; therefore, artificial stirring is not required. Because of their uncomplicated characteristics, they can be used for inexpensive smaller applications. However, the lack of mechanical mixing can lead to issues like sedimentation, uneven nutrient distribution, and inefficient gas exchange, which may impede microalgal growth. Despite these limitations, unstirred ponds are

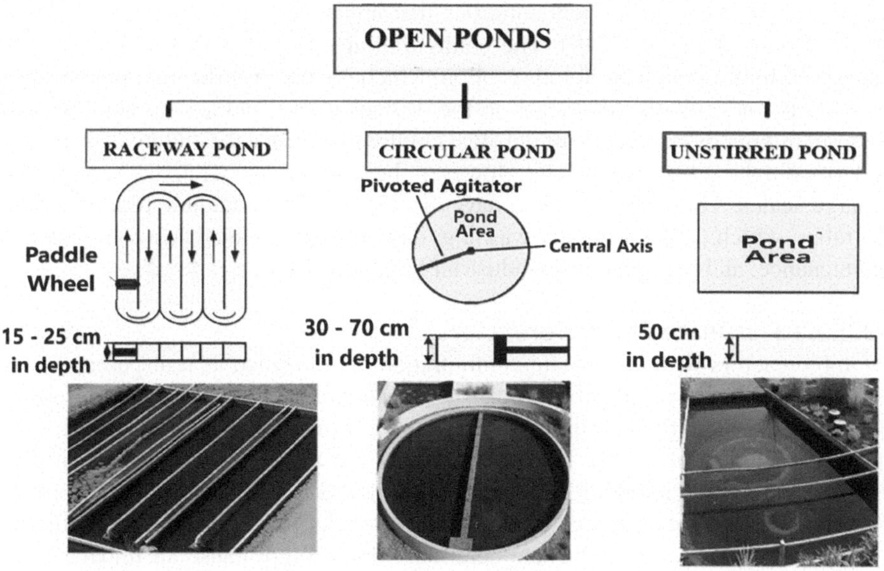

Fig. 13.2 Different types of open pond system

beneficial in certain settings, particularly in areas with high energy costs or limited infrastructure, and can be used to grow resilient microalgae strains that withstand fluctuating conditions (Ledda 2015) (Fig. 13.2).

13.2.3 Photobioreactor Systems

Photobioreactors constitute complete systems specifically engineered for the proper integration of microalgae systems in various wastewater treatment systems. They allow, during the process, the control of parameters such as light, temperature, and CO_2, ensuring effective nutrient removal and biomass production. Because they are closed systems, they do not allow contamination and therefore have superior productivity compared to open ponds (Kumar et al. 2022).

Some of the common designs include tubular, flat plate, column, and bag reactors. Tubular systems are efficient at capturing light, while flat plate designs are compact and space-saving. Column reactors are also space-efficient, whereas bag systems are less expensive to implement for low-scale applications (Smith et al. 2021). However, photobioreactors are more expensive and consume more energy due to their complexity and efficiency, making them suitable for use in areas that require a high degree of precision (García et al. 2020).

13.2.3.1 Tubular Photobioreactors

Tubular photobioreactors are composed of several long transparent tubes that are either arranged in loops, coils, or straight sections. These tubes make it possible for microalgae to thrive in nutrient solutions that are in contact with natural light or artificial ones (Li et al. 2019). The configuration guarantees that a considerable amount of light is available for absorption, which results in sufficient photosynthesis activity and growth. Analogous to the devices above, pumps are employed to perform mixing by circulation of the algal culture to enhance the culture distribution and to avoid the settlement of cells. However, the systems are effective when applied at large scales. They can, however, have drawbacks. These include fouling inside the tubes, which calls for regular washing, increased operational cost as a result of maintenance, and pumping cost (Industrial Plankton 2020).

13.2.3.2 Flat-Plate Photobioreactors

Photobioreactors with a flat-plate configuration are designed as transparent sheets or plates on which microalgal culture grows in a thin layer. This results in uniformly distributed incident light on the culture, thus increasing the efficiency of photosynthesis (Torzillo et al. 2020). Flat-plate systems are also integrated in a relatively compact packaging, allowing use in places where space is limited, like urban areas, combined sewage, or small-scale treatment plants (Hernandez et al. 2021). Still, the systems are faced with inadequate aeration, which might cause an imbalance in nutrient utilization and overheating of the system, particularly in an external environment. Despite these restrictions, they have a high yield and are quite uncomplicated to develop for intermediate-scale production (Rosenberg et al. 2019).

13.2.3.3 Column Photobioreactors

Column photobioreactors are systems with a cylindrical shape that are upright and employ airlift circulation means to mix the algal culture as well as supply carbon dioxide for the process of photosynthesis (Järvinen et al. 2021). Due to their vertical construction features, they occupy relatively less space footprint; therefore, they can be installed in regions with restricted spatial availability. Columns also achieve good gas exchange while ensuring the culture in the column is evenly mixed, which is favorable for algal growth. Because of being enclosed, they are very self-contaminated. However, column PBRs are less applicable in the case of large-scale projects for wastewater treatment, and one of the drawbacks is that they cost more in terms of fixed capital investment than some other designs (Choi et al. 2021).

13.2.3.4 Bag Photobioreactors

Bag photobioreactors are semi-permeable transparent enclosures residing on top of algal biomass, nutrients, and water. They are mostly laid horizontally or hung up to obtain more light. Bag systems are cheap, simple to assemble, and have a portable nature, making them suitable for use in pilot experiments or small-scale applications (Torzillo et al. 2020). They are also not complicated to operate. The disadvantage, however, can be their strength in that the plastic does not last forever (Hernandez et al. 2021). The design may also limit how far light can go through the synthetic materials, reducing how effective the photosynthetic activity can be in comparison to other rigid systems (Rosenberg et al. 2021). But given the limitations, bag PBRs seem to be a reasonable choice for applications and experimental designs that are low-budget (Torzillo et al. 2020) (Fig. 13.3).

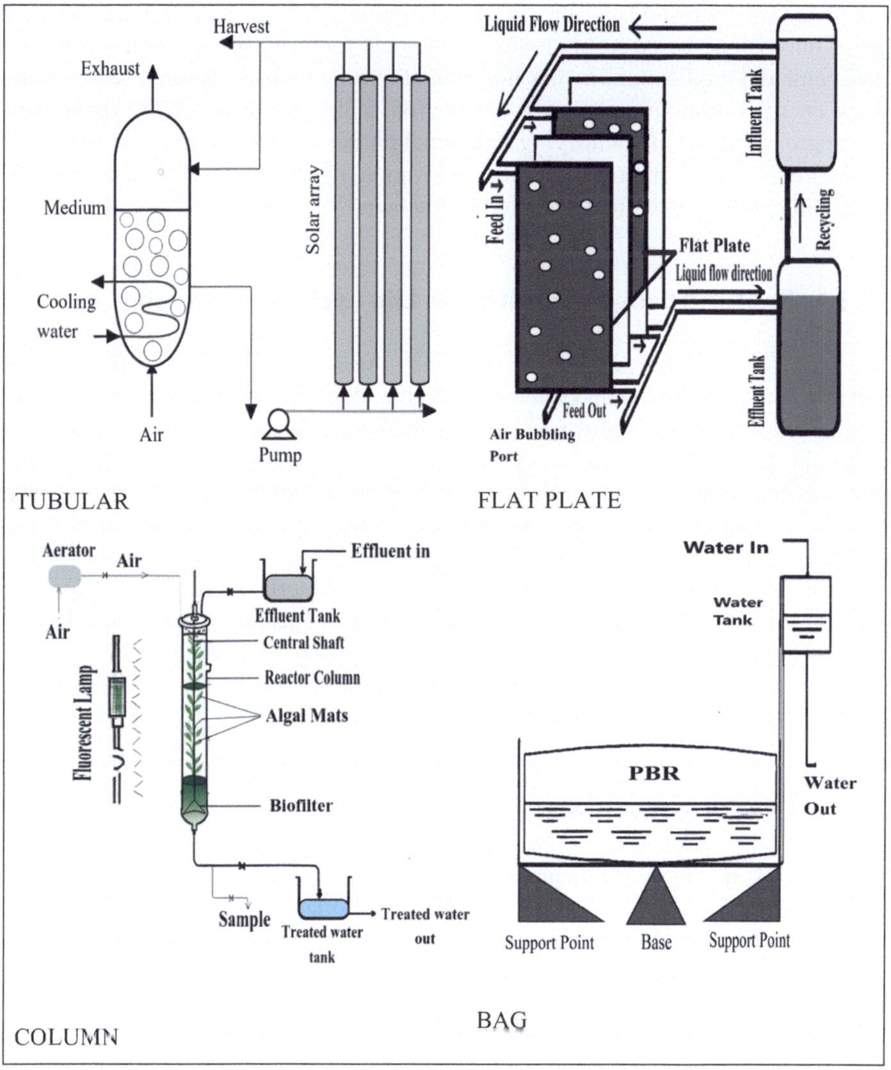

Fig. 13.3 Different types of photobioreactor-based system

13.3 Life Cycle Assessment (LCA)

The life cycle assessment (LCA) of microalgae-based systems illustrates a thorough and extensive assessment of their environmental sustainability, allowing for the identification of areas for improvement and directing the development of more efficient and environmentally friendly processes. This evaluation approach measures the system's inputs (such as energy, water, nutrients, and CO_2) and outputs (such as treated water, algal biomass, and emissions) at each stage of its lifespan, from culture to processing and end-product consumption. LCA ensures a full understanding of microalgae-based technologies' ecological footprint by examining potential environmental implications across the whole lifetime, such as greenhouse gas emissions, energy consumption, resource depletion, and water usage. This comprehensive approach assists in identifying individual phases or processes where efficiency can be improved, whether by technical innovation, operational parameter optimization, or integration with other sustainable practices (Subashchandrabose et al. 2018).

13.3.1 Goal and Scope Definition in Life Cycle Assessment

The goal outlines the purpose of the study, such as evaluating environmental impacts, comparing different systems, or guiding decision-making for sustainability (Heijungs et al. 2021). It also specifies the intended audience, whether it's internal stakeholders, regulators, or consumers (Guinée et al. 2019). The scope defines the details of the study. It includes determining the functional unit, which serves as the basis for comparison (e.g., "per kilogram of product" or "per cubic meter of water treated") (Huppes and Ishikawa 2020). The system boundaries specify which stages of the product's life cycle are included, such as raw material extraction, production, use, and disposal (Jolliet et al. 2018). The scope also clarifies assumptions, limitations, and data availability, ensuring transparency (Kunsch et al. 2021). By setting clear goals and scope, this phase ensures that the LCA remains focused, structured, and relevant throughout the assessment process (Baumann and Tillman 2017).

13.3.2 Life Cycle Inventory

The life cycle inventory (LCI) phase of a life cycle assessment (LCA) is a vital and data-intensive stage that entails comprehensively collecting and quantifying inputs and outputs for a specific system over its entire existence. In the context of microalgae-based systems for wastewater treatment and resource recovery, the LCI phase rigorously documents all aspects of resource use and emissions, laying the groundwork for further impact assessment and interpretation (Arashiro et al. 2022; Josa et al. 2022). During the LCI phase, the system being studied is broken down into its constituent processes, and extensive data is collected for each one. This comprises water, nutrients (nitrogen and phosphorus), CO_2, energy (electricity and heat), and any chemicals required in microalgae cultivation, harvesting, and processing. Outputs are equally detailed, including produced biomass, treated water, air emissions (carbon dioxide, methane, and nitrous oxide), water emissions (residual nutrients and pollutants), and solid trash. For microalgae growth in open pond systems, the LCI would include information on pond construction and maintenance, energy requirements for mixing and aeration, water consumption, and fertilizer inputs (Fagerlund et al. 2022). Centrifugation, flocculation, filtration, and drying are evaluated for resource utilization and efficiency. The inventory would include the electricity and chemical inputs needed, as well as any water recycled or lost during the processes (Fagerlund et al. 2022).

13.3.3 Life Cycle Impact Assessment

The life cycle impact assessment (LCIA) phase is a very significant part of the life cycle assessment process, as it translates or "converts" the amount of data that has been collected for the life cycle inventory (LCI) into measures of environmental damage. This phase involves a systematic determination of the likely environmental consequences of the system inputs and its outputs. LCIA makes evident the impact of microalgae-based wastewater treatment and resource recovery on various environmental dimensions and makes it easy to steer change towards more eco-friendly practices (Arashiro et al. 2022). The first step in LCIA is classification, where relevant inventory data are grouped according to appropriate impact categories. Potential important categories for microalgae-based systems would include (a) Global warming potential (caused by GHG emissions), (b) eutrophication potential (caused by nutrient releases), (c) acidification potential, and (d) ozone depletion potential, and resource depletion (Arashiro et al. 2022). All of the input or output flow specified in the LCI, particularly CO_2 emission, nutrient discharge, energy used, and water consumed, are related to one effect or two effect categories. The next one would be quantitative assessment the amount of loss or change that each of these flows contributes to various categories. For instance, greenhouse gas emissions (CO_2, CH_4, and N_2O) are converted into CO_2-equivalents to assess their impact on global warming potential. Nutrient discharges, especially nitrogen and phosphorus, are analyzed for their role in promoting eutrophication in aquatic ecosystems, which can lead to harmful algal blooms and oxygen depletion (Arashiro et al. 2022).

13.3.4 Interpretation of LCA

The last stage of life cycle assessment (LCA) is what is called the interpretation phase. In this stage, information from the previous phases: LCI and LCIA are analyzed, synthesized, and transformed into practical insights. This phase is vital in order to draw meaningful findings, make well-informed decisions, and identify areas for improving environmental performance (Dasan et al. 2019; Pérez-López et al. 2017). One of the main aims of the interpretation phase is recognizing the relevance or significance of results obtained with respect to LCIA findings by investigating which life cycle stages or processes or inputs contribute most significantly within overall environmental implications found during LCIA. Discovering these hotspots can enable researchers to focus on specific areas for improvement like enhancing energy utilization efficiencies within photobioreactors as well as bettering nutrient recovery efficiency (Onorato and Rösch 2020). This includes ensuring that the functional unit, system boundaries, and impact assessment procedures are consistently used. Sensitivity analysis has the potential of aiding in the determination of key variables and the evaluation of results under different conditions (Branco-Vieira et al. 2020). The algae systems' indirect uncertainties may arise from the changes in the algal growth medium, the surrounding atmosphere, and the composition of the substrates used. Considering these uncertainties from the onset in LCA allows for more elucidations and assurances in interpretations, such that the people involved are able to make considerate choices (Guo and Murphy 2012). For instance, if a recommendation may bear initiatives directed toward enhanced energy utilization such as efficient energy utilization in cultivation methods, advanced harvesting systems incorporate microalgae systems with other renewable sources of energies, (Klein et al. 2017). LCA's interpretation phase is the final stage that integrates results of the LCI and LCIA stages with the aim of confirming that all interpretations are plausible, clear, and useful.

13.3.5 ISO Standards in Life Cycle Assessment

ISO standards (International Organization for Standardization) ensure consistency and reliability in life cycle assessment (LCA) studies, providing a standardized framework for assessing the environmental impacts of products and processes. They guide data collection, impact assessment, and interpretation, making the results credible and comparable (Table 13.2).

13.3.6 Functional Units and System Boundaries

To ensure that LCA gives a complete and correct evaluation of microalgae-based systems' environmental performance, functional units and system boundaries are chosen with care (Tsoy N et al. 2020) (Fig. 13.4). These processes call for an

Table 13.2 ISO Standards for LCA (International Organization for Standardization [ISO])

ISO standard	Description	Key focus areas
ISO 14040	Principles and framework for conducting LCA	LCA phases: goal and scope, inventory, impact assessment, and interpretation
ISO 14044	Detailed guidelines for LCA implementation	Inventory analysis, impact assessment, data quality, and interpretation
ISO 14025	Guidelines for environmental product declarations (EPDs)	Type III declarations, transparency, and product life cycle information
ISO 14046	Water use in LCA (water footprint)	Water assessment, impact quantification, and reporting
ISO 14067	Carbon footprint (greenhouse gases)	Greenhouse gas emissions quantification and communication
ISO 14048	Data documentation and reporting for LCA	Standardized data documentation, traceability, and transparency
ISO 14064	Greenhouse gas emissions and removals	Emissions quantification, monitoring, and reporting

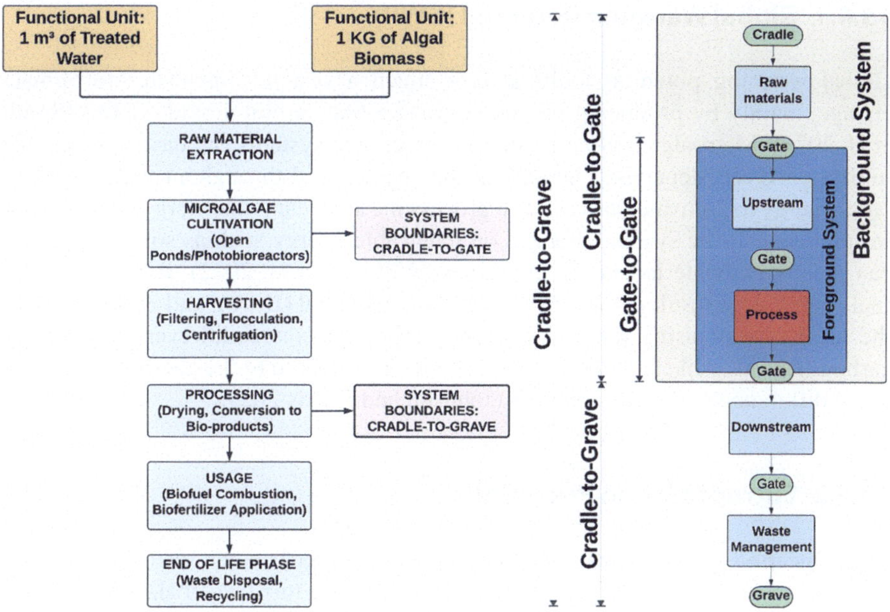

Fig. 13.4 Selection of functional unit and system boundaries

in-depth understanding of the system's operational dynamics as well as of the purpose for which the assessment is carried out (Jolliet et al. 2018). In this way, through careful specification of these parameters, the full complexity and heterogeneity of microalgae-based systems can be captured, hence making better choices possible, aiding in the sustainable development of wastewater treatment and resource recovery (Heijungs et al. 2021). This approach enhances identification of environmental hotspots, optimization of processes, and eco-efficiency technology development as pathways to our more sustainable future (Baumann and Tillman 2017).

13.4 LCA of Microalgae-Based Systems for Wastewater Treatment

LCA (life cycle assessment) as a tool can assist in ascertaining the potential environmental effects associated with any products over their entire life cycle, including extraction, manufacturing, trading, usage, and disposal (Huppes and Ishikawa 2020). In terms of wastewater treatment systems where microalgae are used, LCA allows one to evaluate the level of sustainability of these systems and their benefits and drawbacks (Kunsch et al. 2021). These systems are increasingly recognized as a promising alternative to traditional methods because they not only treat wastewater but also have the potential to recover valuable resources (Tsoy N et al. 2020). To understand their true environmental impact, LCA focuses on several key areas that are vital for assessing sustainability (Jolliet et al. 2018.

13.4.1 Global Warming Potential (GWP)

Global warming potential looks at how much a system contributes to climate change, mainly by producing greenhouse gases like carbon dioxide (CO_2) (Smith et al. 2021). Microalgae-based wastewater treatment systems can have a noticeable impact on GWP, depending largely on the energy used throughout the process. It takes energy to cultivate, harvest, and process the algae, and the energy source is not insignificant. If the system operates on renewable energy such as solar or wind, it can greatly alleviate its carbon emissions (Williams et al. 2022). But using fossil fuel energy does result in more GWP warming potential (Brown and Zhao 2019). In the ideal case of using renewable energy, these systems could even be rendered carbon-neutral or likely carbon-negative, which compared to the alternatives makes a real difference to the environment (Johnson and Harris 2021).

13.4.2 Eutrophication Potential

Eutrophication is the process in which bodies of water become overly dense with nutrients such as nitrogen and phosphorus, which can in turn trigger HABS (Harmful Algal Blooms), oxygen depletion, and aquaculture damage (Clark et al. 2020). Microalgae systems have a clear advantage here because they, on the contrary, tend to absorb unused nutrients during the process of treating wastewater (Li and Zhang 2021). Therefore, by using these nutrients in their synthesis process, microalgae systems counteract eutrophication and enhance the condition of water bodies (Singh and Verma 2022). The efficiency of eutrophication in the system's design is a determining factor when looking for its efficiency in nutrient removal (Kumar and Sharma 2020). Optimally, microalgae systems should be environmentally friendly and acceptable methods of treatment while outdoing traditional means of treatment in decreasing discharges of nutrients.

13.4.3 Acidification Potential

Acidification potential refers to the likelihood of the development of acid rain, which negatively affects the soil and water quality. This is mainly caused by the emissions from burning fossil fuels. While microalgae systems themselves do not produce compounds that are acidifying, the energy required to operate the system may influence its environmental performance (Jones et al. 2021). This is because when the power used in fossil fuels is done, the capability of acidification potential is increased (Brown and Zhao 2019). However, when fossil fuels are avoided and renewable sources are used, the potential is greatly reduced and makes microalgae systems more environmentally friendly (Williams and Harris 2022). This category is not directly related to algae impacts but rather to the methods employed in algae cultivation, especially where the type of energy used is substituted in order to lessen acidifying emissions (Johnson and Harris 2021).

13.4.4 Energy Consumption and Water Usage

Energy consumption and water usage are two major factors that can either make or break the sustainability of microalgae-based wastewater treatment systems. Cultivating and processing microalgae require energy for lighting, aeration, and mechanical stirring to keep the algae growing efficiently (Singh and Kumar 2022). Furthermore, deep water treatment and drying the harvested algae require a lot of energy, which can be a major drawback (Chen and Zhao 2021). However, planting microalgae offers immense offshore potential environmental benefits because, for example, in places where freshwater is scarce: using wastewater that aids in the deepwater treatment of microalgae lowers the supplies even after using a lot of energy during the treatment processes, providing systems with great potential environmental benefits (Davis and Zhang 2020). This massive use of water systems in this case would significantly lower the pressure on water resources availability (Lee and Kim 2021). The emphasis of these issues is placing a great deal of attention patiently on renewable energy sources that help mitigate some of the adverse impacts (Lopez et al. 2022). Furthermore, closed systems like photobioreactors can recycle water within the system, minimizing losses compared to open pond systems.

13.5 Comparative Studies of Microalgae Systems Versus Conventional Treatment (Table 13.3)

13.5.1 Case Studies

These case studies illustrate the application of microalgae-based systems in the treatment of agricultural runoff, municipal, and industrial wastewaters. The majority of the systems were found to be beneficial for the environment through decreasing nutrient loading and greenhouse gas emissions; however, some challenges to these systems arose from consumption of water, upkeep of the system, and high capital costs. The incorporation of solar energy and other renewables could help

Table 13.3 Microalgae systems versus conventional treatment (Arashiro et al. 2020)

	Microalgae systems	Conventional treatment
Energy use	Moderate; renewable energy compatible	High; energy-intensive processes
Nutrient removal	High; supports resource recovery	Effective; relies on chemical additives
Emissions	Low; carbon sequestration potential	High; significant greenhouse gases
Resource Recovery	Produces biofuels, fertilizers, animal feed	Minimal resource recovery
Costs	Higher initial; long-term savings possible	Lower upfront; higher ongoing costs
Applications	Versatile for various wastewater types	Primarily municipal wastewater

solve some of these issues and reduce the energy footprints of these systems allowing for their long-term sustainability.

13.5.1.1 Agricultural Wastewater Treatment in Spain

Chlorella and *Scenedesmus* microalgae were utilized by Spanish researchers for the open pond systems approach to treat agricultural runoff that contains high nutrients. The research emphasized the potential of these microalgae in the removal of 60% nitrogen and 80% phosphorus from wastewater. Life cycle analysis (LCA) showed that this technique significantly reduced the ecological impact as opposed to conventional methods, especially regarding the emissions of greenhouse gases. Besides, the harvested algal biomass was turned into biofertilizers, increasing the value of the system (Lopez et al. 2020). On the other hand, evaporation of water in the open ponds was a problem, and this required more water to be used (Martinez et al. 2021). The research recommended the use of solar energy as a way of further reducing the system's energy requirements (Lopez et al. 2020).

13.5.1.2 Municipal Wastewater Treatment in California

A Californian municipality tested photobioreactors (PBRs) at a WWTP (wastewater treatment plant) aiming at the treatment of secondary effluent using *Spirulina*. This closed system solved the problem of contamination and water loss, enabling the achievement of 95% nitrogen removal efficiency and an 88% phosphorus removal (Wang et al. 2021). In comparison to the previous secondary treatment of this facility using the activated sludge processes, the LCA reduced emissions of greenhouse gases (Jones et al. 2020). Also, this algal biomass was converted to biogas and this contributed to about 30% of the energy requirements of the plant (Davis et al. 2020). Despite its environmental benefits, the system faced challenges such as high initial setup costs and the need for frequent maintenance to manage biofilm buildup. Nonetheless, the study demonstrated the suitability of PBR systems in urban wastewater treatment settings (Chang et al. 2022).

13.5.1.3 Hybrid Microalgae-Wetland System in India

In India, a hybrid system combining microalgae cultivation with a constructed wetland was developed to treat industrial wastewater, and the system utilized *Chlorella vulgaris* and *Scenedesmus* to remove pollutants, while native wetland plants provided additional filtration. This innovative approach achieved a 70% reduction in chemical oxygen demand (COD) and completely removed heavy metals such as lead and cadmium (Rao and Sharma 2020). The LCA revealed moderate energy consumption but significant reductions in soil and water contamination compared to standalone chemical treatments (Patel et al. 2021). Seasonal variations impacted the system's efficiency, requiring operational adjustments, and also this study highlighted the potential of integrating microalgae systems with natural treatment methods in resource-constrained settings (Singh and Verma 2022).

13.5.2 Data Challenges and Limitations

A life cycle assessment of microalgae-based systems for wastewater treatment suffers from several limitations and challenges that affect both the precision and relevance of the results (Gupta and Yadav 2021). One critical problem is the differences witnessed in LCA results which are often caused by variation in system size, location, and more importantly, the type of microalgae used. For example, a small facility in a temperate climate may have completely different environmental and economic consequences when compared to a large facility situated in a sunny area (Kumar et al. 2020). In the same way, the chosen species of microalgae would have different growth rates, nutrient absorption levels, and resource needs which add to the variation again (Singh and Kapoor 2021).

Another significant limitation is the presence of data gaps and uncertainties in the modelling process. Many LCA studies rely on assumptions or generalized data when site-specific or real-world measurements are unavailable. This can lead to inaccurate results or misinterpretations of a system's sustainability (Joshi and Malik 2022). Moreover, the microalgae systems that are used usually involve many different processes such as biomass production, nutrient removal, and even resource recovery, which also make it hard to collect data that is consistent and of high quality across the different stages (Chaudhary et al. 2021). These challenges would be better addressed through more rigorous data collection techniques, better-formulated protocols, and more comprehensive sensitivity analyses that would ensure consistency and ease of comparison of the LCA results (Sharma and Kumar 2021).

13.6 Economic Considerations

Economic factors are highly significant in the assessment of the effectiveness and commercial prospects of industrial microalgae farming (Fig. 13.5). Primarily, setting up the capital of a micro algae production facility which involves photobioreactors, open ponds, and other erection facilities can be very costly. These expected profits, which explain such high capital outlays, are solely dependent on the capacity to manufacture and market such microalgae products (Chisti 2016). Another essential thing to consider when assessing economic practicability is the operational costs. There are two types of activities which use much energy in this process; harvesting and dewatering (Molina Grima et al. 2015). Deploying methods that rely on renewable sources like solar or wind power can minimize these bills. Nevertheless, such strategies normally involve more investments and need technological breakthroughs to work out effectively too (Slade and Bauen 2013). Also, throughout algae farming's production phase, there must be maximum savings on energy input so as to lower operating expenses and also better the general economic profile. The cost for water and fertilizers apart from just energy count much too in terms of an economic feasibility perspective (García et al. 2017).

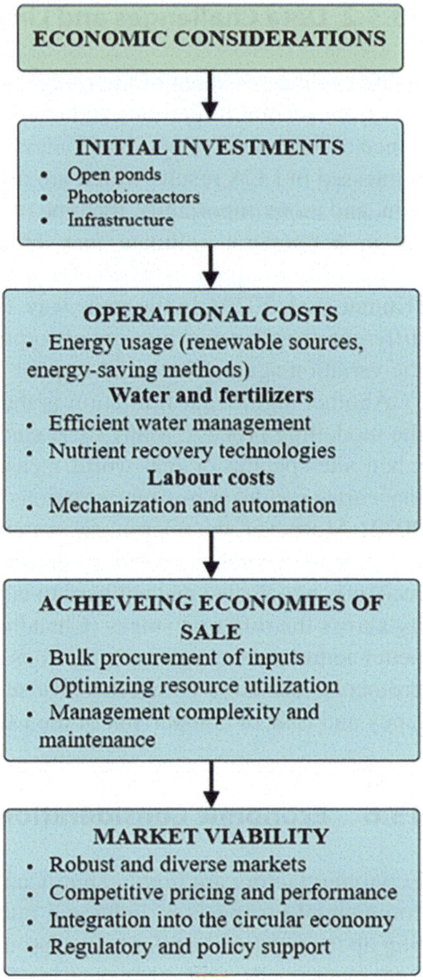

Fig. 13.5 Economic considerations

13.6.1 Integration with Existing Infrastructure

Integration with current infrastructure is critical for the effective implementation and extension of microalgae-based systems, which provide a variety of benefits across multiple industries. By seamlessly integrating microalgae production and processing facilities into existing industrial, agricultural, or municipal infrastructure, these systems can capitalize on synergies, optimize resource utilization, and improve overall sustainability (Brennan and Owende 2016). This allows to somewhat reduce the pressure on development of new areas and new investment in infrastructure, thus having positive effect on capital investment and ecological factors (Christenson and Sims 2016).

In the case of wastewater treatment facilities, microalgae assume an integrating function of nutrients removal, organic matter degradation, and even the pathogen control which actualize an essential aspect of integration (Arashiro et al. 2018). Integrating microalgae growing with wastewater treatment processes offers a win-win option, allowing communities to accomplish effective and ecologically friendly water purification while also producing valuable biomass for bioenergy or biofertilizer development (Gupta et al. 2016). The symbiotic relationship increases the sustainability of wastewater treatment activities and as such helps to advance a circular economy by turning waste into valuable resource. Bringing together algae industrial processes and microalgal productivity together is a chance to enhance efficiency and decrease pollution (Rawat et al. 2016). Further, microalgae growth in aquaculture systems creates opportunities to integrate the harvested nutrient-rich effluent from fish farms to algae growth, thereby creating a resource-efficient and reduced waste producing ecosystem.

To make it more sustainable for agriculture, farming involves the use of microalgal systems that enhance soil fertility as well as crop nutrition and pest control (Garcia et al. 2018). The application of biofertilizers obtained from microalgae can help in enhancing soil health and crop productivity, thereby reducing over reliance on chemical fertilizers for sustainable agriculture (González-Delgado and Kafarov 2018). Besides, a variety of eco-friendly microalgae-based biopesticides can be used instead of chemical pesticides that are environmentally friendly in integrated pest management (IPM) since it reduces unwanted effects associated with pesticide application in the environment (Hamed et al. 2019).

13.7 Conclusion

Microalgae production and wastewater treatment is a rapidly changing trend, fueled by the need for eco-friendly and effective solutions to environmental problems as well as resource constraints. One of the most promising trends is the development of genetically modified strains of microalgae which have high productivity, uptake efficiency for nutrients, and are resistant to environmental stresses. These genetically modified organisms (GMOs) can be made suitable for certain wastewater compositions, make photosynthesis more efficient, and result in increased biomass yield and useful compounds. Improvements in CRISPR technology and other genome-editing methods have hastened the creation of these biotech algae that outperform traditional farming techniques by far. Photobioreactor (PBR) technology is another field that is seeing substantial advancement. Recent PBR designs have been aimed at improving light penetration and dispersion about higher photosynthetic performance of microalgae due to its importance. The new designs use light-emitting diodes (LEDs) and advanced materials that improve both absorption and scattering of light, hence making sure all cells within the reactor receive uniform illumination. Again, the combination of these real-time monitoring and automated control systems in PBRs does enable the manipulation of operating costs by tweaking some parameters such as pH and temperature or nutrient levels to enhance growth conditions. There is also growing enthusiasm for hybrid combination systems which combine algae and bacteria. Seemingly, these systems take the interactions between the microalgae and bacteria into consideration with the aim of improving the depth of treatment of wastewater. For example, some forms of bacteria convert complex organic compounds into simpler ones which algae can easily utilize. Such a symbiotic relationship promotes not only nutrient removal and the breakdown of organic matter but also biomass generation, which can be produced for many purposes including bioenergy and bioproducts. By combining both microorganisms' strengths, these hybrid systems constitute a great all-inclusive approach to wastewater treatment. Harvesting and dewatering technology innovations are essential for scaling up microalgae cultivation and ensuring its economic viability. Conventional methods including centrifugation are energy-intensive as well as expensive, thereby prompting scientists to seek alternative techniques like electrocoagulation, magnetic separation, or application of bio-flocculants for harvesting microalgal biomass. These energy-saving methods are intended to maintain high recovery efficiency while minimizing power consumption and operational costs. For instance, electrocoagulation causes algae cells to form clumps that can be easily removed from the water unlike centrifugation techniques, hence requiring less energy. The combination of Internet of Things (IoT) and artificial intelligence (AI) technologies is transforming the management and optimization of microalgae growing systems. IoT devices may collect massive volumes of data on a variety of operational aspects, which AI algorithms can use to anticipate optimal circumstances, detect anomalies, and recommend corrective actions. The data-oriented approach has improved the efficiency and reliability of microalgae culture systems, making them more adaptable for large-scale operations. Also, the biorefinery concept is a hot topic because it seeks to fully utilize algal biomass in the production of biofuels, bioplastics, drugs, and nutraceuticals, among others.

References

Abinandan S, Shanthakumar S (2015) Challenges and opportunities in application of microalgae (Chlorophyta) for wastewater treatment: a review. Renew Sust Energ Rev 52:123–132

Arashiro LT, Montero N, Ferrer I, Acién FG, Gómez C, & Garfí M (2018) Life cycle assessment of high rate algal ponds for wastewater treatment and resource recovery. Sci Total Environ, 622-623, 1118–1130. doi:https://doi.org/10.1016/j.scitotenv.2017.12.05110.1016/j.scitotenv.2017.12.051

Arashiro LT, Montero N, Ferrer I, Acien FG, Gomez C, Garfi M (2020) Life cycle assessment of high rate algal ponds for wastewater treatment and resource recovery. Sci Total Environ 740:140122. https://doi.org/10.1016/j.scitotenv.2020.140122

Arashiro L, Josa I, Ferrer Martí I, van Hulle S, Rousseau D, Garfi M (2022) Life cycle assessment of microalgae systems for wastewater treatment and bioproducts recovery: natural pigments, biofertilizer and biogas. Sci Total Environ 847:157615

Arun S, Kumar R, Sharma N (2020) Variations of pH during Spirulina platensis growth. ResearchGate

Baumann H, Tillman AM (2017) The hitchhiker's guide to LCA: an orientation in life cycle assessment methodology and applications. Studentlitteratur

Branco-Vieira M, Rangel CM, Fortunato E, Silva A (2020) Environmental impacts of a microalgae-based biodiesel production in Chile. Renew Energy 146:1109–1119

Brennan L, Owende P (2016) Biofuels from microalgae—a review of technologies for production, processing, and extractions of biofuels and co-products. Renew Sust Energ Rev 14(2):557–577

Brown T, Zhao X (2019) The role of energy sources in microalgae-based wastewater treatment systems. Environ Sci Technol 43(8):3245–3256. https://doi.org/10.1021/es3031801

Cai T, Park SY, Li Y (2017) Nutrient recovery from wastewater streams by microalgae: status and prospects. Renew Sust Energ Rev 19:360–369

Chang F, White T, Lee S (2022) Evaluating photobioreactor efficiency in urban wastewater treatment. J Environ Manag 47(3):221–229. https://doi.org/10.1016/j.jenvman.2022.112070

Chaudhary N, Jain P, Shah S (2021) Life cycle assessment of microalgae-based wastewater treatment. Environ Sci Technol 55(10):7245–7253

Chen Y, Zhao J (2021) Energy requirements for microalgae cultivation in wastewater treatment. Renew Energy 145:430–441. https://doi.org/10.1016/j.renene.2019.06.042

Chinnasamy S, Bhatnagar A, Claxton R, Das KC (2015) Biomass and bioenergy production potential of microalgae Scenedesmus sp. and Chlorella sp. recycling nutrients in municipal wastewater. Bioresour Technol 101(6):112–119. https://doi.org/10.1016/j.biortech.2015.09.028

Chisti Y (2016) Microalgae-based carbon capture and utilization systems: a critical review. J Appl Phycol 28(3):1469–1486

Choi J, Park K, Lee Y (2021) Design and optimization of column photobioreactors for microalgae cultivation. J Ind Microbiol 58(3):234–245. https://doi.org/10.1016/j.jid.2021.04.022

Christenson L, Sims R (2016) Production and harvesting of microalgae for wastewater treatment, biofuels, and bioproducts. Biotechnol Adv 29(6):686–702

Clark R, Adams J, Carter M (2020) Eutrophication in aquatic ecosystems: an overview of global impacts. Mar Pollut Bull 112(4):726–732. https://doi.org/10.1016/j.marpolbul.2015.02.042

Dasan Y, Lam MK, Yusup S, Lim JW (2019) Life cycle analysis on fossil energy ratio of algal biodiesel: effects of nitrogen deficiency and oil extraction technology. Sci World J 2015:9

Davis T, Zhang S (2020) The environmental benefits of using wastewater for microalgae cultivation. Environ Pollut 267:115–125. https://doi.org/10.1016/j.envpol.2020.115122

Davis LW, Martinez S, & Taboada B (2020) How effective is energy-efficient housing? Evidence from a field trial in Mexico. J Dev Econ, 143, 102390. https://doi.org/10.1016/j.jdeveco.2019.102390

Fagerlund M, Abrahamsson K, Lagerkvist A (2022) Comparative life cycle assessment of wastewater treatment processes: a Swedish case study. J Clean Prod 330:129855

Fernandez M, Sanchez P, Gonzalez J (2021) Perspective of Spirulina culture with wastewater into a sustainable circular bioeconomy. ResearchGate

García R, Aguilar J, Toro M, Pinto A, & Rodríguez P (2020) A systematic literature review on the use of machine learning in precision livestock farming. Computers and Electronics in Agriculture, 179, 105826. https://doi.org/10.1016/j.compag.2020.105826

García JL, de Vicente M, Galán B, Santos VE (2017) Wastewater treatment using microalgae: how realistic a contribution might it be to significant urban wastewater treatment. Appl Microbiol Biotechnol 101(6):2653–2660

Goh BHH, Yusoff MM, Loh SP (2022) Microalgae Dunaliella as biofuel feedstock and β-carotene production. E3S Web Conf 130:01054. https://doi.org/10.1051/e3sconf/202213001054

González-Delgado J, Kafarov V (2018) Biofertilizers from microalgae for sustainable agriculture: a review of current research and practices. Environ Technol Innov 11:130–145. https://doi.org/10.1016/j.eti.2017.10.005

Guinée JB, Heijungs R, Huppes G (2019) Life cycle assessment: past, present, and future. Environ Sci Technol 53(1):1–8. https://doi.org/10.1021/acs.est.8b06623

Guo M, Murphy RJ (2012) LCA data quality: sensitivity and uncertainty analysis. Sci Total Environ 435:230–243

Gupta PL, Lee SM, Choi HJ (2016) Integration of microalgae cultivation system for wastewater treatment and biofuel production. Renew Sust Energ Rev 58:600–607

Gupta S, Nayak A, Roy C, & Yadav AK (2021) An algal assisted constructed wetland-microbial fuel cell integrated with sand filter for efficient wastewater treatment and electricity production. Chemosphere, 263, 128132. https://doi.org/10.1016/j.chemosphere.2020.128132

Gupta R, Mishra N, Singh G, Mishra S, Lodhiyal N (2024) Microalgae cultivation and value-based products from wastewater: insights and applications. Blue Biotechnol 1:Article number: 20. https://doi.org/10.1186/s44315-024-00019-1

Hamed SM, Abd El-Rhman AA, Abdel-Raouf N, Ibraheem IB (2019) Role of marine macroalgae in plant protection & improvement for sustainable agriculture technology. Bioresour Technol 291:121876

Heijungs R, Huppes G, Guinée J (2021) Environmental impact assessment and sustainability. Springer

Hernandez P, Li L, Rivas D (2021) Compact photobioreactor configurations for urban wastewater treatment. Algal Res 54(5):134–144. https://doi.org/10.1016/j.algal.2021.102039

Huo S, Wang Z, Shen Y (2018) Lipid production from Nannochloropsis under nutrient-deprived conditions. J Appl Phycol 30(3):1023–1030. https://doi.org/10.1007/s10811-018-1411-5

Huppes G, Ishikawa M (2020) A framework for life cycle assessment in environmental management. J Clean Prod 276(7):124123. https://doi.org/10.1016/j.jclepro.2020.124123

Industrial Plankton. (2020, December 10). Photobioreactor advantages and disadvantages. https://industrialplankton.com/2020/12/10/photobioreactor-advantages/

International Organization for Standardization. ISO standards for life cycle assessment (LCA). https://www.iso.org/standards.html

Järvinen M, Lehtinen A, Hänninen L (2021) Column photobioreactors for microalgae cultivation in wastewater treatment. Biochem Eng J 67(4):529–538. https://doi.org/10.1016/j.bej.2020.107013

Johnson M, Harris A (2021) Carbon neutrality in microalgae wastewater treatment systems: challenges and potential. Glob Environ Chang 64(3):115–125. https://doi.org/10.1016/j.gloenvcha.2020.102123

Jolliet O, Margni M, Motoshita M (2018) Life cycle assessment: quantifying the environmental impacts of products and services. Springer

Jones R, Thomas M, Nguyen H (2020) Life cycle assessment of secondary wastewater treatment systems. Environ Sci Technol 54(1):87–95. https://doi.org/10.1021/es4046767

Jones T, Green R, White J (2021) Microalgae systems and energy consumption in wastewater treatment. J Environ Manag 61(5):879–887. https://doi.org/10.1016/j.jenvman.2021.109222

Josa I, Ferrer Martí I, van Hulle S, Rousseau D, Garfí M (2022) Comparative environmental performance of microalgae-based systems for wastewater treatment and resource recovery. Environ Sci Technol 56(22):16294–16303

Joshi D, Malik R (2022) A critical review of LCA in microalgae wastewater treatment. Bioresour Technol 347:126374

Khan A, Gupta M, Sharma T (2019) Heavy metal contaminants removal from wastewater by using Chlorella vulgaris: a review. ResearchGate

Klein BC, Bonomi A, Maciel Filho R (2017) Integration of microalgae production with industrial biofuel facilities: a critical review. Renew Sust Energ Rev 82:1376–1392

Kumar R, Sharma A (2020) Microalgae-based nutrient removal systems: a new frontier in sustainable wastewater treatment. Environ Pollut 267:115–125. https://doi.org/10.1016/j.envpol.2020.115122

Kumar A, Kumar S, Tiwari M (2020) Variability of LCA results in microalgae systems: a global comparison. Sustain Environ Res 30(5):203–215

Kumar R, Saini GK, & Jawed M (2022) Impact of emerging contaminants on biological wastewater treatment process. In Relationship Between Microbes and the Environment for Sustainable Ecosystem Services, Volume 2 (pp. 17–40). Elsevier. https://doi.org/10.1016/B978-0-323-89937-6.00014-0

Kunsch P, Piket M, Van der Goot A (2021) LCA and sustainability in practice. Sustain Rep 6(1):45–56. https://doi.org/10.3390/sus6100456

Ledda L (2015) Unstirred ponds for microalgae production in wastewater treatment and resource recovery. Environ Sci Technol 49(10):1234–1240. https://doi.org/10.1021/es506157g

Lee S, Kim H (2021) Energy and water usage in microalgae wastewater treatment systems. Water Res 213:118–125. https://doi.org/10.1016/j.watres.2022.118139

Lee J, Park Y, Choi H (2021) Valorization of wastewater resources into biofuel and value-added products through microalgae: a review. Front Energy Res 9
Li P, Zhang Z (2021) The role of microalgae in mitigating eutrophication in wastewater treatment. Algal Res 54:102156. https://doi.org/10.1016/j.algal.2020.102156
Li Y, Xu L, Wang D (2019) Nutrient recovery from wastewater using microalgae: potential applications and environmental implications. Sci Total Environ 695:133815. https://doi.org/10.1016/j.scitotenv.2019.133815
Lopez P, Wang L, Chen Y (2020) The role of solar energy in microalgae-based wastewater treatment. Renew Energy 157:205–213. https://doi.org/10.1016/j.renene.2020.03.017
Lopez A, Silva P, Chang F (2022) Sustainable water management through algae-based wastewater treatment systems. Environ Sci Pollut Res 29(4):1905–1915. https://doi.org/10.1007/s11356-021-14172-5
Manning SR et al (2020) Historical use of Spirulina. J Appl Phycol
Martinez R, Johnson K, Green M (2021) Water evaporation issues in open pond microalgae systems for wastewater treatment. Water Res 135:123–130. https://doi.org/10.1016/j.watres.2021.116475
Mata TM, Martins AA, Caetano NS (2010) Microalgae for biodiesel production and other applications: a review. Renew Sust Energ Rev 14(1):217–232
Mehariya S, Al-Najjar MAA, Al-Jabri H, Das P (2023) Utilization of microalgae for urban wastewater treatment and valorization of treated wastewater and biomass for biofertilizer applications. Sustain For 15(22):16019
Molina Grima E, Belarbi EH, Fernández FGA, Medina AR, Chisti Y (2015) Recovery of microalgal biomass and metabolites: process options and economics. Biotechnol Adv 20(7–8):491–515
Onorato C, Rösch C (2020) The EU-funded ABACUS project: detailed LCA on photobioreactors. Sustain For 12(6):2389
Patel R, Kumar A, Soni P (2021) Energy and environmental assessment of microalgae systems for wastewater treatment: a comparative LCA study. J Clean Prod 311:127412
Pérez-López P, López J, González A (2017) Microalgae-based wastewater treatment integrated with industrial CO_2 extraction: life cycle assessment and environmental impacts. J Appl Phycol 29(1):123–135. https://doi.org/10.1007/s10811-016-0987-1
Purba RAP et al (2022) Nutritional properties of microalgae. Food Science and Nutrition
Rao V, Sharma P (2020) Hybrid systems of microalgae cultivation and constructed wetlands for wastewater treatment. Environ Pollut 258:113601
Rawat I, Kumar RR, Mutanda T, Bux F (2016) Biodiesel from microalgae: a critical evaluation from laboratory to large scale production. Appl Energy 103(9):444–467. https://doi.org/10.1016/j.apenergy.2016.02.003
Rosenberg E, Romm M, Ben-Tal A (2019) Photobioreactor technologies for wastewater treatment: current challenges and future prospects. J Environ Eng 145(9):1109–1117. https://doi.org/10.1061/(ASCE)EE.1943-7870.0001492
Rosenberg Z, & Vayig Y (2021) The effect of friction on the shanks of rigid projectiles penetrating concrete targets. Int J Impact Eng, 156, 103926. https://doi.org/10.1016/j.ijimpeng.2021.103926
Sharma S, Kumar M (2021) Data gaps in LCA of microalgae systems: a critical assessment. Sci Total Environ 773:145490
Singh A, Kapoor A (2021) Growth characteristics and nutrient removal efficiency of microalgae. Algal Res 58:102370
Singh R, Kumar A (2022) Microalgae-based nutrient removal systems: a new frontier in sustainable wastewater treatment. Environ Pollut 267:115–125. https://doi.org/10.1016/j.envpol.2020.115122
Singh A, Verma S (2022) Microalgae for wastewater treatment: a sustainable approach. Environ Sustain 14(3):123–135
Slade R, Bauen A (2013) Micro-algae cultivation for biofuels: cost, energy balance, environmental impacts and future prospects. Biomass Bioenergy 53:29–38
Smith A, Brown C, Williams G (2021) Assessing the global warming potential of microalgae systems. Environ Sci Technol 55(3):612–623. https://doi.org/10.1021/acs.est.0c05023

Subashchandrabose SR, Ramakrishnan B, Megharaj M (2018) Environmental sustainability of microalgae-based wastewater treatment: an LCA perspective. Bioresour Technol 247:848–854
Sun X, Khan MN, Bai L (2018) A review of the nutrient removal capacity of microalgae in wastewater treatment. Environ Technol Rev 7(1):16–31
Torzillo G, Tredici M, Posten C (2020) Optimization of flat-plate photobioreactor designs for the large-scale cultivation of microalgae. Bioresour Technol 303(1):122–130. https://doi.org/10.1016/j.biortech.2020.122957
Tsoy N, Steubing B, van der Giesen C et al (2020) Upscaling methods used in ex ante life cycle assessment of emerging technologies: a review. Int J Life Cycle Assess 25, 1680–1692. https://doi.org/10.1007/s11367-020-01796-8
Wang Z, Wang Y, Hu Z (2019) Sustainable cultivation of Nannochloropsis for nutrient removal and biofuel production. Environ Sci Technol 53(1):326–336. https://doi.org/10.1021/acs.est.9b05716
Wang C, Zhang Q, Zhao Y (2020) Systematic sensitivity analysis in life cycle assessment. J Ind Ecol 24(1):114–125
Wang P, Zarei-Baygi A, Sauceda C, Iskander SM, & Smith AL (2021) Long-term surveillance of wastewater SARS-CoV-2 in Los Angeles County. Environmental Science: Water Res & Technol, 7(12), 2282–2294. https://doi.org/10.1039/D1EW00586C
Williams K, Harris A (2022) Renewable energy solutions for microalgae-based wastewater treatment. Renew Sust Energ Rev 70:1189–1201. https://doi.org/10.1016/j.rser.2016.12.015
Williams A, & Thomson M (2022) Net Zero UK – Generation and Energy Storage Requirements for the UK to Become Carbon Neutral. J Energy Power Technol, 04(04), 1–16. https://doi.org/10.21926/jept.2204041
Yadav RK, Singh A, Sharma MP (2018) Comparative study of Scenedesmus microalgae for outdoor application. J Clean Prod 172(6):2647–2654. https://doi.org/10.1016/j.jclepro.2017.11.112

Hydroponics/Aquaponics and Its Co-benefits

14

Ramandeep Saini, Palak Sangal,
Edward Terhemen Akange, and Ayush Madan

Abstract

Hydroponics and aquaponics represent innovative and sustainable agricultural practices that have receiving much attention due to their potential or ability to address current issues in modern farming. Hydroponics involves cultivating plants without soil, by utilizing water solutions rich in nutrient medium whereas aquaponics combines aquaculture with hydroponics and combines aquaculture with hydroponics to form a mutually beneficial environment where fish waste serves as food for plant growth. This chapter examines the co-benefits of hydroponics and aquaponics, emphasizing their ecological, economic, and nutritional advantages. These systems enhance resource efficiency by minimizing water usage compared to traditional soil-based agriculture, making them crucial for water-scarce regions. Furthermore, they offer year-round cultivation, mitigating seasonal limitations and ensuring consistent crop yields. Economically, hydroponics and aquaponics can be implemented in urban settings, reducing transpor-

R. Saini
Chandigarh College of Technology, Chandigarh Group of Colleges,
Landran, Greater Mohali, Punjab, India

P. Sangal
Department of Biological Sciences, Birla Institute of Technology & Science,
Pilani, Rajasthan, India

E. T. Akange
Department of Fisheries and Aquaculture, Joseph Sarwuan Tarka University,
Makurdi, Nigeria

A. Madan (✉)
Department of Biotechnology, School of Research and Technology, People's University,
Bhanpur, Bhopal, Madhya Pradesh, India

Institute of Tropical Aquaculture and Fisheries (AKUATROP), Universiti Malaysia
Terengganu, Kuala Nerus, Terengganu, Malaysia

N. Dwivedi et al. (eds.), *Wastewater Treatment Through Nature-Based Solutions*,
https://doi.org/10.1007/978-981-96-6699-7_14

tation costs and food miles. The nutrient-dense produce from these systems contributes to improved food security and nutrition. As climate change poses challenges to traditional farming, hydroponics and aquaponics emerge as promising alternatives that not only address sustainability concerns but also offer multifaceted benefits for agriculture and food production.

Keywords

Hydroponics · Aquaponics · Sustainable agriculture · Urban farming · Food security

14.1 Introduction

Hydroponics and aquaponics, once niche practices in agriculture, have emerged as dynamic solutions at the forefront of modern food production. The rapidly growing global population and escalating environmental challenges, and these groundbreaking farming techniques provide a new way to approach agriculture (Resh 2022). Hydroponics and aquaponics, beyond their primary role in food production, serve as integral components of nature-based solutions for wastewater treatment (Rakocy et al. 2006). In aquaponics systems, the cultivation of plants is intricately connected to the aquatic environment where fish are reared. The waste produced by fish in the form of ammonia undergoes a natural transformation facilitated by beneficial bacteria. The ammonia undergoes a conversion process, first turning into nitrites and then into nitrates, which are important nutrients for plants. As the plants absorb these nutrients, they act as natural biofilters, purifying the water and creating a symbiotic relationship between fish and plants (Goddek and Delaide 2021).

This natural filtration process not only optimizes nutrient cycling within the closed-loop system but also contributes to the remediation of wastewater. By harnessing the power of microbial communities and the inherent nutrient absorption capabilities of plants, hydroponics and aquaponics offer an eco-friendly solution to wastewater treatment. This aligns with the principles of nature-based solutions, emphasizing the use of natural processes to address environmental challenges (Liu and Chen 2020). Hydroponics and aquaponics play a crucial role in achieving Sustainable Development Goal (SDG) 6 by promoting sustainable water management practices (United Nations 2017a, b, c). The closed-loop design of these systems reduces water usage, while the natural filtration processes help with water purification. By utilizing these technologies, communities can move closer to guaranteeing the access and responsible handling of water resources. The ability of hydroponics and aquaponics to provide a consistent supply of nutritious produce contributes directly to Sustainable Development Goal 2 (United Nations 2017a, b, c). By enhancing food security and offering a solution for year-round cultivation, these practices address the challenges of achieving zero hunger in the face of a growing global population. Hydroponics and aquaponics embody the principles of responsible consumption and production. By optimizing resource use, reducing waste, and minimizing environmental impact, these cultivation methods align with Sustainable Development Goal 12, promoting sustainable practices in agriculture (United Nations 2017a, b, c).

Urban agriculture through hydroponics and aquaponics directly contributes to Sustainable Development Goal 11 by fostering sustainable cities and communities (Benjamin and Mineo 2016). The localized production of fresh, nutrient-rich food reduces the carbon emission from transportation and enhances community resilience through access to locally sourced produce. The innovative and technology-driven nature of hydroponics and aquaponics aligns with Sustainable Development Goal 9 (Ranganathan and Welles 2017). These practices represent advancements in agricultural infrastructure, promoting sustainable and efficient methods that can contribute to the overall development of industries related to food production.

In embracing hydroponics and aquaponics, societies embark on a journey towards achieving multiple Sustainable Development Goals, showcasing the interconnectedness of these practices with broader global aspirations for a more sustainable and fair future.

14.1.1 Definition of Hydroponics and Aquaponics

Hydroponics involves growing plants without soil, by using water solutions rich in nutrient medium to nourish crops directly (Jones 2005). This soilless method provides precise control over nutrient delivery, resulting in accelerated plant growth and increased yields. A technique for growing plants and crops without soil or compost is called hydroponics. Rather, the plants are cultivated in a solution containing water. In this setup, the plant roots are either supported by a non-nutritional medium such as perlite, gravel, or coconut coir, or they are submerged in the nutrient solution (Savvas and Passam 2002).

Aquaponics integrates aquaculture and hydroponics in a mutual benefit environment where fish waste provides essential nutrients for plant growth, and the plants, in turn, filter and purify the water for the fish and creating a closed-loop system where both aquatic life and plants thrive or it is a sustainable and environmentally friendly farming technique, aquaponics combines hydroponics (soilless plant cultivation) with aquaculture (as shown in Fig. 14.1) (Somerville et al. 2015). This method cycles nutrient-rich fish tank water, which contains fish waste through hydroponic grow beds, allowing plants to absorb the necessary nutrients. After being cleaned, the water is put back into the fish tank, starting a self-sustaining cycle that promotes the growth of the fish and plants (Pomoni et al. 2023).

Recent studies highlight the precision and efficiency achieved through hydroponic systems, showcasing their potential for revolutionizing traditional farming practices. The research, spanning various agricultural settings, meticulously assesses the key parameters that characterize hydroponic cultivation and examines their implications for sustainable and high-yield agriculture.

One focal point of the study is the precise control afforded by hydroponic systems over nutrient delivery to plants. The researchers meticulously analyze how this

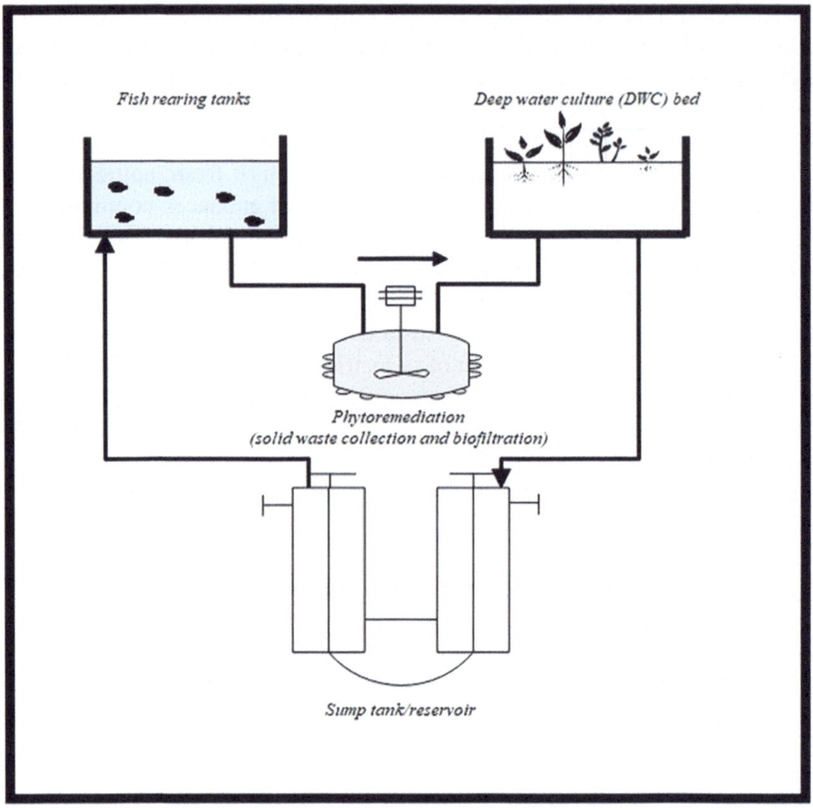

Fig. 14.1 Integration of aquaculture and hydroponics in aquaponics

precise control influences plant growth rates, nutrient absorption and overall crop yields. The findings indicate that the optimized nutrient solutions in hydroponics lead to accelerated growth, allowing for shorter cultivation cycles and increased harvest frequency compared to traditional soil-based methods.

Moreover, Kabir et al. explore the resource efficiency of hydroponic systems, considering factors such as water usage and fertilizer consumption (Magwaza et al. 2020). The study unveils that these systems significantly reduce water requirements, a critical factor in regions facing water scarcity. By recirculating nutrient solutions and minimizing runoff, hydroponics emerges as a sustainable solution that aligns with global goals for responsible water management in agriculture (Lee and Lee 2015).

In addition to resource efficiency, the study investigates the potential reduction in the use of agrochemicals in hydroponics. The findings suggest that the closed-loop nature of hydroponic systems inherently minimizes the need for pesticides and herbicides, contributing to environmentally friendly and ecologically sustainable farming practices.

Furthermore, Kabir et al. explore the adaptability of hydroponics to various crops, including staple foods and high-value horticultural produce. The research underscores the versatility of hydroponic systems, demonstrating their efficacy in cultivating a wide range of crops with diverse nutrient requirements. This adaptability positions hydroponics as a viable option for addressing global food security challenges by diversifying and increasing the resilience of agricultural production (Tzanakakis et al. 2014).

14.1.2 Brief History and Evolution

The roots of hydroponics can be traced back to the Hanging Gardens of Babylon, while aquaponics draws inspiration from ancient agricultural practices in Asia and the Middle East. Over the years, technological advancements have propelled these methods from historical curiosities to cutting-edge agricultural techniques (Okomoda et al. 2023).

In his recent book "Vertical Farming: How shall we feed the three more billion people by 2050?" agricultural historian Dr. F Sabry chronicles the evolution of hydroponics and aquaponics, emphasizing their transformative journey from experimental concepts to integral components of sustainable agriculture.

Meanwhile, the roots of aquaponics draw inspiration from ancient agricultural practices in regions like Asia and the Middle East. Historical accounts reveal that farmers in these areas engaged in integrated fish farming and plant cultivation (Gericke 1937). The practice of using fishponds to fertilize crops is an ancient technique that, over time, evolved into the more sophisticated aquaponic systems we recognize today. The evolution of hydroponics and aquaponics from historical curiosities to cutting-edge techniques owes much to technological advancements. The twentieth century witnessed pioneering efforts by scientists such as William Frederick Gericke, who coined the term "hydroponics" in the 1930s and conducted extensive research on soilless cultivation (Somerville et al. 2015).

Hydroponics, a method of cultivating plants without soil by utilizing water-based nutrients, has been in existence for thousands of years. The first versions of hydroponics were used by ancient societies such as the Aztecs and Babylonians. But the term "hydroponics" was not first used until the 1900s. Early examples of hydroponics can be seen in the Hanging Gardens of Babylon and the Aztec floating gardens. Dr. William F. Gericke popularized hydroponics at the beginning of the twentieth century by carrying out groundbreaking research on soilless plant growth. With advancements in growing media, nutritional solutions, and system designs, the approach developed quickly [Weir 2020.]. Hydroponics is now widely employed in both commercial and household gardening, having become essential in space agriculture. A recently developed method is aquaponics, which is a symbiotic system that combines hydroponics and aquaculture. It establishes a closed-loop environment by using the waste products from fish as plant fertilizers. Even though combined plant and fish cultivation is a relatively new concept, ancient societies such as the Aztecs demonstrated its inception (Wheeler 1986). Due to its efficiency and sustainability, aquaponics became popular in the latter part of the twentieth century. Aquaponics is now widely used, with systems ranging from small-scale household setups to large-scale commercial operations. This is due to rising environmental concerns and technological developments.

The space race in the mid-twentieth century prompted further exploration of hydroponics. NASA's interest in developing systems for growing crops in space led to breakthroughs in controlled-environment agriculture (CEA), as seen in the development of the nutrient film technique (NFT) and other hydroponic methodologies (Smith and Patel 1995). Aquaponics, while rooted in ancient practices, experienced a renaissance in the latter half of the twentieth century. Researchers and aquaculturists began to explore the synergies between fish and plants in closed-loop systems. The concept gained momentum as a sustainable and efficient method of food production (World Economic Forum 2023).

14.1.3 Importance in Modern Agriculture

In the regards of climate change and resource constraints, the importance of hydroponics and aquaponics in modern agriculture cannot be overstated. A report by the World Economic Forum (World Economic Forum 2023) underscores the need for innovative approaches, like hydroponics and aquaponics, to ensure food security amidst changing environmental dynamics (Johnson and Kim 2022).

When compared to conventional farming, hydroponics greatly reduces the amount of water used and the amount of land needed, demonstrating its superior resource utilization. Regardless of outside weather conditions, hydroponic systems offer unmatched control over growth conditions, allowing for the year-round production of high-quality crops. As so, they provide a dependable and steady supply of food, which is especially important in areas with a scarcity of arable land or erratic weather patterns (Davis and Lee 2021). Furthermore, by establishing a controlled environment that lessens the pressures of pests and diseases, hydroponics decreases the usage of pesticides and herbicides. This strategy maintains ecosystem balance while also protecting human health. A harmonious fusion of hydroponics and aquaculture, aquaponics represents sustainable food production. In a closed-loop system, combining fish farming with plant growing maximizes resource efficiency and reduces waste (Thompson and Roberts 2020). A symbiotic relationship is developed when fish waste, which is often regarded as a pollution, is converted into nutrient-rich water for plant growth. This circular economy strategy lessens the ecological impact of agriculture while also conserving water. Because aquaponics efficiently recycles water, it is especially beneficial in areas where freshwater resources are limited. Additionally, it provides a varied food source by including both plants and fish, enhancing nutritional diversity and food security (Garcia and Patel 2023).

As we navigate a future with unpredictable weather patterns and diminishing arable land, hydroponics and aquaponics offer a lifeline for sustainable, high-yield farming. This chapter explores not only their definitions and historical underpinnings but also deals with their important role in shaping the future of global agriculture. Furthermore, one of the critical aspects where hydroponics and aquaponics shine is in their ability to address water-related challenges, particularly through nature-based solutions (Miller and Wong 2022). Aquaponics, as a closed-loop system, serves as a natural water filtration system. The mutual relationship between fish and plants allows for the conversion of fish waste into valuable nutrients for plant growth. This natural nutrient cycling not only fosters plant development but also acts as a biofilter, purifying the water for the aquatic ecosystem. The result is a sustainable solution for wastewater treatment within the context of food production. However, hydroponics, with its soilless cultivation method, minimizes water usage compared to old soil-based farming. The precise control over nutrient delivery ensures that water is used efficiently. In regions grappling with water scarcity, hydroponic systems offer a lifeline by optimizing resource utilization and contributing to the broader goal of sustainable water management (Dash et al. 2024).

Beyond addressing water challenges, hydroponics and aquaponics play an important role in enhancing the overall resilience of agriculture and increased food production (as shown in Table 14.1).

14.1.3.1 Climate-Resilient Cultivation

Hydroponic systems, by providing a controlled environment, enable farmers to mitigate the impact of unpredictable weather patterns. This climate resilience is especially valuable in the face of extreme weather events, allowing for consistent crop production regardless of external conditions (Karim and Osama 2022).

14.1.3.2 Year-Round Crop Availability

Hydroponics facilitates year-round cultivation, decoupling food production from traditional growing seasons. This perpetual availability of crops contributes to food security by reducing the vulnerability associated with seasonal fluctuations and climate-induced disruptions (Chatterjee et al. 2020).

14.1.3.3 Increasing Localized Food Production

Hydroponics and aquaponics support the trend of localized food production, aligning with the principles of sustainability and reducing dependence on long-distance transportation.

14.1.3.4 Urban Agriculture and Vertical Farming

In urban settings, hydroponics enables vertical farming, transforming unused spaces into productive agricultural zones. This urban agriculture model significantly shortens the food supply chain, reducing carbon emissions associated with long route transportation and contributing to a more sustainable, localized food system (Gasperi 2017).

14.1.3.5 Reduction of Food Miles

By enabling agriculture in urban environments and utilizing vertical spaces, hydroponics and aquaponics contribute to a reduction in food miles. This reduction is a

Table 14.1 Different ways in which hydroponics and aquaponics are crucial to advancing modern agriculture

Aspects	Hydroponics	Aquaponics	Importance in modern agriculture	References
Optimized resource use	Efficient use of water, nutrients, and space with precise control	Recycles fish waste as plant nutrients, reducing reliance on external inputs	Promotes sustainability by reducing resource consumption and waste	Niu and Masabni (2022)
Climate resilience	Allows cultivation in controlled environments, independent of weather	Similar to hydroponics with the added benefit of fish production	Ensures stable food production amidst climate change and extreme weather	Karim and Osama (2022)
Year-round production	Enables continuous cultivation, regardless of season	Supports continuous crop and fish production	Enhances food security by providing consistent food supply year-round	Chatterjee et al. (2020)
Urban agriculture potential	Ideal for vertical farming, converting urban spaces into productive zones	Integrates fish farming into urban agriculture	Shortens food supply chains, reduces carbon footprint, and increases local food availability	Gasperi (2017)
Reduction of food miles	Facilitates local food production, reducing transportation needs	Localizes production with fish and plants in the same system	Decreases environmental impact associated with long-distance food transport	Nowysz et al. (2022)
Data-driven decision-making	Inherently supports data collection and precise control of growing conditions	Allows monitoring of both plant and fish health, optimizing overall system performance	Increases agricultural efficiency and productivity through technology integration	Alipio et al. (2017)
Sustainable food systems	Reduces reliance on traditional farming inputs, promotes sustainability	Integrates fish farming with plant production, creating a closed-loop system	Contributes to the development of sustainable and resilient food systems	Sharma et al. (2023)

vital component of sustainable agriculture, lessening the environmental impact associated with transporting food over long distances (Nowysz et al. 2022).

14.1.3.6 Technological Innovation and Precision Agriculture

Hydroponics, being a technology-driven cultivation method, represents a paradigm shift toward precision agriculture.

14.1.3.7 Optimized Resource Use

The precision in nutrient delivery in hydroponic systems leads to optimized resource use. This includes not only water but also fertilizers and other inputs. This efficiency is crucial in the context of resource constraints and supports the sustainable use of agricultural inputs (Niu and Masabni 2022).

14.1.3.8 Data-Driven Decision-Making

Hydroponic systems are inherently conducive to data collection and analysis. Farmers can make well-informed decisions and optimize plant growth and resource utilization with this data-driven approach.

In summary, hydroponics and aquaponics are groundbreaking agricultural technologies that address the complexities of modern food production. By optimizing resource utilization, increasing yields, and minimizing environmental impact, these systems offer promising solutions to the global challenges of feeding a growing population while preserving our planet.

14.2 Hydroponics and Aquaponics Unveiled

14.2.1 Hydroponics: Cultivating Without Boundaries

14.2.1.1 Beyond Soil: Redefining Plant Growth

A cutting-edge technique for growing plants without traditional soil is hydroponics. Rather than this, plants get their vital nutrients from a nutrient-rich water solution, which gives them exact control over the conditions under which they grow. Due to its many benefits, such as higher growth rates, water efficiency, and the capacity to grow plants in areas with poor soil quality, this technique has become more popular (Rodríguez-Delfín 2011).

In hydroponics, plants are typically placed in inert growing mediums such as perlite, coconut coir, or rock wool, providing dual role allowing easy access to the nutrient solution and allows the optimization of nutrient uptake, as plants can absorb nutrients directly from the water, eliminating the need for nutrient absorption through soil microbes.

Growers can customize the nutrition of their plants to meet their unique requirements by adjusting the quantity and content of nutrients in the water solution, leading to more resilient and productive harvests. This level of precision is especially beneficial for growing crops in areas with nutrient-deficient soils.

Advantages of Hydroponics

Hydroponics has garnered widespread acclaim due to its multifaceted advantages. One of its primary merits lies in the substantial increase in growth rates observed in hydroponically cultivated plants. By directly delivering nutrients to the roots, plants can focus their energy on growth rather than extensive root development (Rajendran et al. 2024).

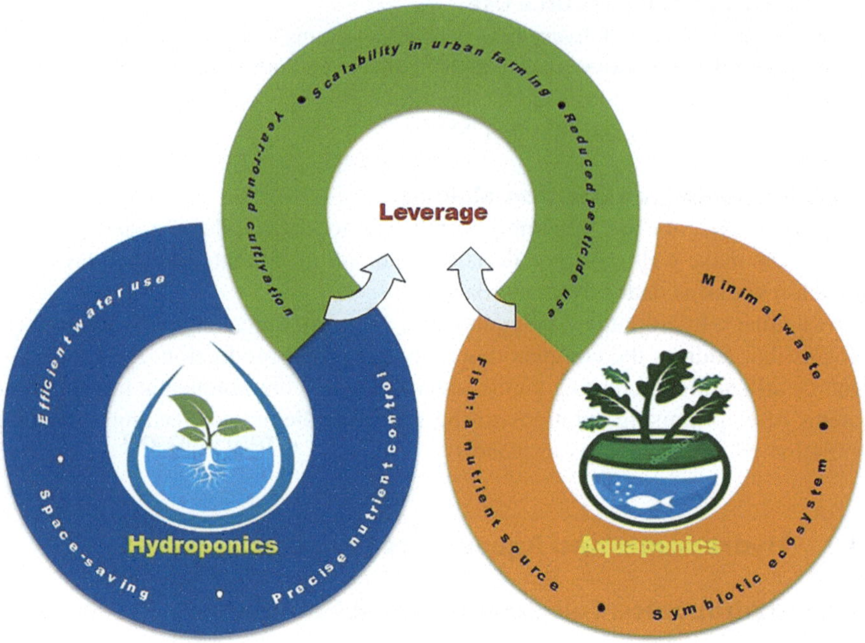

Fig. 14.2 Advantages of hydroponics and aquaponics

Additionally, hydroponics stands out for its unparalleled water efficiency. The closed-loop system allows for the recycling of water, minimizing wastage and addressing the global concern of water scarcity (as shown in Fig. 14.2).

Growing Beyond Geographical Constraints

An inherent advantage of hydroponics is its ability to liberate plant cultivation from the limitations imposed by soil quality and geography. In hydroponic systems, plants are strategically positioned in inert growing mediums such as perlite, coconut coir, or rock wool. These mediums not only provide structural support but also facilitate easy access to the nutrient solution (Lim et al. 2024).

This approach allows growers to cultivate plants in locations with poor or even no soil, such as arid regions or urban environments with limited space for traditional agriculture. By breaking free from geographical constraints, hydroponics emerges as a versatile solution for urban farming and sustainable agriculture in diverse ecosystems.

Precision Nutrition

The heart of hydroponics lies in its precision nutrition capabilities. Growers exercise meticulous control and insights over the composition and concentration of nutrients in the water solution. This level of precision enables the tailoring of a plant's diet to its specific needs, fostering healthier and more productive crops (Cho et al. 2017).

The elimination of soil as an intermediary in nutrient absorption streamlines the process, allowing plants to directly uptake essential elements from the water. This not only enhances nutrient absorption efficiency but also reduces the risk of nutrient imbalance or deficiency, promoting optimal plant health.

Overcoming Nutrient-Deficient Soils

Hydroponics proves particularly advantageous in regions with nutrient-deficient soils. By providing a regulated environment where nutrient levels can be finely tuned, hydroponic systems offer a strategic solution to the challenges posed by soils lacking essential nutrients. This is particularly relevant in areas where traditional agriculture might struggle due to poor soil quality (Lennard and Goddek 2019).

The method empowers growers to create an environment where plants receive precisely what they need, ensuring robust growth and mitigating the agricultural limitations imposed by soil quality variations.

In essence, hydroponics represents a paradigm shift in plant cultivation, emancipating agriculture from the confines of traditional soil-based methods. With its capacity for rapid growth, resource efficiency, and adaptability to diverse environments, hydroponics stands at the forefront of redefining the future of global agriculture.

Nutrients in the Flow: Water Solutions at the Helm

Water is the lifeblood of hydroponic systems, serving as the conduit through which essential nutrients are delivered to plants. With hydroponic solutions, all the nutrients required for plant growth are dissolved in water in a well-regulated mixture. Nutrient solutions that combine macronutrients (nitrogen, phosphorus, and potassium) with micronutrients (iron, manganese, and zinc) are frequently utilized.

The recirculation of nutrient solutions in hydroponic systems enhances water efficiency, making this method more environmentally sustainable than traditional soil-based cultivation. This setup minimizes nutrient wastage and reduces the overall water requirements for plant growth.

Regular monitoring and adjustment of these parameters, such as pH and electrical conductivity (EC), are crucial to maintaining optimal growing conditions. This active management allows growers to respond to the changing needs of plants during different growth stages.

14.2.2 Aquaponics: The Symbiotic Symphony

14.2.2.1 Dance of the Elements: Aquaculture Meets Hydroponics

Aquaponics represents a harmonious combination of aquaculture and hydroponics. In this closed-loop system, fish waste acts as a natural source of nutrients for plants, while the plants filter and purify the water for the fish. This mutual relationship creates a self-sustaining ecosystem where both fish and plants thrive (Maucieri et al. 2018).

The nitrogen cycle is essential for aquaponics. In this process, fish excrete ammonia, which is then converted by beneficial bacteria into nitrites, and then into nitrates. These nitrates serve as nutrients for the plants. As the plants absorb these nutrients, the water is filtered and returned to the fish tank. This closed-loop system is not only environmentally friendly but also resource-efficient, as it reduces the need for external fertilizers and minimizes water usage. The synergy between aquaculture and hydroponics creates a balanced and sustainable method of food production.

14.2.2.2 Harmony in a Tank: Fish and Plants in Symbiosis

In aquaponics, the compatibility between fish and plants is crucial for the success of the system. Certain fish species, such as tilapia, carp, and catfish, are well suited for aquaponic environments due to their waste production and adaptability to confined spaces.

The selection of plant varieties is equally important, with leafy greens, herbs, and certain fruiting plants thriving in aquaponic setups. The plant beds serve not only as a nutrient absorption system but also as a biofilter, promoting a healthy environment for the fish. This setup creates a balanced ecosystem where each element contributes to the overall well-being of the system.

Maintaining water quality is a continuous process in aquaponics, as any imbalance can disrupt the delicate symbiosis between fish and plants. Regular monitoring of water parameters, such as ammonia, nitrite, nitrate, pH, and temperature, ensures a stable and productive aquaponic system (Proksch et al. 2023).

14.3 Ecological Marvels

14.3.1 Water Wizards: The Art of Reduced Consumption

14.3.1.1 Hydroponics' Elixir: Navigating Water Scarcity

Hydroponics emerges as a water-efficient agricultural marvel, addressing the challenges posed by water scarcity. In traditional soil-based cultivation, a significant amount of water is lost through runoff and evaporation. Hydroponic systems, on the other hand, recirculate water, minimizing wastage and optimizing resource use.

Hydroponics' closed-loop design allows for precise control over the water flow to the plants. In contrast to traditional farming, nutrient-rich water solutions decrease the overall water demand as plants absorb them efficiently. This is especially crucial in regions with limited water resources, as hydroponics offers a sustainable method for cultivating crops without detrimental effects on the environment.

14.3.2 Seasons on Mute: Year-Round Cultivation's Evergreen Promise

14.3.2.1 Breaking the Chains: Overcoming Nature's Calendar

Year-round cultivation, facilitated by advanced agricultural practices, challenges the traditional constraints imposed by seasonal changes. Through techniques like hydroponics and controlled-environment agriculture (CEA), farmers can manipulate growing conditions to extend production beyond the natural growing season.

Hydroponics enables farmers to provide consistent and optimized conditions for plant growth, decoupling cultivation from external factors such as climate and weather patterns. Controlled environments, including greenhouse systems, further amplify this capability by regulating temperature, humidity, and light (Proksch et al. 2023).

This liberation from the limitations of nature's calendar enhances food security by ensuring a continuous and predictable food supply. Moreover, it reduces the need for long route transportation of perishable goods, contributing to lower carbon footprints in the agricultural supply chain.

14.4 Economic Alchemy

14.4.1 Urban Oasis: Hydroponics and Aquaponics in Cityscapes

14.4.1.1 Vertical Dreams: Reducing Food Miles in Urban Farms

In densely populated urban areas, hydroponics and aquaponics emerge as transformative technologies, turning vertical spaces into thriving agricultural zones. In urban settings, vertical farming involves growing crops in stacked layers or vertical towers to make effective use of restricted space.

This vertical approach significantly reduces "food miles," the distance food travels from the point of production to consumption. With urban farms strategically located within or near cities, the need for long-distance transportation is minimized, leading to lower carbon emissions and fresher produce for urban consumers (Bugbee 1999).

The integration of hydroponics and aquaponics into vertical farming systems enhances resource efficiency, as these methods require less water compared to traditional or old soil-based agriculture. Additionally, the controlled environments in urban farms provide opportunities for year-round cultivation, contributing to a more stable and reliable food supply.

14.4.2 Coins from Cultivation: Economic Blossoms

14.4.2.1 Tending to Profitability: The Entrepreneurial Spirit

The adoption of hydroponics and aquaponics has not only transformed agricultural practices but has also created economic opportunities for entrepreneurs. Tending to profitability, individuals and businesses have embraced the entrepreneurial spirit within the realm of soilless cultivation.

Entrepreneurs in this space explore various business models, including commercial hydroponic farms, aquaponic systems for local food production, and supplying specialized equipment and nutrients. As the demand for fresh, locally grown produce increases, entrepreneurs find innovative ways to capitalize on this growing market.

The economic benefits extend beyond food production to technology development, consulting services, and education. Entrepreneurs play a crucial role in advancing the field, driving innovation, and promoting the widespread adoption of hydroponics and aquaponics in diverse economic landscapes.

14.5 Nutritional Renaissance

14.5.1 Beyond the Plate: Nutrient-Rich Produce Takes Center Stage

14.5.1.1 Cultivating Health: A Nutritional Revolution

Hydroponics and aquaponics play an important role in the nutritional revolution by providing a platform for cultivating exceptionally nutrient-rich produce. In traditional agriculture, soil quality and composition impact the nutritional content of crops. However, hydroponic and aquaponic systems allow for precise control over the nutrient composition of the growing medium, resulting in plants with enhanced nutritional profiles.

Research indicates that hydroponically and aquaponically grown fruits and vegetables can contain higher levels of essential vitamins and minerals compared to their soil-grown counterparts. The controlled environment in these systems ensures that plants receive the optimal balance of nutrients, promoting the production of healthier and more nutrient-dense crops (Langenfeld et al. 2022).

This nutritional revolution is particularly significant in addressing modern dietary challenges, where access to fresh and nutrient-rich food is crucial for promoting overall health and preventing nutritional deficiencies.

14.5.2 Food Security's Fresh Face: Hydroponics and Aquaponics' Contribution

The nutrient-rich produce from hydroponics and aquaponics contributes significantly to food security by providing a reliable source of fresh and healthy food. These cultivation methods enable the year-round production of crops, reducing dependence on seasonal fluctuations and external environmental conditions.

The controlled environments in hydroponic and aquaponic systems mitigate the impact of climate-related challenges, ensuring a stable supply of nutritious food even in regions with harsh climates or limited arable land. This fresh face of food security aligns with the goal of ensuring that communities have access to a diverse and consistent supply of essential nutrients.

The efficiency of these systems in utilizing water resources also adds to their contribution to food security, particularly in regions where water scarcity is a concern. By maximizing water usage and minimizing waste, hydroponics and aquaponics represent sustainable solutions for enhancing food security on a global scale.

14.6 Conclusion

The challenges of modern agriculture can be effectively addressed through the implementation of hydroponics and aquaponics. These cutting-edge farming techniques provide a compelling solution to the escalating issues of environmental degradation, water scarcity, and food security. Hydroponics and aquaponics demonstrate their potential to transform agriculture by increasing crop yields, reducing environmental impact, and optimizing resource efficiency. By cultivating crops without soil, hydroponics allows for precise control over growing conditions, resulting in higher productivity and reduced water usage. Aquaponics, which merges aquaculture and hydroponics, establishes a closed-loop system that embodies sustainability and resource efficiency. This symbiotic relationship between fish and plants not only conserves water but also yields nutrient-rich food. Moreover, these systems offer promising solutions for urban agriculture by minimizing food transportation distances and bolstering community resilience. While challenges and opportunities exist, it is essential to integrate technology and research to fully realize the potential of these systems. Embracing hydroponics and aquaponics could pave the way for a more sustainable, resilient, and food-secure future. In conclusion, these innovative practices are instrumental in transforming agriculture from a resource-intensive industry to a sustainable and efficient model capable of nourishing a growing global population.

References

Alipio MI, Cruz AEMD, Doria JDA, Fruto RMS (2017) A smart hydroponics farming system using exact inference in Bayesian network. In: 2017 IEEE 6th global conference on consumer electronics (GCCE). IEEE, pp 1–5

Benjamin AE, Mineo C (2016) Urban agriculture: a review of the literature. Urban Agric Region Food Syst 1(1):1–14. https://doi.org/10.2139/ssrn.2914573

Bugbee B (1999) 284 towards efficient nutrient management in recirculating hydroponic culture. HortScience 34(3):491C

Chatterjee A, Debnath S, Pal H (2020) Implication of urban agriculture and vertical farming for future sustainability. In: Urban horticulture-Necessity of the future. IntechOpen

Cho WJ, Kim HJ, Jung DH, Kang CI, Choi GL, Son JE (2017) An embedded system for automated hydroponic nutrient solution management. Trans ASABE 60(4):1083–1096

Dash S, Jemla Naik D, Chinnu VS (2024) Climate crisis and agricultural response: climate resilient crops for sustainability in food production systems. J Exp Agric Int 46(6):440–458

Davis RA, Lee HJ (2021) Integrating aquaponics into modern agriculture: benefits and challenges. Sustain Farming J 18(2):112–123
Garcia ER, Patel AM (2023) Nature-based solutions in modern agriculture: the role of hydroponics and aquaponics. Environ Sustain J 15(2):67–79
Gasperi D (2017) Urban horticulture: reducing food miles to improve cities microclimate and environmental sustainability. University of Bologna, Bologna
Gericke WF (1937) Hydroponics: a new era in gardening. The California Garden 4(2):14–17
Goddek S, Delaide B (2021) Aquaponics: a sustainable solution for wastewater treatment. Springer. https://doi.org/10.1007/978-3-030-54684-8
Johnson LK, Kim ST (2022) Water efficiency and land use in hydroponics: a comparative analysis. J Control Environ Agric 12(3):45–58
Jones JB (2005) Hydroponics: a practical guide for the soilless grower. CRC Press
Karim SR, Osama R (2022) Sustainable nutrient-rich food production during COVID-19 pandemic through year-round vegetable farming using hydroponic technique. Chem Proc 10(1):69
Langenfeld NJ, Pinto DF, Faust JE, Heins R, Bugbee B (2022) Principles of nutrient and water management for indoor agriculture. Sustain For 14(16):10204
Lee S, Lee J (2015) Beneficial bacteria and fungi in hydroponic systems: types and characteristics of hydroponic food production methods. Sci Hortic 195:206–215. https://doi.org/10.1016/j.scienta.2015.09.011
Lennard W, Goddek S (2019) Aquaponics: the basics. In: Aquaponics food production systems, vol 113. Springer, Cham
Lim D, Keerthi K, Perumbilavil S, Suchand Sandeep CS, Antony MM, Matham MV (2024) A real-time on-site precision nutrient monitoring system for hydroponic cultivation utilizing LIBS. Chem Biol Technol Agric 11(1):111
Liu J, Chen Y (2020) The role of aquaponics in sustainable agriculture and water management. Sustain For 12(21):8799. https://doi.org/10.3390/su12218799
Magwaza ST, Magwaza LS, Odindo AO, Mditshwa A, Buckley C (2020) Partially treated domestic wastewater as a nutrient source for tomatoes (Lycopersicum solanum) grown in a hydroponic system: effect on nutrient absorption and yield. Heliyon 6(e05745). https://doi.org/10.1016/j.heliyon.2020.e05745
Maucieri C, Nicoletto C, Junge R, Schmautz Z, Sambo P, Borin M (2018) Hydroponic systems and water management in aquaponics: a review. Ital J Agron 13(1):1–11
Miller CL, Wong RK (2022) Efficient water management through hydroponics: addressing water scarcity and resource optimization. Water Resour Manag Rev 24(3):134–145
Niu G, Masabni J (2022) Hydroponics. In: Plant factory basics, applications and advances. Academic, pp 153–166
Nowysz A, Mazur Ł, Vaverková MD, Koda E, Winkler J (2022) Urban agriculture as an alternative source of food and water security in today's sustainable cities. Int J Environ Res Public Health 19(23):15597. https://doi.org/10.3390/ijerph192315597. PMID: 36497678; PMCID: PMC9739535
Okomoda VT, Oladimeji SA, Solomon SG, Olufeagba SO, Ogah SI, Ikhwanuddin M (2023) Aquaponics production system: a review of historical perspective, opportunities, and challenges of its adoption. Food Sci Nutr 11(3):1157–1165
Pomoni DI, Koukou MK, Vrachopoulos MG, Vasiliadis L (2023) A review of hydroponics and conventional agriculture based on energy and water consumption, environmental impact, and land use. Energies 16(4):1690. https://doi.org/10.3390/en16041690
Proksch G, Horn E, Lee G (2023) Urban integration of aquaponics: advancing integrated food systems for the circular city. In: Urban and regional agriculture. Academic, pp 403–430
Rajendran S, Domalachenpa T, Arora H, Li P, Sharma A, Rajauria G (2024) Hydroponics: exploring innovative sustainable technologies and applications across crop production, with emphasis on potato mini-tuber cultivation. Heliyon 10:e26823
Rakocy JE, Masser MP, Losordo TM (2006) Aquaponics—integrating fish and plant culture, SRAC publication no. 454. Southern Regional Aquaculture Center

Ranganathan J, Welles S (2017) Advances in hydroponics and aquaponics: a review of technological innovations. J Agric Food Chem 65(31):6352–6360. https://doi.org/10.1021/acs.jafc.7b02764

Resh HM (2022) Hydroponic food production: A definitive guidebook for the advanced home gardener and the commercial hydroponic grower, 7th edn. CRC Press

Rodríguez-Delfín A (2011) Advances of hydroponics in Latin America. In: II International symposium on soilless culture and hydroponics, vol 947, pp 23–32

Savvas D, Passam HC (2002) Hydroponics: soilless culture for the production of quality vegetables and plant propagation. Elsevier

Sharma S, Lishika B, Shahi A, Kaushal S (2023) Hydroponics: the potential to enhance sustainable food production in non-arable areas. Curr J Appl Sci Technol 42(39):13–23

Smith JA, Patel MR (1995) The renaissance of aquaponics: modern applications of ancient techniques. Aquac Res 26(6):789–797

Somerville C, Cohen M, Pantanella E, Stankus A, Lovatelli A, Love DC, Fry JP, Li X, Hill ES, Genello L, Semmens K, Thompson RE (2015) An international survey of aquaponics practitioners. PLoS One 10(7):e0129600. https://doi.org/10.1371/journal.pone.0129600

Thompson GM, Roberts DC (2020) Aquaponics as a solution for food security and water conservation. Glob Food Secur Rev 29(4):89–101

Tzanakakis V, Koo-Oshima S, Haddad M, Apostolidis N, Angelakis A, Angelakis A, Rose J (2014) The history of land application and hydroponic systems for wastewater treatment and reuse. In: Evolution of sanitation and wastewater technologies through the centuries, vol 1. IWA Publishing, London, p 457

United Nations (2017a) Sustainable development goal 6: clean water and sanitation. United Nations. https://www.un.org/sustainabledevelopment/water-and-sanitation/

United Nations (2017b) Sustainable development goal 2: zero hunger. United Nations. https://www.un.org/sustainabledevelopment/hunger/

United Nations (2017c) Sustainable development goal 12: responsible consumption and production. United Nations. https://www.un.org/sustainabledevelopment/sustainable-consumption-production/

Weir R (2020) Historical insights into aquaponics and its modern applications. J Sustain Agric 36(4):221–234

Wheeler R (1986) Hydroponics in space: innovations and challenges. NASA technical paper, 1526, pp 1–12

World Economic Forum (2023) The future of food security. World Economic Forum

Enzymatic and Microbial-Based Systems for Bioremediation of Wastewater 15

Suhaily Suhaimi, Nurfadhila Nasya Ramlee, and Noor Hidayah Abd Rahman

Abstract

Environmental pollution is a never-ending global issue and with the recent progress in analytical chemistry, diverse types of organic and inorganic compounds from various industries are found in the environment which further impose a threat to the ecosystem and human health. Therefore, appropriate methodology is required to treat these contaminants and revive the polluted area. Wastewater treatment via physical, chemical, and biological are among the conventional methods available. However, the recent advancement found that enzymatic and microbial-based bioremediation is considered emerging and sustainable technology. Enzymes possess strong catalytic activity under mild reaction conditions with minimal energy consumption, efficient reaction rate at low concentration of contaminants, and reduce the generation of toxic by-products. However, microbes can metabolize carbon and nitrogen as the energy sources concomitant with decomposition of organic matter. These advantages make the bioremediation much greener, safer, and energy effective. Thus, enzymes and microbes have become important tools for practical application and rapidly gaining interest for the utilization in the environment. Moreover, the increasing use of biological components and the need for low reaction cost and mild reaction conditions are among the main forces behind the implementation of enzymes and microbial-based bioremediation. In this chapter, understanding the bioremediation process using various types of enzymes and microbes to treat different types of contaminants, integration of methods for wastewater treatment will be described in detail.

S. Suhaimi · N. N. Ramlee (✉) · N. H. A. Rahman
School of Chemical and Energy Engineering, Faculty of Engineering, UniversitiTeknologi Malaysia, Skudai, Johor, Malaysia

N. Dwivedi et al. (eds.), *Wastewater Treatment Through Nature-Based Solutions*,
https://doi.org/10.1007/978-981-96-6699-7_15

Keywords

Bioremediation · Enzymes · Microbes · Emerging contaminant · Wastewater treatment · Organic matter

15.1 Introduction

The main causes of environmental pollution are industrialization, modern agriculture, rapid urbanization, and other human-caused activities. Numerous hazardous contaminants are released into the ecosystems as a result of these operations, endangering the health of local plants, animals, and people (Weralupitiya et al. 2021). According to Vasilachi et al. (2021), more than 700 contaminants grouped into 200 categories such as dyes, pharmaceutical products, pesticides, and polycyclic aromatic hydrocarbons are considered as the most harmful due to their carcinogenicity, non-biodegradability, and persistence in the environment.

Dyes are color-giving molecules that can be classified into different types based on their chemical structure which are employed in food, textile, leather, and paint industries. Untreated waste containing dyes may produce toxic and mutagenic contaminants and cause various harmful diseases to living beings (Su et al. 2020). On the other hand, pharmaceutical products especially antibiotics have been accumulated in the environment through industrial production processes, human and animal excretion, hospital effluents, as well as inappropriate disposal (Mukhopadhyay et al. 2022). The consumption of antibiotics has exceeded 100, 000 tons annually and the presence of untreated antibiotics in the environment can contribute to the increase in antimicrobial resistance population which caused ineffective common antibiotic treatment (Danner et al. 2019). Pesticides are used to prevent illnesses spread by vectors. However, the use of pesticides does not only harm target organisms but also affects non-target organisms with fish being the most notable one. Fish consume algae, plankton, and other aquatic plants contaminated with pesticides which result in the damage of tissues and organs of the fish. Hence, fish mirror the state of water quality (Pathak et al. 2022). Meanwhile, polycyclic aromatic hydrocarbons (PAH) are typically produced from incomplete oil or coal burning. Specifically, petroleum-based polycyclic aromatic hydrocarbons appear to float on the surface of water and form a rigid film that prevents sunlight penetration and photosynthesis (Saravanan et al. 2021).

Thus, different physical, chemical, and biological methods have been used to reduce the toxic contaminants in the environment. Among them, the biological methods called bioremediations are considered environmentally safer, lower energy consumption, and can convert toxic contaminants into non-toxic metabolites in comparison to the other two methods (Saravanan et al. 2021). Generally, bioremediation involves the use of microorganisms and their enzymes for the degradation of pollutants. The main process of degrading and detoxifying pollutants is through intracellular accumulation or enzymatic transformation.

Several species of bacteria, fungi, and algae have been discovered to demonstrate bioremediation ability. There is also a significant role portrayed by plants and their corresponding bacteria to degrade toxic contaminants found in water. Plant species such as *Rhus vernicifera* (Chinese lacquer tree) and *Amoracia rusticana* (horseradish) can secrete oxidative enzymes for biotransformation. Furthermore, in the presence of hazardous compounds in water, microbes can grow at extreme temperatures in the range of −127 °C to 648 °C (Bala et al. 2022). The microbes also able to metabolize organic contaminants by consuming carbon sources via both aerobic and anaerobic processes. Due to their adaptability, they are an excellent choice for water decontamination (Bala et al. 2022). In this chapter, major enzymatic-based (oxidoreductases and hydrolases) and microbial-based bioremediation (bacteria and fungi) involved in bioremediation of wastewater were discussed. Besides, the innovation and integration of methods in the water treatment also was described in detail.

15.2 Role of Enzymes in Bioremediation of Wastewater

15.2.1 Oxidoreductases

Oxidoreductases have been used in the degradation of various pollutants including natural and synthetic compounds. The oxidoreductases are typically secreted by various species (bacteria, fungi, plants) and undergo oxidation of pollutant compounds by electron transfer during the enzymatic reaction. The transfer of the electron in a compound results in the release of chloride ions, carbon dioxide, and methanol. In addition, the presence of reactive diffusible redox mediator in the enzymes could dramatically enhance the reaction rate and increase the range of substrates that could be degraded by the oxidoreductase enzymes. A literature survey shows that most enzymatic remediation mainly uses three classes of oxidoreductase; oxygenase, laccase, and peroxidase (Sharma et al. 2018). This section mainly focuses on the sources, types or class, and mechanisms of oxidoreductases.

15.2.1.1 Oxygenase

Oxygenase (EC 1.13 or EC 1.14) participates in the aerobatic degradation of aromatic compounds and catalyzes the cleavage of aromatic rings by the addition of one or two molecules of oxygen as the primary step. A wide range of substrates that are persistent in the environment including compounds that contain chlorine (herbicides, fungicides, and pesticides) can be broken down through the dehalogenation process. The oxygenase can be classified into two sub-classes based on the number of oxygen molecules involved during the reaction; monooxygenase and dioxygenase (Cheng et al. 2022).

Monooxygenase is involved in the degradation of aromatic compounds via the addition of only one oxygen molecule to enhance the reactivity and insolubility. The monooxygenase participates in several mechanisms during the degradation process such as dehalogenation. Desulfurization, denitrification, and hydroxylation. Based on the cofactor used, monooxygenase can be classified into two. The first class is flavin-dependent monooxygenase (FMO) in which the flavin is used as an electronic medium to bind to the enzyme and circulates between oxidation state and reduction state for electron transfer. This type of monooxygenase usually uses nicotinamide adenine dinucleotide phosphate (NADPH) as a hydrogen donor to catalyze the oxygen-containing compounds into small molecules (Deng et al. 2022). Oxidation of aromatic organic compound, indole among indigo-producing plant to indoxyl was observed using FMO cloned from *Polygonum tinctorium* (Inoue et al. 2021).

The second class of monooxygenase is P450 monooxygenase which has more than 200 subfamilies that can be found in prokaryotes and eukaryotes. Unlike FMO, P450 monooxygenase has an iron-containing porphyrin group to oxidize the target contaminants. For instance, cytochrome P450 monooxygenase (CYP) is one of the families used in many industries to oxidize the contaminant release. The CYP is able to catalyze a wide range of reaction such as N/S-oxidation, N/O/S-dealkylation, M/S-oxidation, hydroxylation, and C-C bond cleavage (Yan et al. 2022). CYP1A2 from *Arabidopsis thaliana* plant managed to degrade herbicide isoproturon and chlortoluron for weed control (Ehad et al. 2020; Jang et al. 2020).

Another type of oxygenase is dioxygenase which performs degradation of aromatic pollutants by the addition of two oxygen molecules in the ring. The dioxygenase can be classified into two based on the mode action: (1) aromatic ring hydroxylation dioxygenase (ARHD) which performs the degradation by adding two oxygen molecules into the ring of target contaminant and (2) aromatic ring cleavage dioxygenase (ARCD) which cleaves the aromatic ring of the target pollutants. A new enzyme from *Bacillus* sp. for the degradation of polycyclic aromatic hydrocarbons has been reported by Abdelhaleem et al. (2019). In the study, enzyme protocatechol 3,4- dioxygenase and catechol 1,2- dioxygenase play important role in the oxidation of polycyclic aromatic hydrocarbons (PAH) compound named anthracene to dihydrodiol. In addition, dihydroxylation also occurs during the reaction which resulted in the cleavage of the PAH structure and subsequent intermediates of tricarboxylic acid are formed. Then, thiol oxidation together with the cleavage of C-C and C-S bonds was catalyzed by deoxygenation. It can be concluded that the newly discovered enzymes are able to convert hydrophobic compounds into hydrophilic compounds which promote their metabolism by other enzymes (Abdelhaleem et al. 2019).

15.2.1.2 Laccase

Laccase (EC 1.10.3.2) is a multi-copper oxidase enzyme that catalyzes one-electron (e-) oxidations to produce free radicals by the synergistic effect of four copper ions in the laccase molecules. The catalytic site of a laccase can be categorized into three groups: one copper atom of type-1 (T1), one copper atom of type-2 (T2), and two copper atoms of type-3 (T3). The T1 catalytic site is the first stage of substrate oxidation, while T2 and T3 act as a place for the reduction in molecular oxygen to water (Brugnari et al. 2021). Since its first extraction, laccase has been widely expressed from a different sources of fungi (e.g., *Trametes versicolor*, *Phanerochaete chrysosporium*, and *Coprinus cinereus*), bacteria (e.g., *Bacillus, Streptomyces*, and

Pseudomonas), plants (e.g., *Rhus vernicifera*), and insects (e.g., *Manduca Sexta* and *Bombyx mori*) (Arregui et al. 2019).

Due to the discovery of laccase from different sources, it presented remarkable biotechnological potential as the enzyme can oxidize a broad range of substrates that are vastly used in several industrial sectors. For instance, laccase can degrade phenolic compounds, amino groups that contain phenols, lignin, and aryl groups containing diamines. These compounds can be found in various pharmaceuticals, pesticides, dyes, pulps, and textiles products (Nunes and Kunamneni 2018). A study by Kumar et al. (2022) found that laccase managed to remove phenols and aromatic amines pollutant compounds in water via the oxidation of the compounds to free radicals, and subsequently undergo polymerization along with partial precipitation. Besides, laccase was able to reduce dioxygen molecules of pollutants in the water by the removal of electrons from the organic substrate (Paraschiv et al. 2011).

Despite having wide range of application, the use of laccase is limited due to the presence of electron-withdrawing and bulky substituent of the pollutant. Hence, a free radical mediator is typically added to a reaction system to enhance its enzymatic activity. The mediator would undergo oxidation by laccase to produce high-activity radicals and transfer the electron to destroy the structure of targeted pollutants (Bankole et al. 2021). Laccase from strain *Coriolopsis gallica* enhanced the degradation of carbazole, *N*-ethyl carbazole, fluorine, and dibenzothiophene in the presence of 1-hydroxybenzotriazole (1-HBT) and 2.2-azinobis (3-ethylbenzthiazoline)-6-sulfonic acid (ABTS) as the mediators (Bressler et al. 2000). In the application of dye decolorization, laccase from *Trametes* sp. showed high degradation of Remazol brilliant blue, malachite green, and Eriochrome black in the presence of violuric acid and acetosyringone as the mediators (Wang et al. 2018). The fourier transform infrared spectrosopy (FTIR) spectrum of Remazol brilliant blue after the treatment using laccase stain *Arthrographis kalrae* showed new peaks ($-NH_2$, $-CH_3$) and deformation of NH_3 in amino acids indicating that new intermediate products were formed after the enzymatic activity. Interestingly, the intermediates produced are non-toxic and environmentally safe (Yadav et al. 2021).

15.2.1.3 Peroxidase

Peroxidase (EC 1.11.1.X) is a large family of enzymes that can be classified into two broad categories; heme peroxidase (divided into animal peroxidase and non-animal peroxidase) and non-containing heme peroxidase (Fig. 15.1). Heme is a complex between iron ion (Fe^{3+}) and the molecule protoporphyrin IX. About 74% of the peroxidase of the protein sequence were identified to have heme and mainly found in plants, fungi, bacteria, and animals. Besides, 87% of known heme peroxidases are reported as non-animal origin which are categorized into three classes namely class I, II, and III. Class I can be widely found in prokaryotes and eukaryotes intracellular such as ascorbate peroxidase, yeast cytochrome c peroxidase, and bacterial catalase peroxidase. Peroxidases in Class I play an important role in the prevention of oxidative stress in bacteria and they are necessary in scavenging and eliminating hydrogen peroxide (Twala et al. 2020). Compared to other classes of

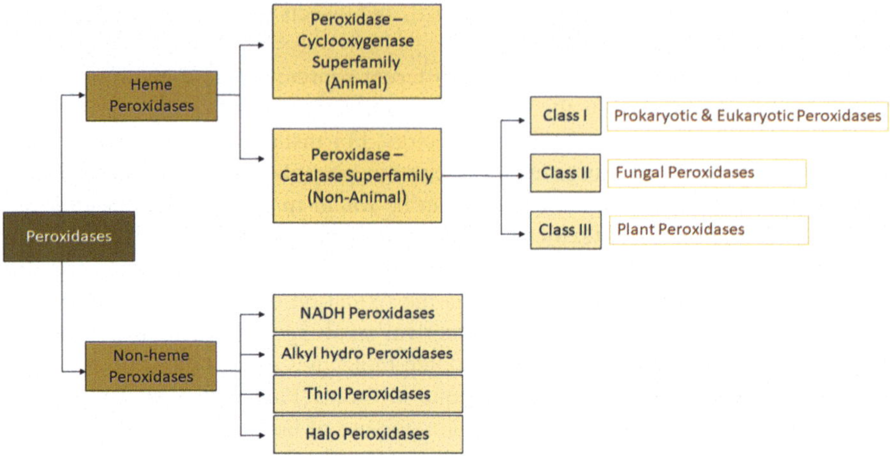

Fig. 15.1 Schematic classification of peroxidases. (Adapted from Pandey et al. 2017)

peroxidase, Class I is typically involve within the cell and rarely participates in bioremediations.

Class II heme peroxidase includes lignin peroxidase (LiP), manganese peroxidase (MnP), and versatile peroxidase (VP). The peroxidases in Class II have the potential for degradation of plant-based lignin. The oxidation of non-phenolic lignin compound by LiP involves the cleavage of side chain, ring-opening, demethoxylations, and oxidative dechlorinations that generally comprise one oxidation and two reduction steps. Similarly, the MnP has the same reaction mechanism as LiP during the peroxidative cycle except the MnP acts differently on lignin and phenolic lignin. Mn^{2+} is oxidized to Mn^{3+} by the MnP. This is followed by the formation of a Mn^{3+} and organic acid complex that act as a redox mediator to attack phenolic lignin structures and break the aromatic rings of lignin particles. For instance, MnP from *Cerrena unicolor* BBP6 was found to have great potential for various compound including denim decolorization, fluorine, phenanthrene, and polycyclic aromatic hydrocarbons (Zhang et al. 2020a). As for the degradation mechanism of VP, the oxidation ability shares a hybrid characteristic of those by LiP and MnP. Due to its specific bifunctional ability, the VP has obtained research interest for the biotechnological application. The VP from *Physisporinus vitreus* oxidized both phenolic and non-phenolic as well as cleaved β-*O*-4 and 5–5′ dimer lignin model compounds (Kong et al. 2017).

Class III peroxidase can be widely found in plant with the most representative member of this class being horseradish peroxidase (HRP) isolated from root of horseradish (*Armoracia rusticana*) and soybean peroxidase (SBP) isolated from seed hulls of the soybean (*Glycine max*) (Fernandes et al. 2020). Unlike Class I that is lack of disulfide bonds, calcium, and endoplasmic reticulum signal, Class II and Class III peroxidases contain four disulfide bridges and calcium ions in their structure (Twala et al. 2020). Moreover, peroxidases in Class III are known as enzymes-catalyzed H_2O_2 oxidation system which can be used as an alternative of Fenton process. Thus, peroxidase/H_2O_2 system has been used for the degradation of various pollutants including phenol, polychlorinated biphenyls, and dyes.

Despite that, the peroxidase/H_2O_2 system is hampered by slow degradation process and low degradation efficiency; hence, electron shuttle is introduced to the system to overcome these problems. Wang et al. (2022a) conducted an experiment to compare the degradation efficiency of sulfamerazine by HRP/ H_2O_2 in the presence and absence of iodine ions as the electron shuttle. It was found that the degradation of sulfamerazine in the presence of the iodine ions increased to 90.1% within 10 min in comparison to the absence of the iodine ions which was less than 10% degradation. This shows that the presence of electron shuttle in the system definitely enhances the degradation efficiency. Table 15.1 summarizes the previously reported studies on the application of oxidoreductases for bioremediations.

15.2.2 Hydrolases

Hydrolases is a class of enzyme that catalyzes the hydrolytic cleavage of several chemical bonds such as carbon–oxygen (C–O), carbon–nitrogen (C–N), carbon–carbon (C–C), phosphorus–nitrogen (P–N) bonds, and facilitates the breakdown of large molecules into small fragments (Ekta et al. 2022). Hydrolases enzymes have been extensively employed in industrial application such as in food and beverages, biofuels, detergents, as well as in wastewater treatment. This section mainly focuses on the sources of hydrolase enzymes, mechanisms, existing problems, and improvement methods to enhance catalytic activity and efficiency of the hydrolase enzymes for bioremediation of wastewater, with a purpose to give an understanding of the challenge and applicability in large-scale operations.

15.2.2.1 Lipase

The ability of lipase to catalyze wide range of catalytic reaction either in aqueous or non-aqueous medium is accompanied by high specificity and selectivity advancing its application in wastewater treatment. Lipase is classified in triacylglycerol acylhydrolase group and identified with EC 3.1.1.3 and possesses the ability to hydrolyze triacylglycerol to diacylglycerol, monoacylglycerol, glycerol, and fatty acids. The efficiency of lipase activity depends on the chain length, location, and degree of unsaturation of fatty acid present in the triacylglycerol (Nimkande and Bafana 2022). The versatility of lipase in wide range of industrial application ascribed to its capability in performing interesterification, transesterification, esterification, aminolysis, acidolysis, and alcoholysis (Ramlee et al. 2022a, b; Ramlee et al. 2021).

Lipase is one of the most explored enzymes by researchers due to its unique structure that can undergo open and close conformational changes via interfacial activation. For example, *Candida rugosa* lipase is equipped with a mobile polypeptide chain called lid with amphipathic properties that can rotate almost 90° from the closed to the open conformation (Tejo et al. 2004; Wu et al. 2015). During contact with hydrophobic environment, lipase favors an open, active conformation. The lid is extending away of the binding site, exposing the hydrophobic side of the catalytic

Table 15.1 Application of oxidoreductases in bioremediation for water treatment

Class	Enzyme	Source	Pollutant	Compound	Degradation (%)	References
Oxygenase	Monooxygenase	*Rhodococcus* sp. B2	Herbicide	Pretilachlor	81.6	Liu et al. (2021)
	Monooxygenase	*Microbacterium* sp. CJ77	Antibiotic	Sulfonamide	90.0	Kim et al. (2019)
	Dioxygenase	*Pseudomonas aeruginosa*	Dyes	Brilliant green Aniline blue Malachite green	40.1 75.9 76.5	Lu et al. (2020)
	Dioxygenase	*Buttiauxella* sp. S19-1	Pesticide/explosive	2,4,6-trinitrotoluene	>90	Xu et al. (2016)
	Monooxygenase & Dioxygenase	*Bacillus subtilis* EB1	Pesticide	Phenanthrene	93.0	Ganesh Kumar et al. (2023)
Laccase	Laccase	*Bacillus subtilis*	Antibiotics	Tetracycline Ampicillin	< 90.0	Zhang et al. (2020b)
	Laccase	*Myrothecium verrucaria ITCC-8447*	Lignin	Lignocellulose	64.7	Agrawal et al. (2019)
	Laccase	*Beltraniopsis* sp. ET-17	Dyes	Congo red-brilliant blue mixture	53.6	Hoyos et al. (2023)
	Laccase	*Trametes versicolor*	Dyes	Reactive blue 19 Acid Orange 7	62.0 80.0	Gao et al. (2022)
	Laccase	*Pleurotus dryinus*	Herbicide Herbicide Fungicide	Metolachlor Atrazine Kresoxim-methyl	100.0 100.0 91.0	Vaithyanathan et al. (2022)
	Laccase	*Parvibaculum lavamentivorans*	Petroleum	Hydrocarbon >C17	83.0	Diefenbach et al. (2024)

Peroxidase	Lignin peroxidase	*Fusarium* sp. ZH-H2	Lignin	Lignocellulose	85.9	Zhang et al. (2020b)
	Lignin peroxidase	*Pseudomonas fluorescens LiP-RL5*	Dyes	Nigrosine Bromophenol blue Coomassie brilliant blue Methylene blue	57.2 81.6 51.4 88.2	Kiran et al. (2020)
	Manganese peroxidase	*Bacillus velezensis* strain Al-Dhabi 140	Antibiotic	Tetracycline	99.2	Al-Dhabi et al. (2021)
	Manganese peroxidase	*Aspergillus flavus*	Dyes	Direct red 31 Acid black 234	100.0 92.0	Kalsoom et al. (2022)
	Versatile peroxidase	*Pleurotus eryngii*	Dyes	Evans blue Amido black 10B Guinea green	79.0 99.0 99.0	Ilić Đurđić et al. (2020)
	Versatile peroxidase	*Lentinus squarrosulus*	Lignin	Lignocellulose	11.0–17.0	Ravichandran et al. (2019)
	Horseradish peroxidase	Commercial	Dyes	Reactive black 5 Malachite Green	70.0 83.0	Jankowska et al. (2021)
	Horseradish peroxidase	Commercial	Phenol	2,4-dichlorophenol	66.0	Miłek (2024)
	Soybean peroxidase	Commercial	Fungicide	Mercaptobenzothiazole Meloxicam Caffeic acid	99.4 98.8 88.2	Morsi et al. (2021)
	Soybean peroxidase	Soybean seedcoat	Aromatic amine	Poly-substituted mono-cyclic 4-chloro-o-toluidine 4,4′-methylenebis-(2-chloroaniline)	> 95.0	Mukherjee et al. (2020)
	New discovery peroxidase	*Bacillus aryabhattai* B8W22	Phenol	Phenol	95.0	Elmetwalli et al. (2023)
	New discovery peroxidase	*Aspergillus terreus GS28*	Dye	Direct Blue-1	98.4	Singh and Dwivedi (2020)

pocket, thus enhancing the accessibility of substrate towards the catalytic triad (Ser209, His449, and Glu341) (Grochulski et al. 1994). The catalytic action occurs via nucleophilic attack. The catalytic mechanism of lipase begins by the binding of lipids or substrates on the active sites, then activation of nucleophilic serine residues through hydrogen donors to nearby histidine residues followed by nucleophilic attacks by serine residues on substrates to form tetrahedral intermediate conformations. Histidine then donates protons to release the alcohol component from the substrate to form intermediate covalent bonds (acyl enzymes). The acidic component of the substrate is esterified with a serine residue nearby and then water enters to carry out a nucleophilic attack on the C carbonyl atom in the covalent intermediates. Then histidine residue donates protons taken from water molecules to oxygen atoms in the active serine residue, subsequently the enzyme ester bond with the broken acyl component and the product in the form of acyl (acid) is released (Subroto et al. 2020). In contrast, during closed, inactive conformation, the hydrophilic side of the lid is exposed to the solvent, covers the substrate binding site and reduces the catalytic activity of lipase (Vanleeuw et al. 2019).

The employment of lipases to treat lipid-rich wastewater receives remarkable attention worldwide due to its capabilities in solving environmental problems such as clogged sewer, and oil pollution that affect the aquatic life (Nimkande and Bafana 2022). The application of lipases in bioremediation processes has been reported by previous researcher including treatment of hydrocarbon organic pollutants (Theerachat et al. 2017), sludge digestion (Rashvanlou et al. 2021), improving sludge dewatering (Kang et al. 2023), and fat, oil, and grease (FOG) digestion (Baena et al. 2022) in the wastewater treatment plant. The removal efficiency of the pollutant is strongly dependent on the types of lipases, composition of wastewater, as well as environmental and physicochemical conditions (temperature, pH, presence of organic solvents, metals, salts) (Nimkande and Bafana 2022).

The wastewater treatment through lipase hydrolysis is cost-effective due to mild reaction condition and environment-friendly condition. Thus, isolation, characterization, and identification of microbial and fungal species that can produce lipase with excellent activity is a pioneer step prior its application in wastewater treatment. Another important thing is the determination of the microbial or fungal sources because lipases produced from different sources differ in their substrate range, selectivity, and tolerance to environmental parameters. Ariaeenejad et al. (2023) discovered a novel and robust stable lipase (PersiLipase1) from metagenomic data of tannery wastewater for effective bio-degradation of several types of oily wastewater. PersiLipase1 presented remarkable performance resulted in 90.76% efficiency in hydrolysis of oil and grease, and 100% efficiency for hydrolysis of olive oil and sheep fat. Utilizing of free bacteria and enzymes as biocatalysts sometimes present some drawbacks such as limitation in the multiple reuses, low stability in the field application, and unwanted by-products formations. The immobilization of enzymes or microorganisms is a solution for solving the problem.

In addition, the bioremediation of wastewater using lipases could be executed using whole-cell biocatalyst. The whole-cell biocatalyst can survive in wastewater by metabolizing an organic pollutant as their growth medium, while secreting lipase extracellularly to the medium. Previous studies have reported the isolation of microbes that have the ability to produce lipase from lipid-rich wastewater or natural environment. For example, Hassan et al. (2018) had successfully isolated and screened marine bacteria for lipase production. Subsequently, the highest lipolytic hydrolysis (20 mm) was identified as *Bacillus cereus* HSS and the immobilized system presented remarkable performance (2.8-fold of lipolytic activity) as compared to free lipase. On top of that, the removal efficiencies of biochemical oxygen demand (BOD), total suspended solids, and oil and grease were 87.63, 90, and 94.70%, respectively. Besides, Ktata et al. (2020) have isolated novel *Aeribacillus pallidus* from geothermal oil-field thermostable which is categorized as haloalkaline, and possessed remarkable biochemical characteristics including high tolerance in 50% (v/v) of solvent, high pH stability, as well as outstanding performance in the presence of detergents and organic solvents. In addition, the isolated microbes have the ability to hydrolyze oil wastewater up to 96% at 50 °C for 1 h.

Nevertheless, the efficiency of whole-cell biocatalyst is far from achieving the discharge standard set by the authority. This is because lipases from different sources differ in their substrate range, selectivity, and tolerance to environmental parameters. Thus, Mohebrad et al. (2022) improved the treatment system for dairy wastewater by immobilization of *Pseudomonas aeruginosa* onto sodium alginate and incorporating direct electric current during treatment process to stimulate the microorganisms and enhance lipase and protease production. The maximum production of protease (193.81 IU/L) and lipase (47.89 IU/L) was recorded in pH = 6.0 and 8.0 mA of induced current, ultimately resulted in 95.67% of COD removal rates.

It may be noted that enzymes are susceptible to variable environmental and physicochemical conditions. Hence, their pollutant removal efficiency depends on the wastewater composition and prevailing conditions. Integrated approach consisting of identifying and preventing oil spill sources, separation from waste before treatment, and biological treatment using lipase can result in effective lipid waste management (Nimkande and Bafana 2022). The economic aspect of lipase treatment such as high-cost need to be considered especially in industrial scale of wastewater treatment plant. However, the advantages of lipase treatment for oily wastewater outweigh its limitations especially in mild operating condition which can reduce energy consumption and its ability to work in aqueous solution eliminate the separation process between oil and aqueous phase.

15.2.2.2 Organophosphorus Hydrolase

Organophosphorus hydrolase (EC 3.1.8.1) also known as a phosphotriesterase is a promising enzyme for bioremediation of organophosphorus compounds (Wang and Sun 2021; Efremenko and Sergeeva 2001). The action of organophosphorus hydrolase is through hydrolysis of organophosphorus pesticides and nerve agents into less or non-toxic compound (Grimsley et al. 1997). For example, organophosphorus hydrolase is able to catalyze the hydrolysis of methyl parathion in to *p*-nitrophenol and dialkyl-thio/oxophosphate in the presence of water (Mehta et al. 2019). Organophosphorus hydrolase is a metal dependent enzyme that contains two Zn^{2+} and Co^{2+} ions per subunit and the presence of metal ions significantly affects its catalytic activity (Efremenko and Sergeeva 2001). The ability of organophosphorus hydrolase in hydrolysis of wide range of organophosphorus compounds such as paraoxon, methyl paraoxon, methyl parathion, coroxon, coumaphos, dichlorvos, and chlorpyrifos, suggesting its great potential in the bioremediation of organophosphorus pesticides (Santillan et al. 2020a).

The use of organophosphorus pesticides is so indispensable to increase food production and to stabilize food prices, making it impossible to be completely prohibited at present (Wang and Sun 2021). Organophosphorus pesticides are generally degraded and used as the source of phosphate by many bacteria, such as *Pseudomonas* sp., *Agrobacter* sp., *Arthrobacter* sp., *Flavobacterium* sp., and others (Wu et al. 2019). The biodegradation of organophosphorus contaminants by microorganisms involves complex reactions and many enzymes. The reactions include oxidation, reduction, hydrolysis, dealkylation, hydroxylation, alkylation, and ring cleavage. Generally, the first step of the degradation is hydrolysis followed by further degradation of the less toxic products. However, the main enzyme responsible for the degradation of organophosphate contaminants is organophosphorus hydrolase which hydrolyzes phosphoester bond in orthophosphates such as P—O, P—F, and P—S bond (Efremenko and Sergeeva 2001). Enzymatic catalyst is becoming a first choice in bioremediation technique due to its efficiency, and clean strategy in degrading the pollutant, while chemical degradation can result in toxic emission.

The most important factor that determine the biodegradation ability of the microorganisms towards organophosphorus compounds is the isolation source. Usually, microorganisms isolated from contaminated area possessed the gene that are responsible in the expression of organophosphorus hydrolase. For example, Haque et al. (2020) reported that *Leuconostocmesenteroides* WCP307 strain was isolated from chlorpyrifos contaminated kimchi contain two organophosphorus hydrolase genes (opdA and opdE). The bacteria strain presented remarkable catalytic activity in biodegradation of organophosphorus insecticides such as adusafos, comnaphos, diazinon, dyfonate, ethoprophos, fenamiphos, methylparathion, and parathion insecticides. In addition, Lourthuraj et al. (2022) isolated chlorpyrifos degrading bacteria from the pesticide contaminated environment and presented excellent catalytic properties in biodegradation by utilizing pesticide as the source of carbon and energy. The degrading bacteria were identified as *Enterobacter aerogenes* CP2 and *Streptococcus pyogenes* CP11 which expressed high organophosphate hydrolase activity in intracellular, followed by outer membrane and extracellular sample.

Immobilization technique could be implemented to facilitate catalyst recovery to be reused in repeated reaction cycle. Haque et al. (2020) synthesized dual-enzyme immobilization employing organophosphorus hydrolase and aryloxyphenoxypropionate hydrolase via encapsulation in magnetic ZIF-8 in aqueous solution that simultaneously removed up to 69.30 and 59.60% of chlorpyrifos (insecticide) and quizalofop-*P*-ethyl (herbicide) in 90 min. Furthermore, the catalyst could be easily collected by magnetic separation to be reused for repeated biodegradation process.

Apart from that, bioremediation of organophosphorus pollutant could be executed through in situ technique by introducing the microorganisms that have the capability to expresses organophosphorus hydrolase. Wu et al. (2019) conducted simultaneous bioremediation of soybean wastewater, removal of organophosphorus pollutant, and improvement in soil fertility by inoculating *Rhodobacter sphaeroides* into the target polluted area. Soybean wastewater contains organic materials that can provide carbon sources and energy to the bacteria, and can provide the best condition for the induction and secretion of the organophosphorus acid hydrolase.

The great advantage of organophosphorus hydrolase is its ability to hydrolyze wide range of substrates, especially organophosphorus derived-compound. In general, the efforts on the improvement in organophosphorus hydrolase are focus on immobilization technique and genetic engineering, with an aim to enhance the substrate spectrum and improve catalytic efficiency with a final purpose to detoxification of organophosphorus insecticides. On top of that, biotechnological advances play an important role in enhancing its catalytic activity and stability in order to be fitted in large-scale bioremediation process which will be discussed in the next section.

15.2.2.3 Cutinase

Cutinases (EC 3.1.1.74) belong to the α/β-serine hydrolase superfamily and classified as esterases which can perform depolymerization by breaking down ester bond in the polymer and have Ser His Asp Glu residues similar to that of lipase, forming a catalytic triad (Anbalagan et al. 2021; Chen et al. 2018). These properties enable cutinase resembles the activity of lipase such as hydrolysis of *p*-nitrophenyl (*p*NP) acetate, *p*NP butyrate, and *p*NP caproate (Yang et al. 2013). Cutinase was derived from phytopathogenic fungi and possessed the capability to hydrolyze the polymer from plant's cutin (Ueda et al. 2021; Zhu and Wei 2019). Cutinase can catalytically degrade high-molecular-weight synthetic polyesters, low-molecular weight soluble esters, and short- and long-chain triacylglycerols. Previous studies have reported that cutinases were capable of degrading some emerging contaminants such as phthalates, malathion, polyethylene terephthalate (PET), polycaprolactone (PCL), and poly(butylene succinate) (PBS) (Yang et al. 2013; Sahu et al. 2023).

Apart from hydrolysis reaction, cutinase can also perform esterification and transesterification reactions for biosynthesis of product such as polyamides and polyester (Liang and Zou 2023). The catalytic mechanism of cutinase begins with deprotonatation of Ser136 by His204 base to allow its nucleophilic attack on the ester bond. Subsequently, His204 donates a proton to the leaving group and then activates a water molecule to allow the hydrolysis of the acyl enzyme. Then, Asp191 alters the pKa of the His204 to allow to act as an effective base in the reaction. Ultimately, Gln137 and Ser58 form the oxyanion hole to stabilize the transition state and release the product (Nyon et al. 2009; Martinez et al. 1994; Martinez et al. 1992).

Previous studies have reported several microorganism species that have the capability in expressing cutinase. For instance, Zhang et al. (2021) conducted a comparison study and found that among six types of microbes, *Thermobifida fusca*, *Thermobifida cellulosilytica*, *Humicola insolens*, *Fusarium solani*, *Meiothermus taiwanensis*, and *Thermobifida alba*, the last species presented remarkable expression yield, temperature tolerance, and good catalytic activity in decolorization of molasses wastewater. Subsequently, *Thermobifida alba* was subjected to the covalent immobilization procedure employing chitosan as a carrier and glutaraldehyde as a cross-linker, which resulted in 79.30–81.20% of decolorization efficiency. Besides, *Fusarium solani* cutinase showed exceptional capability in degrading biodegradable polymers as compared to esterase and arylesterase (Santos-Beneit et al. 2023). In order to enhance protein expression, *Fusarium solani* cutinase was engineered and heterologous expressed in *Escherichia coli*. As a result, the engineered microbes can grow on a PBS as sole carbon source of the culture media and can degrade plastic via microbial fermentation.

The capability of microbes to degrade pollutant begins by colonizing the surfaces of insoluble substrates, secreting the extracellular enzymes, and finally breakdown of the pollutant into simpler compound as a source of growth (Wei et al. 2022). Hence, the source of microbes is among the main factor in determining its catalytic activity. Din et al. (2023) have isolated mesophilic microorganisms identified as *Stenotrophomonas maltophilia* PRS8 from soil of landfill that can grow in a saline medium containing polyethylene terephthalate as the sole carbon source. Then, the degradation of polyethylene terephthalate using an extracellular medium showed the maximum degradation rate as compared to degradation using microbial cells, and a purified cutinase-like enzyme, thus reducing the cost and time during laborious process of enzyme purification. Besides, the optimization of the degradation reaction is crucial in obtaining the maximum degradation rate. Anbalagan et al. (2021) optimized the hydrolysis of polyethylene terephthalate at the pH of 8.5 and 80 °C which resulted up to 13.68 μmol/mL/h of maximum hydrolysis rate.

The ability of microbes to maintain normal metabolic activity under various stressful conditions, such as low pH and an oligotrophic medium, nominates its potential in plastic degradation. Nevertheless, the degradation rate is very low. Thus, Yan et al. (2021) constructed a synergistic process through combination of microbial (*Microbacterium oleivorans* JWG-G2) and enzymatic catalysis (cutinase) for the degradation of plastic pollutant such as bis(hydroxyethyl) terephthalate and poly(ethylene terephthalate). In addition, integration of advanced identification such as genome sequencing analysis helped in predicting the presence of extracellular hydrolases (three types of carboxylesterases, esterase and a lipase) that could be expressed by *Microbacterium oleivorans* JWG-G2.

Even though cutinase has been proven as powerful biocatalyst in the degradation of various pollutants, including emerging contaminants, the application of cutinase in reclamation of contaminants in wastewater treatment plant remains challenges due to high cost and limitation of study in large-scale reaction. Thus, new strategy must be developed to synthesize functional, stable, reusable, and regenerable biocatalysts that can deliver the task efficiently.

15.2.2.4 PETase

In 2016, Yoshida et al. (2016) discover poly(ethylene terephthalate) (PET) degrading microorganism from plastic-bottle recycling factory which embarks the research on its catalytic performance in treating plastic pollution. PET-degrading enzymes also known as PETase have cutinase-like activity and was expressed by *Ideonella sakaiensis*. PETase have the capability to hydrolyze PET to bis(2-hydroxylethyl) TPA (BHET), mono(2-hydroxyethyl) TPA (MHET), ethylene glycol (EG), and TPA(Chen et al. 2018; Mohanan et al. 2020). The mechanism of PETase starts with the binding of PET to the active site via hydrophobic interactions. Then proton transfer of Ser 131 to His208, accompanied by a nucleophilic attack by Ser 131 on the substrate, forms a tetrahedral conformation. Consecutively, deprotonation of an active site water molecule by His208 occurs while hydroxide attacks the acylated Ser131 intermediate, leading to breaking its bond to the substrate. Finally, His208 transfers the water proton to Ser131, and releases MHET that leads to enzyme regeneration (Chen et al. 2018; Jerves et al. 2021).

Even though the discovery of PETase leads to a major breakthrough in addressing plastic pollution, the catalytic activity of native PETase is still low, restricting its potential in large-scale bioremediation process. Thus, Chen et al. (2020) aimed to improve the degradation efficiency of PETase by displaying PETase on the surface of yeast (*Pichia pastoris*). As a result, the turnover rate increases up to 36fold. On the other hand, the applicability of PETase in large-scale reactions is restricted due to a lack of homogeneity between the experimental protocols used in the published literature. Hence, Arnal et al. (2023) assessed the performance of different types of PETase by developing a standard protocol regarding PETase hydrolysis under operating conditions that was relevant for larger scale reaction. Among four types of PETase (LCC^{ICCG}, FAST-PETase, HotPETase, and $PES\text{-}H1^{L92F/Q94Y}$), LCC^{ICCG} outperforms the other enzymes, converting 98% of PET into the monomeric products TPA and EG in 24 h, followed by $PES\text{-}H1^{L92F/Q94Y}$ with 80% of PET depolymerization, while the remaining PETases presented low depolymerization rates and not suitable for large-scale application.

The improvement in PETase mainly focuses on genetic engineering to increase the expression of the protein while facilitating the purification process of the enzyme. *Yarrowia lipolytica* has been reported as the best host for cloning due to its performance in heterologous protein expression, its ability to metabolize PET as a carbon source, and its great adaptability in the medium contained with intermediate degradation products. Thus, Kosiorowska et al. (2022) expressed a codon-optimized PETase gene in the yeast *Yarrowia lipolytica f*or its ability to degrade PET directly in the culture and observed that the addition of olive oil as a supplement helps in maximizing PET hydrolysis, while the addition of salt reduces the hydrolysis rate. In this context, while genetic engineering results in improvement in the hydrolysis activity of the enzyme, medium composition also plays a pivotal role in determining the rate of pollutant degradation.

The polluted soil, water, or land is the potential source for the plastic-degrading microorganisms. Cao et al. (2023) isolated plastic-degrading microorganisms named as *Comamonas testosteroni* CNB-1 from wastewater of *p*-chloronitrobenzene production factory that exhibited the capability on utilizing PET degradation intermediates, such as terephthalic acid (TPA), ethylene glycol (EG), and bis(2-hydroxyethyl)-TPA (BHET), for growth. In order to improve the performance of plastic-degrading microorganisms, his research group mutated 10 amino acid sites of *Is*PETase (DuraPETase) to improve the hydrolytic activity of DuraPETase in hydrolyzing PET to TPA and EG. After mutation, they performed heterologous expression of the gene into *Comamonas testosterone* as the host. The genetically engineered species successfully mineralized PET by consuming the intermediates compound, instead of breaking it down into smaller molecules.

The research on PETases provides a great breakthrough in the area of genetic engineering through improvement in the protein expression and its catalytic activity. It is worth noting that the degradation of plastic performed by PETases was carried out in laboratory scale. Further research needs to be explored in order to evaluate the performance of PETase in industrial application, such as the development of bioreactor in wastewater treatment plant, development of a standard protocol, as well as the optimization of reaction condition. The application of hydrolases enzymes and its treatment method in wastewater treatment is presented in Table 15.2.

15.3 Role of Microorganisms for Bioremediation of Wastewater

15.3.1 Bacterial Bioremediation

Several bacteria are widely used for bioremediation especially for the treatment of wastewater including *Pseudomonas*, *Bacillus subtilis*, and endophytes. For example, a study conducted by Das et al. (2012) used these bacteria to treat wastewater effluent in the pharmaceutical sector from different places and collected it in different forms: effluent treatment plant (ETP), condensate online water, cooling tower water, condensate water, sewage treatment plant, multiple-effect evaporator (MEE), boiler blowdown water, processed salts, and MEE feed water. In another previous study, a bacterial consortium of *Aspergillus niger, Nitrosomonas, Bacillus* sp. *(B. licheniformis, B. subtilis, B. pumilus, B. megaterium*), and *Pseudomonas fluorescens* were inoculated into wastewater effluent. After the treatment, sulfates, total dissolved solids (TDS), and total suspended solids (TSS) were remarkably decreased as compared to control sample without a treatment (Chakraborty et al. 2012). Meanwhile, chemical oxygen demand (COD) demonstrated minor changes during treatment and showed a consistent level in all samples. The results show that various types of bacterial strains are excellent candidates for the degradation of TSS and sulfate but are unable to reduce COD in water.

Two types of bacteria have been identified as the most common bacterial strains in heavy metal bioremediation: endophytic bacteria and plant growth promoting rhizobacteria. For endophytic bacteria, they will colonize the subepidermal layer of plant tissues, which helps protect plant cells from heavy metals in stressful conditions. These bacteria such as *Bacillus, Pseudomonas,* and *Rahnella* diminish the phytotoxicity of heavy metal by altering their phytoavailability and show high resistance to manganese, cadmium and lead (Abo-Alkasem et al. 2023). Plant growth promoting rhizobacteria (PGPR) are comprised of a group of free living, symbiotic, or endophytic bacteria like *Erwinia, Klebsiella, Enterobacter, Pseudomonas, Gluconacetobacter*, and *Flavobacterium*. These bacteria can improve plant growth

Table 15.2 The application of hydrolases enzymes and its treatment method in wastewater treatment

Types of hydrolases	Species	Types of wastewater	Degradation method	Result	References
Lipase	*Aeribacillus pallidus*	Oily wastewater	Free lipase	Oil removal efficiency: 96.11%	Ktata et al. (2020)
	Bacillus stearothermophilus	Oil layer biodegradation	Lipase adsorption On $CaCO_3$, Celite 545, and silica gel	COD removal: 99.79% Lipid removal: 99.92%	Ben Bacha et al. (2022)
	Pseudomonas aeruginosa	Dairy wastewater	Whole-cell immobilization on calcium alginate in a microbial electrochemical system	COD removal rates: 95.67%	Mohebrad et al. (2022)
	Aeribacillus pallidus	Oily wastewater	Free lipase	Removal rates: 96%	Ktata et al. (2020)
	Eversa transform 2.0	Oily wastewater	Polydopamine-assisted one-step immobilization of lipase on α-alumina membrane	Permeance performance: 90% Reduction in pure water permeance after oil fouling: 35%	Mulinari et al. (2023)
	Candida lipolytica	Phthalic acid esters	Combination of molecular blotting and immobilized enzyme technology	Degradation rate: 94.76%	Sun et al. (2022)

(continued)

Table 15.2 (continued)

Types of hydrolases	Species	Types of wastewater	Degradation method	Result	References
Organophosphorus hydrolase	*Enterobacter aerogenes* CP2 and *Streptococcus pyogenes* CP11	Chlorpyrifos	Whole cell	Removal efficiency: 77% and 74.20%	Lourthuraj et al. (2022)
	Arthrobacter sp. HM01	Paraoxon, chlorpyrifos, methyl parathion, malathion	Genetic engineering	Catalytic efficiency: (up to tenfold) Stability: (up to twofold)	Mali et al. (2022)
	Sphingomonas sp. and *Brevundimonas* sp.	Chlorpyrifos	Cell immobilization	Degradation efficiency: 75.4%	Santillan et al. (2020b)
	Pseudomonas sp. WBC-3 *Pseudomonas* sp. J-2	Chlorpyrifos and quizalofop-*P*-ethyl	Encapsulation organophosphorus hydrolase and aryloxyphenoxypropionate hydrolase in magnetic ZIF-8	Removal efficiency: 69.30% & 59.60% Enhanced tolerance to broad pH range, high temperatures, and various chemical solvents	Li et al. (2023)
	Pseudomonas diminuta	Organophosphorus pesticides	Enzyme immobilization onto mesoporous silica nanoparticles coated with a zwitterionic polymer containing short hydrophobic chains	2.4-fold improvement in k_{cat}/K_m, and about 3.8-fold lower in K_m	Wang et al. (2022b)
Cutinase	*Thermobifida alba*	Molasses wastewater	Immobilization on modified chitosan	Pigments removal: 79.30–81.20%	Zhang et al. (2021)
	Thermobifida fusca	BHEP and PET	Integration of M*icrobacterium oleivorans* JWG-G2 and cutinase	Degradation degree: 2.79 and 2.26	Yan et al. (2021)
	Stenotrophomonas maltophilia PRS8	PET	Extracellular medium	Depolymerization percentage: 22.50%	Din et al. (2023)
	Penicillium crustosum	PET	Purified enzymes	Hydrolysis rate: 13.68 μmol/mL/h	Anbalagan et al. (2021)
	Phylloplane yeast (Pseudozyma and Cryptococcus)	Cutin	Free cutinase	Cutin monomer: 3.20 and 1.0 μg	Ueda et al. (2021)

PETase	*Ideonella sakaiensis*	Microplastic	Immobilization of the enzyme onto silica nanoparticles	2.5-fold enhanced stability and 6.2-fold increased activity	Zurier and Goddard (2022)
	Ideonella sakaiensis	PET	Genetically engineering the curli of an *Escherichia coli* cell with PETase	Degradation efficiency: 9.10%	Zhu et al. (2022)
	Comamonas testosteroni CNB-1	PET microplastics	DuraPETase was heterologous expressed and optimized in *Comamonas testosteroni* CNB-1.	Degradation efficiency: 9.00%	Cao et al. (2023)
	Ideonella sakaiensis	PET	PETase on the surface of yeast (*Pichia pastoris*)	Turnover rate: 36-fold	Chen et al. (2020)
	Ideonella sakaiensis	PET	Overexpressed the PETase from I. *sakaiensis*in the yeast *Yarrowia lipolytica* and optimized the PET degradation process by medium supplementation	TPA release: 66%	Kosiorowska et al. (2022)

in heavy metal-contaminated soils (Seth 2012), mitigate the toxicity of heavy metals (Nadeem et al. 2010), and produce phytohormones and siderophores and phosphate solubilization (Ullah et al. 2015). Furthermore, Actinobacteria including *Streptomyces, Rhodococcus, Corynebacterium*, and *Amycolatopsis* are other bacteria species that can tolerate and remediate heavy metals like Cd(II), Ni(II), Hg(II), and Co(II) (Alvarez et al. 2017; Oyetibo et al. 2010).

15.3.2 Fungal Bioremediation

Fungi species like *Aspergillus niveus, Aspergillus niger*, and *Aspergillus fumigatus* also have potential for bioremediation, even though there are drawbacks in some strains due to a prolonged cycle of spore production. Ascomycetes are a well-known species that can reduce COD from industrial wastewater and effluent released by distilleries. Moreover, some fungal strains like *Stachybotrys Allescheriella, Pleurotus*, and *Phlebia* have been investigated to possess sequestration systems and metals (D'Annibale et al. 2007). These processes could improve their tolerance toward heavy metals and biotransformation to reduce the toxicity of heavy metals. In addition, hyphal and high biomass growth lead to an advantage for this type of microorganism because it allows simple harvesting process together with the adsorption of heavy metals (Aly et al. 2011). For instance, in a review study stated that *Aspergillus, Rhizopus,* Cephalosporium, and *Penicillium* are among the most explored fungal genera for the potential capability for the removal of heavy metals such as zinc and lead from soils and aqueous solutions (Abo-Alkasem et al. 2023). Additionally, another study also examined the impact of white rot fungus called *Bjerkandera adusta* MUT 2295 on waste sample mixtures derived from humic and tannic acids. They reported that the elimination of COD was 61% in the tannic acid effluent; however, the removal of BOD was 75% and 89% in the mixture of humic and tannic acids (Bardi et al. 2017).

15.4 Mechanisms of Microbial Action in Bioremediation of Wastewater

Naturally, bacteria have the ability to degrade metals without chemical intervention in large-scale wastewater sources, yet their efficacy can be compromised by heavy metal resistance. However, heavy metals are non-degrading and reach the environment through natural and anthropogenic activities, which cause ecological, nutritional, genetic, and evolutionary impacts on ecosystems. Regulations and standards by Food and Agriculture Organization are stipulated to maximize the permissible limit of toxic metals in irrigation water. Therefore, many techniques and strategies have been implemented to alleviate heavy metals in water and minimize their harmful effects. Bioremediation is a process that utilizes and depends on biological mechanisms to degrade or detoxify pollutants to an innoxious state and ultimately mineralize them into water, carbon dioxide, and nitrogen using live or dead biomass (Kapahi and Sachdeva 2019).

Microorganisms, especially bacteria, are recognized as the best candidates for bioremediation because of their abundance, small size, diversity, and unique capacity to grow and propagate under specific conditions with environmental resilience. By using bacteria heavy metals in wastewater can be adsorbed on the bacteria cell surface through bio-adsorption or bioaccumulation. Several bacterial species such as *Pseudomonas*, *Enterobacter*, *Flavobacterium*, and *Bacillus* have been tested for their biosorption abilities toward heavy metals. They showed great biosorption capability because of their active chemisorption sites and high surface-to-volume ratios (Mosa et al. 2016). However, the efficiency of these techniques depends on the way biomass binds with pollutants, either through active or passive processes. These bacteria have evolved diverse, discrete mechanisms adapt to and interact with environments rich in heavy metals. Different mechanisms are responsible for transporting heavy metals to the bacterial cells, including attachment on the membrane surface by bio-adsorption and intracellular sequestration by accumulation, which are discussed in this section.

15.4.1 Biosorption

Biosorption is comprised of several processes in which alive or dead biomass eliminates heavy metals or any pollutants from solution. This process is a fast and simple way to retrieve pollutants from an aqueous solution. Biosorption involves non-directed active or passive physiochemical interaction between inorganic and organic metals and cellular substances to survive toxicity (Gupta et al. 2000). The main mechanisms involved in biosorption processes are physisorption, chemisorption, surface adsorption, ion exchange, and surface complexation. Physisorption majorly occurred in the presence of Van der Waals forces, while chemisorption or "activated" adsorption happened when there was an attraction between the adsorbent and the adsorbate.

Furthermore, surface adsorption involves an "exchange" reaction of electrical attraction between positively charged metal ions in the medium and negatively charged ligands that exist on the cell wall (Ledin 2000). Moreover, heavy metal binding comprises a two-stage process that involves interaction between metal ions and reactive groups on the bacterial cell surface, followed by metal deposition. Overall, the combination of all these mechanisms leads to complete metal adsorption on the microbial surface (Brown et al. 2000). For a successful adsorption process, there are many metals binding cellular radicals on the bacterial cell wall that undergo ionization, which turns the bacterial cell surface negatively charged and viable for cationic metal attachment, for example, carbonyl, phosphate, hydroxyl sulfate, imine, amide, and phosphonate. Table 15.3 tabulates three different mechanisms such as physical adsorption, ion exchange, and complexation for the biosorption technique in liquid medium for heavy metal remediation.

Table 15.3 Mechanisms in adsorption techniques for heavy metal remediation

Mechanisms	Description	Process	Examples
Physical adsorption/ physisorption	Transfer ions from liquid phase to solid phase by Van der Waals forces or electrostatic adsorption	(i) Bulk transfer to peripheral layer (ii) Massive transmission through random motion of atoms by solid diffusion (iii) Attachment or adhesion to the exterior surface of bacterial cells (Nasser Sahmoune and Louhab 2010)	(i) Biosorption of copper on *Zoogloea ramigera* through Van der Waals forces or electrostatic adsorption arising between bacterial cell surface and charged solute particles (Ahalya et al. 2003) (ii) Dead dried biomass of *Aeromonas hydrophila* comprised of amine, carboxyl, and hydroxyl groups responsible for Pb (II) sorption through pores (Selatnia et al. 2004)
Ion exchange	Binding of metal cations to a vacant site which adsorbed by exchange with polysaccharides counter ions on the outer membrane and cell wall of bacteria	(i) The ionization process depends on their ionization pattern, and different kinds and number of sites that exist on the cell surface (ii) Amine group is positive charge when protonated, while change to neutral when deprotonated (iii) Carboxyl, phosphate, and sulfate groups are neutral after the addition of proton and turns to negative charge in deprotonated environment (Naja et al. 2010)	(i) *Micrococcus luteus* and *Pseudomonas pseudo-alkali genes* successfully adsorbed high amount of copper and lead. The adsorption rate increased as pH increased from 2 to 6 (Leung et al. 2000) (ii) *Pseudomonas fluorescens* 4F39 is reported to undergo nickel biosorption by ion exchange technique (López et al. 2000)
Complexion	A major biosorption process that formed a complex poly-atomic molecules over cell periphery	(i) Complexion of poly-atomic molecules with positive, negative and neutral charge between reactive radicals on cell wall and metal ions in solution normally in alkaline condition	(i) Complexation of several heavy metals such as cadmium, zinc, copper, and mercury in *Pseudomonas syringae*(Leung et al. 2000) (ii) Carboxyl, phosphoryl, hydroxyl, amino group on extracellular matrix of *Shewanella putrefaciens* formed a complex with Cr(VI) (An et al. 2022)

15.4.2 Bioaccumulation

Bioaccumulation is a toxicokinetic process in which the accretion of metal within a cell organism is higher than that of losing it. Bacterial cells in normal conditions can resist high levels of chemicals up to a certain level; however, beyond this level, the chemicals become pollutants, which are toxic and endanger the bacterial cell. The sensitivity of bacteria to chemicals or pollutants is highly dependent on the chemicals and types of bacterial cells (Mishra and Malik 2013). The pathway for the accumulation of non-metabolic metals within bacterial cells is similar to that for metabolic metals. Initially, the metal ions bind to the reactive radicals on the outer surface of the bacterial cell, and subsequently, bind to the internal region of the bacterial cell through an energy-independent mechanism. On the other hand, metabolic metal ions like potassium, sodium, and magnesium are important for heavy metal transport through the cell membrane for intracellular accumulation. One of the main mechanisms for heavy metal bioaccumulation is intracellular accumulation which has three stages including intracellular sequestration, reduction in heavy metal ions and methylation of metals. This section will be focused on intracellular accumulation of microorganism for heavy metals bioremediation.

15.4.2.1 Intracellular Transport

In bacterial cells, heavy metals are transported into the periplasmic space via several methods, such as channels, secondary carrier proteins, and primary active transporters that exist on the cell membrane. The channels are comprised of α-helical proteins that assist in the passive diffusion of heavy metals across the cell membrane following the concentration gradient. These channels are considered the major intrinsic proteins of the superfamily and are found in various bacterial species including *Streptomyces, Pseudomonas*, *E. coli,* and *Serratia* to transport mercury and arsenic (Diep et al. 2018). Moreover, the translocation of heavy metals is also assisted by the presence of β-barrel proteins and porins at the outer layer membrane of gram-negative bacteria.

Meanwhile, the secondary carrier proteins involved in heavy metal accumulation are uniporters, symporters, and antiporters. The movement of these proteins depends on the proton motive force for metal shifts into the bacterial cells. Similarly, the uptake of Co, Ni, and As in *Helicobacter pylori* (Rahul and Wilson 2000), *Staphylococcus aureus* (Deng et al. 2013), and *Rhodopseudomonas palustris* (Gogada et al. 2015). There are multi-domain tertiary proteins that are embedded in the plasma membrane of bacterial cells called primary active transporter proteins. These proteins comprise a trans-membrane portion and a cytosolic ATPase coupling component that act to transport heavy metals opposite to the concentration gradient by means of energy resulting from the action of ATP reserves in cells. In summary, bioaccumulation is a process when microorganisms metabolically transport heavy metals into their intracellular space using importer complexes by developing a translocation pathway through the lipid bilayer (Fig. 15.2).

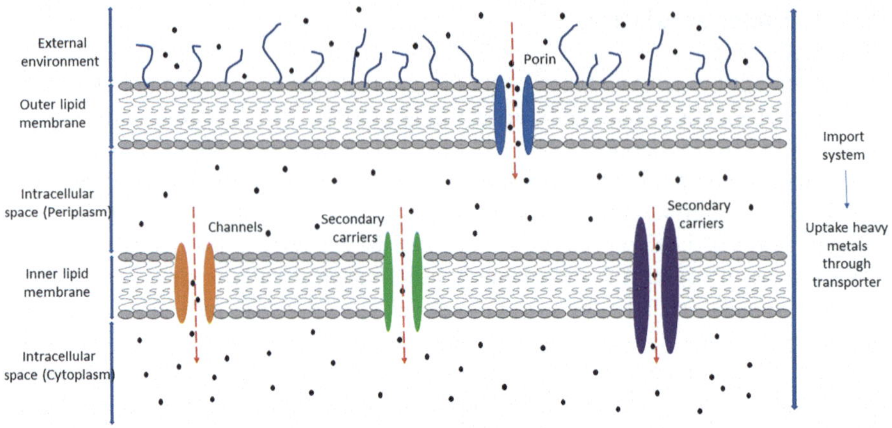

Fig. 15.2 Schematic diagram of intracellular transport in bioaccumulation of heavy metal

15.4.2.2 Intracellular Sequestration

Intracellular sequestration is a process for the formation of metal–ion composites by interacting with various cellular cytoplasmic components. Once the heavy metal enters the intracellular region of bacterial cells, it is attached, translocated, or embedded in cellular organelles and bound to ligands or compounds through the chelation process. The chelation process occurred in the cytoplasmic region with the aid of metallothionein peptides rich in cysteine and polyphosphate, the chelating molecules produced in most halophilic bacteria and archaea species. Several metal-sequestrating proteins exist in bacterial cells, including copper, zinc-storing metallothioneins, and the iron-storage protein ferritin (Klein and Lewinson 2011). The presence of metal in the cytoplasm leads to high metal stress, which induces high gene expression and overproduction of metallothionein in bacterial cells, resulting in metal binding and sequestration (Bae et al. 2000). Intracellular sequestration has been applied for effluent treatment using specific bacterial strains with the assistance of glutathione, which demonstrated a considerable amount of intracellular cadmium, aluminum, and lead ion sequestration in *Rhizobium leguminosa, Plectonema boryanum, Anabaena cylindrica,* and *Bacillus* species (Bae et al. 2000; Jroundi et al. 2020; Lima et al. 2006).

15.4.2.3 Reduction of Heavy Metal Ions

The reduction in heavy metal ions makes bacterial cells transform the oxidation state of metal ions to reduce their toxicity. This could be due to the use of metals and metalloids as electron donors or acceptors by bacteria for energy generation. Furthermore, the oxidized metals are used as terminal electron acceptors during anaerobic respiration in bacteria. Many anaerobic and facultative anaerobic bacteria species use oxidized ferum (III), manganese (IV) or chromium (VI) as terminal electron acceptors (Thamdrup 2000). For example, bacterial strains such as *Bacillus, Pseudomonas putida*, and *Pedomicrobium* reduce manganese-Mn (III and IV) and then oxidize it before being used as terminal electron acceptors in the electron transport chain (Mohanty et al. 2018).

15.4.2.4 Methylation of Metals

Another mechanism involved in the bioaccumulation of heavy metals by bacteria is the methylation of metals. This process occurs in the cytosol after heavy metals are transported from the outer membrane of the bacterial cells. Methylation is one of the fundamental biochemical pathways of bacterial cells, which is required for methylating various heavy metals, metalloids, and organometallic compounds to methyl derivatives that are easily volatile and evaporate (Gadd 2008). Intracellular methylation can occur through three different pathways involving methyl-cobalamin, S-adenosylmethionine, and *N*-methyltetrahydrofolate. Methyl-cobalamin is used by sulfate or iron reducing bacteria to methylate mercury, while arsenic is methylated via S-adenosylmethionine (Mason 2012). Commonly, the methylated metals produced by the methylation process are explosive. Previously, biomethylation of selenium led to the formation of volatile dimethyl selenide, and also volatilization of arsenic produced gaseous arsines, which can be found in polluted topsoil containing bacteria (Ramasamy et al. 2007).

15.5 Innovation and Integration of Methods for Bioremediation of Wastewater

The use of biological catalysts (enzymes and microbes) for remediation is proven to be useful since (1) they are natural green catalysts that are environmentally safe and (2) have high activity toward various types of pollutants. Due to the advancement in technology, these biocatalysts are becoming important tools for numerous applications and rapidly gaining interest. However, they are not well utilized at the industrial scale as they are intolerant and unstable at unfavorable conditions which lead to the inevitable loss of activity. Besides, they have poor long-term stability, consequently increasing the overall production and operational cost. Over the years, researchers have been continuously working to identify feasible approaches to overcome the drawbacks. Immobilization, genetic engineering, combination of methods between biological and physical/chemical are some of the innovations and integration of methods that are being extensively studied to overcome the limitations of biocatalysts. Figure 15.3 summarizes the innovation and integration of methods for bioremediation of wastewater.

15.5.1 Immobilization

The concept of immobilization has been recognized as a classical strategy to tackle the limitations of using enzymes and microbes (Lou et al. 2021). Immobilization offers a few techniques including adsorption, entrapment, cross-linking, and covalent attachment which not only improve stability but also offer re-utilization of the biocatalysts several times (Suhaimi et al. 2023). This is because the biocatalysts are attached on/in a support which makes it possible to separate them from the reaction

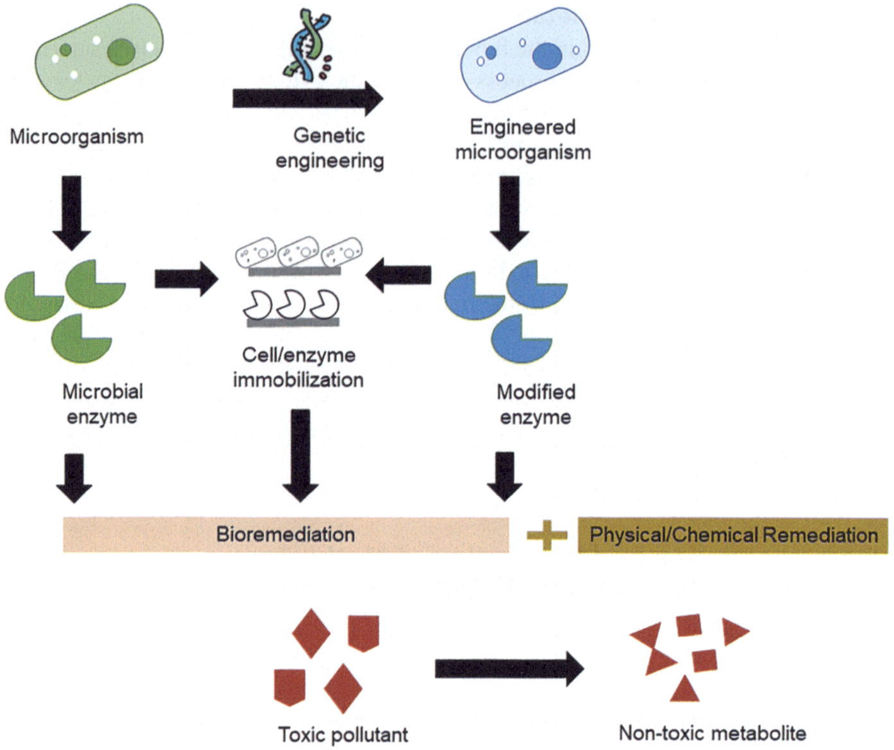

Fig. 15.3 Innovation and integration of methods for the bioremediation of wastewater adapted and modified (Saravanan et al. 2021)

mixture. The recyclability feature of the immobilization system is important to establish a cost-effective system.

A study by Imam et al. (2021) found that 100% degradation of anthracene was achieved using immobilized laccase from *Trametes maxima* on rice straw biochar. Besides, the immobilized laccase managed to be reused for six cycles (24 h per cycle). In another study, laccase immobilized peroxidase mimicking magnetic metal-organic framework (MMOF) was used to investigate the degradation of industrial dyes (methylene blue and crystal violet) in batch and continuous modes (Ladole et al. 2020). Both modes successfully achieved 90–95% degradation efficiency and sustained 89% of enzyme activity after 10 cycles for a total of 25 days. The application of a hollow fiber membrane bioreactor for the immobilization of whole-cell *P. restrictum* by Fakhri et al. (2021) found that the system has led to significant changes in the antibiotic (sulfonamide, erythromycin, and tetracycline) degradation efficiency (more than 90%). The fungi used in the system also managed to reduce the sorption of antibiotics on the membrane which significantly assisted in the reduction in membrane clogging (Fakhri et al. 2021). Other than that, the use of *Aspergillus tabacinus* LZ-immobilized on polyurethane foam in augment anaerobic activated sludge for the treatment of municipal wastewater was able to degrade

77.4% of metronidazole and increased the chemical oxygen demand to 83.30% (Yu et al. 2023).

In addition, immobilization procedure can improve the stability of the enzymes as compared to its free form. Wang et al. (2022b) evaluated that mesoporous silica nanoparticles coated with zwitterionic polymer coating can be used to modify the microenvironment of enzymes, leading to the stabilization of the immobilized enzyme while maintaining its catalytic activity during hydrolysis of methyl parathion. Thus, in this context, the types of support materials play an important role in determining the biochemical characteristic of immobilized enzymes. On top of that, immobilization procedure also has been presented as an efficient method to enhance some enzyme properties, such as activity, stability, selectivity, specificity, and resistance to inhibitors. Sun et al. (2022) integrated molecular imprinting technology and immobilized enzyme technology to synthesize nanotube-based immobilized enzyme for degradation of phthalic acid esters. The immobilized enzymes successfully degrade 94.76% of phthalic acid esters at the optimum conditions (pH = 9 and 50 °C) and can be reused up to 10 cycles associated with 65.95% of degradation rate. Therefore, the use of immobilized lipase is advantageous in many aspects, especially in term of its operational stability and recyclability.

The reusability of enzymes for repeated reaction cycle can be achieved through immobilization of cutinase by cross-linking with glutaraldehyde on magnetic nanoparticles (Muley et al. 2019) and immobilization via entrapment method into magnetic genipin-cross-linked chitosan beads (Wang et al. 2021). Nevertheless, conventional immobilization method is time consuming as it involves tedious enzyme purification step and use lots of chemicals. The new paradigm of enzyme immobilization can be achieved via exploitation of synthetic biology to engineer biocatalysts by expressing the target enzymes on microbial cell surface (Shi et al. 2018). Zhu and Wei (2019) expressed the enzyme *Fusarium solanipisi* cutinase on the cell surface of Baker's yeast *Saccharomyces cerevisiae* and displayed an amazing performance by degrading 800 μg/L of propylparaben, butylparaben, and benzylparaben, either as a single compound or as a mixture, within 48 h and maintaining 93% of its initial catalytic activity after six consecutive degradation cycle.

Moreover, Mulinari et al. (2023) developed new technique that requires single-step immobilization which simultaneously conducted with polydopamine polymerization. Single-step immobilization method gave similar results with conventional two-step immobilization method especially in enzyme loading and hydrolytic activity, while outperformed the permeance performance of the latter method by 90%. The integration of enzyme technology and materials science plays a major role in developing good immobilization technique in order to achieve high catalytic activity, stability, and reusability of the enzymes.

15.5.2 Genetic Engineering

Another way to increase the usage of biocatalysts at the industrial scale level is via genetic engineering. This can be achieved by either genetic experimentation to pick a natural form or by mutagenesis of a gene in vitro and expression in a host cell. For instance, a newly discovered thermostable, solvent-tolerant lipase from *Aeribacillus pallidus* by Ktata et al. (2019) reported that, the enzyme displayed a resistance at high temperatures (up to 70 °C) and had prominent stability in a wide range of organic solvents. In addition, the new lipase was able to degrade 96.1% of oil from the effluent of a canned tuna production industry. In another study, Harzallah et al. (2023) reported isolated *bacilli* strains from oil refinery wastewater. The genes encoding phenol hydrolase and catechol-2,3-dioxygenase have been detected in the strain and showed outstanding performance in degrading phenol at concentrations up to 1200 mg/L via meta pathway.

Liu et al. (2020) cloned enzyme laccase, MnP, and LiP from *Aspergillus* sp. TS-A and they were expressed in *Pichia pistoris* GS115 for the decolorization of bromophenol blue. The three recombinant enzymes managed to achieve more than 65% of decolorization. The recombinant enzymes also were found to exhibit strong activities over a wide range of pH and temperature in comparison to the wild-type enzymes. Certain type of enzymes requires redox mediator to perform efficient biocatalytic reaction of pollutants which would be inconvenient for on-field application. Hence, genetic cloning would offer opportunities to enhance the degradation system (Ganesh Kumar et al. 2023). Chiadò et al. (2021) combined computational simulation and molecular biology to generate engineered laccase from *Trametes versicolor*, expressed in yeast to increase the ability to process large molecules of polycyclic aromatic hydrocarbons (PAH) in the absence of mediator. The mutant laccase was reported to have a larger binding pocket to accommodate the large PAH molecules, resulting in an enhanced of activity without any mediator.

Furthermore, the integration of recombinant DNA and protein engineering to produce recombinant lipase can improve its yield and efficiency, as well as enhance its tolerance in solvent or harsh environment. Hence, there is hope in a breakthrough of the recombinant DNA technology to improve biochemical characteristics of enzymes. Ko et al. (2023) developed a mutant lipase CALB1422 by employing protein-specific secretion system in *Saccharomyces cerevisiae* and immobilized the mutant on a porous methacrylic resin. Subsequently, mutant was subjected for ester synthesis from food waste oil in the presence of water. The mutant presented 2.1-fold increase in ester conversion yield against commercial lipase (Novozyme 45).

Genetic engineering can be a great tool to improve the stereoselectivity, catalytic efficiency, and stability of the enzyme. The recombinant organophosphorus hydrolase (opdH) obtained from *Arthrobacter* sp. HM01 revealed excellent performance especially in catalyzing a broad-range of organophosphorus pesticides in organic-solvent. On top of that, the recombinant enzymes exhibited exceptional increment in catalytic efficiency (up to tenfold) and stability (up to twofold) as compared to the native enzyme (Mali et al. 2022). Even though genetic engineering can improve the catalytic efficiency and stability of the enzymes, its applicability in the large-scale operation is questionable due to catalyst recovery issue.

15.5.3 Integration of Methods

In general, the remediation of wastewater can be classified into three ways: physical, chemical, and biological-based technologies. Indeed, biological wastewater treatment is an environmentally friendly procedure (Sharma et al. 2018). However, the procedure is less effective in removing certain type of contaminants and usually requires a lot of maintenance (Su et al. 2016). On the other hand, physical and chemical are a promising procedure but suffer several significant disadvantages such as production of toxic by-products, high energy demand, high investment, regeneration of sludge, and risk of membrane fouling in filtration process. Hence, researchers have taken the initiative to explore hybrid treatment process to overcome the limitations of the individual methods (Su et al. 2016). In this section, the physical and chemical methods are integrated with biological methods to form one system that are effective for the removal of various pollutant. In addition, the integrated methods can significantly breakdown the pollutant molecules compared to individual method. Table 15.4 summarizes the integrated methods for wastewater treatment.

Table 15.4 Integrated method for wastewater treatment

Classification	Combined process	Type of wastewater	Removal (%)	References
Chemical and biological	Electrochemical oxidation and constructed wetland	Cooling tower effluent	Benzotriazole: 100 COD: 76 TOC: 48	Saha et al. (2020)
	Electrochemical oxidation and membrane bioreactor	Synthetic	Naphthalene: 100 Phenanthrene: 100	Treviño-Reséndez and Nacheva (2021)
	Anaerobic baffled reactor and down-flow hanging sponge reactor	Synthetic dyeing	Hellozol reactive black: 58 COD: 90	Nguyen et al. (2020)
	Ozonation and biofilm	Synthetic	Tetracycline hydrochloride: 97	Su et al. (2020)
Physical and biological	Ultrasound and bioreactor	Synthetic	Ibuprofen: 99 COD: 45	Abdelhay et al. (2020)
	Ultrasound-assisted alkaline solution and enzymatic hydrolysis	Oil palm front	Lignin: 47 Glucose: 72–90	Ong et al. (2021)
Biological and biological	Different microbial strains	Textile effluents	Methylene green: 74 Basic violet: 75	Shamim Hossain et al. (2022)
	Different enzymes	Trace organic contaminants	Acetaminophen: 97 Amlodipine: 98 Levothyroxine: 99 Trimethoprim: 99	Alokpa et al. (2022)

The application of chemical-biological processes is the most popular combination methods especially for dye decolorization (Su et al. 2016). According to Thanavel et al. (2020), the combined process using biological component (*Aeromonas hydrophila* SK16) and chemical component (hydrogen peroxide under UV radiation) is able to enhance the degradation of Remazol Yellow RR dye in comparison to single method. In another study, a couple system combined photocatalysis with *Bacillus cereus* was adopted for tetracycline degradation (Ji et al. 2024). It was reported that under the dual action, variety types of tetracycline intermediates were produced and finally mineralized to carbon dioxide and water. This study also found that long processing cycle and high cost associated with separate photocatalytic and microbial degradation were solved in this one system. A study by Abdelhay et al. (2020) shows an example of physical-biological methods in which sequential ultrasound-assisted activated sludge bioreactor was used to degrade ibuprofen. In the study, ultrasound was used as a pre-treatment that mainly was associated with acoustic cavitation of the ibuprofen into smaller intermediates. The intermediates were then catalyzed by the activated sludge into non-toxic products.

Combined biological-biological process is especially useful in the degradation of various contaminant in the same sample. For instance, bacterial strains of *Bacillus pseudomycoides* and *Acinetobacter haemolyticus* managed to degrade two different dyes: methylene green and basic violet found in a textile effluent (Shamim Hossain et al. 2022). The hydrolysis of ten types of trace-organic contaminants using a mixture of hydrolases (laccases, amylase, lipase cellulase and protease) in one-pot system was found efficient and cost effective as compared to single enzyme (Alokpa et al. 2022). Moreover, integration of genetic engineering and enzyme technology creates an amazing discovery in solving plastic pollution. For instance, the depolymerization of highly crystalline postconsumer polyethylene terephthalate (PET) has been investigated under ambient condition by genetically engineered the curli of *Escherichia coli* with PETase enzyme. The effects of key reaction parameters, including pH, temperature, plastic substrate mass load, and surfactant addition, were studied which resulted in 9.10% of degradation efficiency in 7 days (Zhu et al. 2022). Another study conducted by Zurier and Goddard (2022) reported that a direct immobilization of PETase could be achieved by genetically fused PETase with silica binding peptide to enable the adsorption of modified PETase onto mesoporous silica nanoparticles. This strategy minimizes the cost and steps required in immobilization procedure, while enhancing up to 2.5-fold increment of stability and 6.2-fold increment of activity of the engineered-PETase as compared to the free PETase.

15.6 Conclusion

Rapid industrialization and urbanization have facilitated the discharge of toxic effluent to the water bodies, polluting the environment and disturbing the ecological balance in nature. Conventional chemical or physical wastewater treatments are expensive, consume vast amount of energy, and are inefficient in treating emerging pollutant. Thus, bioremediation using biological methods is an efficient process to mitigate the pollutants, in conjunction with water reclamation. The selection of biological methods is primarily dependent on the types of contaminants. Enzyme-based bioremediation has been proven in catalyzing wide range of contaminants efficiently and effectively into a safe and non-toxic product. Meanwhile, microbial-based bioremediation such as bacteria and fungi is commonly used to treat heavy metals and involve complex and multi-step mechanisms during the process. Despite the efficiency of biological-based treatment, it lacks stability, low catalytic activity, and cannot be recycle for repeated used. On top of that, microorganisms and enzymes are susceptible to variable changes in environmental and physicochemical conditions. Thus, employing biotechnological advancements including immobilization methods and genetic engineering has become a great prospect in bioremediation process.

References

Abdelhaleem HAR, Zein HS, Azeiz A, Sharaf AN, Abdelhadi AA (2019) Identification and characterization of novel bacterial polyaromatic hydrocarbon-degrading enzymes as potential tools for cleaning up hydrocarbon pollutants from different environmental sources. Environ Toxicol Pharmacol 67:108–116. https://doi.org/10.1016/j.etap.2019.02.009

Abdelhay A, Allafi A, Albsoul A (2020) Optimization of ibuprofen degradation in water using high frequency ultrasound-assisted biological reactor. Water Sci Technol 81:2250–2259. https://doi.org/10.2166/wst.2020.291

Abo-Alkasem MI, Hassan NMH, Abo Elsoud MM (2023) Microbial bioremediation as a tool for the removal of heavy metals. Bull Natl Res Cent 47. https://doi.org/10.1186/s42269-023-01006-z

Agrawal K, Bhardwaj N, Kumar B, Chaturvedi V, Verma P (2019) Process optimization, purification and characterization of alkaline stable white laccase from *Myrothecium verrucaria* ITCC-8447 and its application in delignification of agroresidues. Int J Biol Macromol 125:1042–1055. https://doi.org/10.1016/j.ijbiomac.2018.12.108

Ahalya N, Ramachandra TV, Kanamadi RD (2003) Biosorption of heavy metals. Res J Chem Environ 7:71–78

Al-Dhabi NA, Esmail GA, Valan Arasu M (2021) Effective degradation of tetracycline by manganese peroxidase producing *Bacillus velezensis* strain Al-Dhabi 140 from Saudi Arabia using fibrous-bed reactor. Chemosphere 268:128726. https://doi.org/10.1016/j.chemosphere.2020.128726

Alokpa K, Lafortune F, Cabana H (2022) Application of laccase and hydrolases for trace organic contaminants removal from contaminated water. Environ Adv 8:100243. https://doi.org/10.1016/j.envadv.2022.100243

Alvarez A, Saez JM, Davila Costa JS et al (2017) Actinobacteria: current research and perspectives for bioremediation of pesticides and heavy metals. Chemosphere 166:41–62. https://doi.org/10.1016/j.chemosphere.2016.09.070

Aly AH, Debbab A, Proksch P (2011) Fungal endophytes: unique plant inhabitants with great promises. Appl Microbiol Biotechnol 90:1829–1845. https://doi.org/10.1007/s00253-011-3270-y

An H, Tian T, Wang Z, Jin R, Zhou J (2022) Role of extracellular polymeric substances in the immobilization of hexavalent chromium by *Shewanella putrefaciens* CN32 unsaturated biofilms. Sci Total Environ 810:151184. https://doi.org/10.1016/j.scitotenv.2021.151184

Anbalagan S, Raghava H, Venkatakrishnan R et al (2021) Hydrolytic degradation of polyethylene terephthalate by cutinase enzyme derived from fungal biomass–molecular characterization. Biointerface Res Appl Chem 12:653–667. https://doi.org/10.33263/briac121.653667

Ariaeenejad S, Kavousi K, Han J-L, Ding X-Z, Hosseini Salekdeh G (2023) Efficiency of an alkaline, thermostable, detergent compatible, and organic solvent tolerant lipase with hydrolytic potential in biotreatment of wastewater. Sci Total Environ 866:161066. https://doi.org/10.1016/j.scitotenv.2022.161066

Arnal G, Anglade J, Gavalda S et al (2023) Assessment of four engineered PET degrading enzymes considering large-scale industrial applications. ACS Catal 13:13156–13166. https://doi.org/10.1021/acscatal.3c02922

Arregui L, Ayala M, Gómez-Gil X, Gutiérrez-Soto G, Hernández-Luna CE, Herrera de los Santos M, Levin L, Rojo-Domínguez A, Romero- Martínez D, Mario CNS, Mauricio A. Trujillo-Roldán, Norma A, Valdez-Cruz (2019) Laccases: structure, function, and potential application in water bioremediation. Microb Cell Fact 18:200. https://doi.org/10.1186/s12934-019-1248-0

Bae W, Chen W, Mulchandani A, Mehra RK (2000) Enhanced bioaccumulation of heavy metals by bacterial cells displaying synthetic phytochelatins. Biotechnol Bioeng 70:518–524. https://doi.org/10.1002/1097-0290(20001205)70:5<518::AID-BIT6>3.0.CO;2-5

Baena A, Orjuela A, Rakshit SK, Clark JH (2022) Enzymatic hydrolysis of waste fats, oils and greases (FOGs): status, prospective, and process intensification alternatives. Chem Eng Process Process Intensif 175:108930. https://doi.org/10.1016/j.cep.2022.108930

Bala S, Garg D, Thirumalesh BV et al (2022) Recent strategies for bioremediation of emerging pollutants: a review for a green and sustainable environment. Toxics 10:1–24. https://doi.org/10.3390/toxics10080484

Bankole PO, Semple KT, Jeon BH, Govindwar SP (2021) Impact of redox-mediators in the degradation of olsalazine by marine-derived fungus, *Aspergillus aculeatus* strain bpo2: response surface methodology, laccase stability and kinetics. Ecotoxicol Environ Saf 208:111742. https://doi.org/10.1016/j.ecoenv.2020.111742

Bardi A, Yuan Q, Tigini V et al (2017) Recalcitrant compounds removal in raw leachate and synthetic effluents using the white-rot fungus *Bjerkanderaadusta*. Water 9. https://doi.org/10.3390/w9110824

Ben Bacha A, Alonazi M, Alanazi H, Alharbi MG, Jallouli R, Karray A (2022) Biochemical study of *Bacillus stearothermophilus* immobilized lipase for oily wastewater treatment. Processes 10. https://doi.org/10.3390/pr10112220

Bressler DC, Fedorak PM, Pickard MA (2000) Oxidation of carbazole, N-ethylcarbazole, fluorene, and dibenzothiophene by the laccase of *Coriolopsis gallica*. Biotechnol Lett 22:1119–1125. https://doi.org/10.1023/A:1005633212866

Brown PA, Gill SA, Allen SJ (2000) Metal removal from wastewater using peat. Water Res 34:3907–3916. https://doi.org/10.1016/S0043-1354(00)00152-4

Brugnari T, Braga DM, dos Santos CSA et al (2021) Laccases as green and versatile biocatalysts: from lab to enzyme market—an overview. Bioresour Bioprocess 8:1–29. https://doi.org/10.1186/s40643-021-00484-1

Cao Z, Xia W, Wu S et al (2023) Bioengineering *Comamonas testosteroni* CNB-1: a robust whole-cell biocatalyst for efficient PET microplastic degradation. Bioresour Bioprocess 10:94. https://doi.org/10.1186/s40643-023-00715-7

Chakraborty S, Drioli E, Giorno L (2012) Development of a two separate phase submerged biocatalytic membrane reactor for the production of fatty acids and glycerol from residual vegetable oil streams. Biomass Bioenergy 46:574–583. https://doi.org/10.1016/j.biombioe.2012.07.004

Chen C-C, Han X, Ko T-P, Liu W, Guo R-T (2018) Structural studies reveal the molecular mechanism of PETase. FEBS J 285:3717–3723. https://doi.org/10.1111/febs.14612

Chen Z, Wang Y, Cheng Y et al (2020) Efficient biodegradation of highly crystallized polyethylene terephthalate through cell surface display of bacterial PETase. Sci Total Environ 709:136138. https://doi.org/10.1016/j.scitotenv.2019.136138

Cheng M, Chen D, Parales RE, Jiang J (2022) Oxygenases as powerful weapons in the microbial degradation of pesticides. Ann Rev Microbiol 76:325–348. https://doi.org/10.1146/annurev-micro-041320-091758

Chiadò A, Bosco F, Bardelli M et al (2021) Rational engineering of the lccβ *T. versicolor* laccase for the mediator-less oxidation of large polycyclic aromatic hydrocarbons. Comput Struct Biotechnol J 19:2213–2222. https://doi.org/10.1016/j.csbj.2021.03.017

D'Annibale A, Leonardi V, Federici E, Baldi F, Zecchini F, Petruccioli M (2007) Leaching and microbial treatment of a soil contaminated by sulphide ore ashes and aromatic hydrocarbons. Appl Microbiol Biotechnol 74:1135–1144. https://doi.org/10.1007/s00253-006-0749-z

Danner MC, Robertson A, Behrends V, Reiss J (2019) Antibiotic pollution in surface fresh waters: occurrence and effects. Sci Total Environ 664:793–804. https://doi.org/10.1016/j.scitotenv.2019.01.406

Das MP, Bashwant M, Kumar K, Das J (2012) Control of pharmaceutical effluent parameters through bioremediation. J Chem Pharm Res 4:1061

Deng X, He J, He N (2013) Comparative study on Ni^{2+}-affinity transport of nickel/cobalt permeases (NiCoTs) and the potential of recombinant Escherichia coli for Ni^{2+} bioaccumulation. Bioresour Technol 130:69–74. https://doi.org/10.1016/j.biortech.2012.11.133

Deng Y, Zhou Q, Wu Y, Chen X (2022) Properties and mechanisms of Flavin-dependent monooxygenases and their applications in natural product synthesis. Int J Mol Sci 23:1–25

Diefenbach T, Sumetzberger-Hasinger M, Braunschmid V et al (2024) Laccase-mediated degradation of petroleum hydrocarbons in historically contaminated soil. Chemosphere 348:140733. https://doi.org/10.1016/j.chemosphere.2023.140733

Diep P, Mahadevan R, Yakunin AF (2018) Heavy metal removal by bioaccumulation using genetically engineered microorganisms. Front Bioeng Biotechnol 6. https://doi.org/10.3389/fbioe.2018.00157

Din SU, Kalsoom, Satti SM et al (2023) The purification and characterization of a cutinase-like enzyme with activity on polyethylene terephthalate (PET) from a newly isolated bacterium *Stenotrophomonas maltophilia* PRS8 at a mesophilic temperature. Appl Sci. https://doi.org/10.3390/app13063686

Efremenko E, Sergeeva V (2001) Organophosphate hydrolase – an enzyme catalyzing degradation of phosphorus-containing toxins and pesticides. Russ Chem Bull 50:1826–1832. https://doi.org/10.1023/A:1014377912147

Ehad A, Hegazy AK, Gobouri AA, Amr E (2020) Impact of transgenic *Arabidopsis thaliana* plants on herbicide Isoproturon phytoremediation through expressing human cytochrome P450-1A2. Biology 9:1–14

Ekta S, Ameya DB, Sushama MG (2022) Hydrolases: the most diverse class of enzymes. In: Sajjad H, Adnan H, Angel C (eds) Hydrolases. IntechOpen, Rijeka, p Ch. 1

Elmetwalli A, Allam NG, Hassan MG et al (2023) Evaluation of *Bacillus aryabhattai*B8W22 peroxidase for phenol removal in wastewater effluents. BMC Microbiol 23:1–13. https://doi.org/10.1186/s12866-023-02850-9

Fakhri H, Shahi A, Ovez S, Aydin S (2021) Bioaugmentation with immobilized endophytic *Penicillium restrictum* to improve quorum quenching activity for biofouling control in an aerobic hollow-fiber membrane bioreactor treating antibiotic-containing wastewater. Ecotoxicol Environ Saf 210:111831. https://doi.org/10.1016/j.ecoenv.2020.111831

Fernandes M, Diego HS, Rosana OH, Mauricio VA, Everton S, Agenor FI (2020) Obtaining soybean peroxidase from soybean hulls and its application for detoxification of 2,4-dichlorophenol contaminated water. J Environ Chem Eng, 8: 3. https://doi.org/10.1016/j.jece.2020.103786

Gadd G (2008) Accumulation and transformation of metals by microorganisms. In: Biotechnology, a multi-volume comprehensive treatise, volume 10: special processes, pp 225–264

Ganesh Kumar A, Manisha D, Nivedha Rajan N et al (2023) Biodegradation of phenanthrene by piezotolerant *Bacillus subtilis* EB1 and genomic insights for bioremediation. Mar Pollut Bull 194:115151. https://doi.org/10.1016/j.marpolbul.2023.115151

Gao Y, Wang M, Shah K, Singh Kalra S, Rome LH, Mahendra S (2022) Decolorization and detoxification of synthetic dye compounds by laccase immobilized in vault nanoparticles. Bioresour Technol 351:127040. https://doi.org/10.1016/j.biortech.2022.127040

Gogada R, Singh SS, Lunavat SK et al (2015) Engineered *Deinococcusradiodurans* R1 with NiCoT genes for bioremoval of trace cobalt from spent decontamination solutions of nuclear power reactors. Appl Microbiol Biotechnol 99:9203–9213. https://doi.org/10.1007/s00253-015-6761-4

Grimsley JK, Scholtz JM, Pace CN, Wild JR (1997) Organophosphorus hydrolase is a remarkably stable enzyme that unfolds through a homodimeric intermediate. Biochemistry 36:14366–14374. https://doi.org/10.1021/bi971596e

Grochulski P, Li Y, Schrag JD, Cygler M (1994) Two conformational states of *Candida rugosa* lipase. Protein Sci 3:82–91

Gupta R, Ahuja P, Khan S, Saxena RK, Mohapatra H (2000) Microbial biosorbents: meeting challenges of heavy metal pollution in aqueous solutions. Curr Sci 78:967–973

Haque AM, Hwang CE, Kim SC et al (2020) Biodegradation of organophosphorus insecticides by two organophosphorus hydrolase genes (opdA and opdE) from isolated *Leuconostoc mesenteroides* WCP307 of kimchi origin. Process Biochem 94:340–348. https://doi.org/10.1016/j.procbio.2020.04.026

Harzallah B, Grama SB, Bousseboua H, Jouanneau Y, Yang J, Li J (2023) Isolation and characterization of Indigenous *Bacilli* strains from an oil refinery wastewater with potential applications for phenol/cresol bioremediation. J Environ Manag 332:117322. https://doi.org/10.1016/j.jenvman.2023.117322

Hassan SWM, Abd El Latif HH, Ali SM (2018) Production of cold-active lipase by free and immobilized marine *Bacillus cereus* HSS: application in wastewater treatment. Front Microbiol 9:2377

Hoyos LV, Chaves A, Grandezz D et al (2023) Systematic screening strategy for fungal laccase activity of endophytes from *Otoba gracilipes* with bioremediation potential. Fungal Biol 127:1298–1311. https://doi.org/10.1016/j.funbio.2023.08.003

Ilić Đurđić K, Ostafe R, Đurđević Đelmaš A et al (2020) Saturation mutagenesis to improve the degradation of azo dyes by versatile peroxidase and application in form of VP-coated yeast cell walls. Enzym Microb Technol 136:109509. https://doi.org/10.1016/j.enzmictec.2020.109509

Imam A, Suman SK, Singh R, Vempatapu BP, Ray A, Kanaujia PK (2021) Application of laccase immobilized rice straw biochar for anthracene degradation. Environ Pollut 268:115827. https://doi.org/10.1016/j.envpol.2020.115827

Inoue S, Morita R, Minami Y (2021) An indigo-producing plant, *Polygonum tinctorium*, possesses a flavin-containing monooxygenase capable of oxidizing indole. Biochem Biophys Res Commun 534:199–205. https://doi.org/10.1016/j.bbrc.2020.11.112

Jang J, Khanom S, Moon Y, Shin S, Lee OR (2020) PgCYP76B93 docks on phenylurea herbicides and its expression enhances chlorotoluron tolerance in Arabidopsis. Appl Biol Chem 63:1–11. https://doi.org/10.1186/s13765-020-00498-x

Jankowska K, Zdarta J, Grzywaczyk A et al (2021) Horseradish peroxidase immobilised onto electrospun fibres and its application in decolourisation of dyes from model sea water. Process Biochem 102:10–21. https://doi.org/10.1016/j.procbio.2020.11.015

Jerves C, Neves RPP, Ramos MJ, da Silva S, Fernandes PA (2021) Reaction mechanism of the PET degrading enzyme PETase studied with DFT/MM molecular dynamics simulations. ACS Catal 11:11626–11638. https://doi.org/10.1021/acscatal.1c03700

Ji L, Li K, Gao M et al (2024) Elucidating tetracycline degradation in a coupled system of photocatalysis combined with *Bacillus cereus*. J Water Process Eng 57:2–11. https://doi.org/10.1016/j.jwpe.2023.104616

Jroundi F, Martinez-Ruiz F, Merroun ML, Gonzalez-Muñoz MT (2020) Exploring bacterial community composition in Mediterranean deep-sea sediments and their role in heavy metal accumulation. Sci Total Environ 712:135660. https://doi.org/10.1016/j.scitotenv.2019.135660

Kalsoom U, Ahsan Z, Bhatti HN et al (2022) Iron oxide nanoparticles immobilized *Aspergillus flavus* manganese peroxidase with improved biocatalytic, kinetic, thermodynamic, and dye degradation potentialities. Process Biochem 117:117–133. https://doi.org/10.1016/j.procbio.2022.04.002

Kang X, Li C, Ding W et al (2023) Optimization of operating conditions in the biological enzymes for efficient waste activated sludge dewatering. Process Saf Environ Prot 170:545–552. https://doi.org/10.1016/j.psep.2022.12.046

Kapahi M, Sachdeva S (2019) Bioremediation options for heavy metal pollution. J Health Pollut 9:191203

Kim DW, Thawng CN, Lee K, Wellington EMH, Cha CJ (2019) A novel sulfonamide resistance mechanism by two-component flavin-dependent monooxygenase system in sulfonamide-degrading actinobacteria. Environ Int 127:206–215. https://doi.org/10.1016/j.envint.2019.03.046

Kiran, Rathour RK, Bhatia RK, Rana DS, Bhatt AK, Thakur N (2020) Fabrication of thermostable and reusable nanobiocatalyst for dye decolourization by immobilization of lignin peroxidase on graphene oxide functionalized $MnFe_2O_4$ superparamagnetic nanoparticles. Bioresour Technol 317:124020. https://doi.org/10.1016/j.biortech.2020.124020

Klein JS, Lewinson O (2011) Bacterial ATP-driven transporters of transition metals: physiological roles, mechanisms of action, and roles in bacterial virulence†. Metallomics 3:1098–1108. https://doi.org/10.1039/c1mt00073j

Ko H, Kim M-J, Kim H-J et al (2023) Efficient valorization of food waste oils to renewable biodiesel by a *Candida antarctica* lipase B mutant that catalyzes the ester synthesis reaction in the presence of water. J Clean Prod 428:139336. https://doi.org/10.1016/j.jclepro.2023.139336

Kong W, Xiao F, Lei W, Ahmad A, Jingli Z, Fuying M, Xiaoyu Z, Hongbo Y (2017) A novel and efficient fungal delignification strategy based on versatile peroxidase for lignocellulose bioconversion. Biotechnol Biofuels, 10: 218. https://doi.org/10.1186/s13068-017-0906-x

Kosiorowska KE, Moreno AD, Iglesias R, Leluk K, Mirończuk AM (2022) Production of PETase by engineered *Yarrowia lipolytica* for efficient poly(ethylene terephthalate) biodegradation. Sci Total Environ 846:157358. https://doi.org/10.1016/j.scitotenv.2022.157358

Ktata A, Krayem N, Aloulou A et al (2019) Purification, biochemical and molecular study of lipase producing from a newly thermoalkaliphilic*Aeribacillus pallidus* for oily wastewater treatment. J Infect Dis 220:89–99. https://doi.org/10.1093/jb/mvz083

Ktata A, Krayem N, Aloulou A et al (2020) Purification, biochemical and molecular study of lipase producing from a newly thermoalkaliphilic *Aeribacillus pallidus* for oily wastewater treatment. J Biochem 167:89–99. https://doi.org/10.1093/jb/mvz083

Kumar VV, Venkataraman S, Kumar PS et al (2022) Laccase production by *Pleurotus ostreatus* using cassava waste and its application in remediation of phenolic and polycyclic aromatic hydrocarbon-contaminated lignocellulosic biorefinery wastewater. Environ Pollut 309:119729. https://doi.org/10.1016/j.envpol.2022.119729

Ladole MR, Pokale PB, Patil SS, Belokar PG, Pandit AB (2020) Laccase immobilized peroxidase mimicking magnetic metal organic frameworks for industrial dye degradation. Bioresour Technol 317:124035. https://doi.org/10.1016/j.biortech.2020.124035

Ledin M (2000) Accumulation of metals by microorganisms – processes and importance for soil systems. Earth Sci Rev 51:1–31. https://doi.org/10.1016/S0012-8252(00)00008-8

Leung WC, Wong MF, Chua H, Lo W, Yu PHF, Leung CK (2000) Removal and recovery of heavy metals by bacteria isolated from activated sludge treating industrial effluents and municipal wastewater. Water Sci Technol 41:233–240. https://doi.org/10.2166/wst.2000.0277

Li M, Li L, Sun Y, Ma H, Zhang H, Li F (2023) Facile synthesis of dual-hydrolase encapsulated magnetic ZIF-8 composite for efficient removal of multi-pesticides induced pollution in water. Chemosphere 314:137673. https://doi.org/10.1016/j.chemosphere.2022.137673

Liang X, Zou H (2023) Biotechnological application of cutinase: a powerful tool in synthetic biology. SynBio. https://doi.org/10.3390/synbio1010004

Lima AIG, Corticeiro SC, de Almeida Paula Figueira EM (2006) Glutathione-mediated cadmium sequestration in *Rhizobium leguminosarum*. Enzym Microb Technol 39:763–769. https://doi.org/10.1016/j.enzmictec.2005.12.009

Liu S, Xu X, Kang Y, Xiao Y, Liu H (2020) Degradation and detoxification of azo dyes with recombinant ligninolytic enzymes from *Aspergillus* sp. with secretory overexpression in Pichia pastoris: degradation and detoxification of dye. R Soc Open Sci 7:1–15. https://doi.org/10.1098/rsos.200688

Liu H, Yuan M, Liu A, Ren L, Zhu G, Sun L (2021) A bifunctional enzyme belonging to cytochrome P450 family involved in the O-dealkylation and N-dealkoxymethylation toward chloroacetanilide herbicides in *Rhodococcus* sp. B2. Microb Cell Factories 20:1–13. https://doi.org/10.1186/s12934-021-01544-z

López A, Lázaro N, Priego JM, Marqués AM (2000) Effect of pH on the biosorption of nickel and other heavy metals by *Pseudomonas fluorescens* 4F39. J Ind Microbiol Biotechnol 24:146–151. https://doi.org/10.1038/sj.jim.2900793

Lou WY, Fernández-Lucas J, Ge J, Wu C (2021) Editorial: enzyme or whole cell immobilization for efficient biocatalysis: focusing on novel supporting platforms and immobilization techniques. Front Bioeng Biotechnol 9:1–4. https://doi.org/10.3389/fbioe.2021.620292

Lourthuraj AA, Hatshan MR, Hussein DS (2022) Biocatalytic degradation of organophosphate pesticide from the wastewater and hydrolytic enzyme properties of consortium isolated from the pesticide contaminated water. Environ Res 205:112553. https://doi.org/10.1016/j.envres.2021.112553

Lu J, Zhang C, Leong HY, Show PL, Lu F, Lu Z (2020) Overproduction of lipoxygenase from *Pseudomonas aeruginosa* in Escherichia coli by auto-induction expression and its application in triphenylmethane dyes degradation. J Biosci Bioeng 129:327–332. https://doi.org/10.1016/j.jbiosc.2019.09.006

Mali H, Shah C, Rudakiya DM, Patel DH, Trivedi U, Subramanian RB (2022) A novel organophosphate hydrolase from *Arthrobacter* sp. HM01: characterization and applications. Bioresour Technol 349:126870. https://doi.org/10.1016/j.biortech.2022.126870

Martinez C, De Geus P, Lauwereys M, Matthyssens G, Cambillau C (1992) *Fusarium solani* cutinase is a lipolytic enzyme with a catalytic serine accessible to solvent. Nature 356:615–618. https://doi.org/10.1038/356615a0

Martinez C, Nicolas A, van Tilbeurgh H et al (1994) Cutinase, a lipolytic enzyme with a preformed oxyanion hole. Biochemistry 33:83–89. https://doi.org/10.1021/bi00167a011

Mason R (2012) The methylation of metals and metalloids in aquatic systems. In: Methylation-from DNA, RNA and histones to diseases and treatments, pp 271–302

Mehta J, Dhaka S, Paul AK, Dayananda S, Deep A (2019) Organophosphate hydrolase conjugated UiO-66-NH_2 MOF based highly sensitive optical detection of methyl parathion. Environ Res 174:46–53. https://doi.org/10.1016/j.envres.2019.04.018

Miłek J (2024) The energies of activation and deactivation of 2,4-Dichlorophenol degradation by horseradish peroxidase immobilized on the modified nanofibrous membrane. Appl Sci 14:2423. https://doi.org/10.3390/app14062423

Mishra A, Malik A (2013) Recent advances in microbial metal bioaccumulation. Crit Rev Environ Sci Technol 43:1162–1222. https://doi.org/10.1080/10934529.2011.627044

Mohanan N, Montazer Z, Sharma PK, Levin DB (2020) Microbial and enzymatic degradation of synthetic plastics. Front Microbiol 11:580709. https://doi.org/10.3389/fmicb.2020.580709

Mohanty S, Ghosh S, Bal B, Das AP (2018) A review of biotechnology processes applied for manganese recovery from wastes. Rev Environ Sci Biotechnol 17:791–811. https://doi.org/10.1007/s11157-018-9482-1

Mohebrad B, Ghods G, Rezaee A (2022) Dairy wastewater treatment using immobilized bacteria on calcium alginate in a microbial electrochemical system. J Water Process Eng 46:102609. https://doi.org/10.1016/j.jwpe.2022.102609

Morsi R, Al-Maqdi KA, Bilal M et al (2021) Immobilized soybean peroxidase hybrid biocatalysts for efficient degradation of various emerging pollutants. Biomol Ther 11:1–19. https://doi.org/10.3390/biom11060904

Mosa KA, Saadoun I, Kumar K, Helmy M, Dhankher OP (2016) Potential biotechnological strategies for the cleanup of heavy metals and metalloids. Front Plant Sci 7:1–14

Mukherjee D, Taylor KE, Biswas N (2020) Soybean peroxidase-catalyzed oligomerization of arylamines in water: optimization, kinetics, products and cost. J Environ Chem Eng 8:103871. https://doi.org/10.1016/j.jece.2020.103871

Mukhopadhyay A, Duttagupta S, Mukherjee A (2022) Emerging organic contaminants in global community drinking water sources and supply: a review of occurrence, processes and remediation. J Environ Chem Eng 10:107560. https://doi.org/10.1016/j.jece.2022.107560

Muley AB, Chaudhari SA, Bankar SB, Singhal RS (2019) Stabilization of cutinase by covalent attachment on magnetic nanoparticles and improvement of its catalytic activity by ultrasonication. Ultrason Sonochem 55:174–185. https://doi.org/10.1016/j.ultsonch.2019.02.019

Mulinari J, Ambrosi A, Feng Y et al (2023) Polydopamine-assisted one-step immobilization of lipase on α-alumina membrane for fouling control in the treatment of oily wastewater. Chem Eng J 459:141516. https://doi.org/10.1016/j.cej.2023.141516

Nadeem SM, Zahir ZA, Naveed M, Asghar HN, Arshad M (2010) Rhizobacteria capable of producing ACC-deaminase may mitigate salt stress in wheat. Soil Sci Soc Am J 74:533–542. https://doi.org/10.2136/sssaj2008.0240

Naja GM, Murphy V, Volesky B (2010) Biosorption, Metals. In: Encyclopedia of industrial biotechnology, pp 1–29

Nasser Sahmoune M, Louhab K (2010) Kinetic analysis of trivalent chromium biosorption rimosus biomass. Arab J Sci Eng 35:2–2

Nguyen TH, Watari T, Hatamoto M, Sutani D, Setiadi T, Yamaguchi T (2020) Evaluation of a combined anaerobic baffled reactor–downflow hanging sponge biosystem for treatment of synthetic dyeing wastewater. Environ Technol Innov 19:100913. https://doi.org/10.1016/j.eti.2020.100913

Nimkande VD, Bafana A (2022) A review on the utility of microbial lipases in wastewater treatment. J Water Process Eng 46:102591. https://doi.org/10.1016/j.jwpe.2022.102591

Nunes CS, Kunamneni A (2018) Laccases-properties and applications. In: Enzymes in human and animal nutrition: principles and perspectives. Elsevier

Nyon MP, Rice DW, Berrisford JM et al (2009) Catalysis by *Glomerella cingulata* Cutinase requires conformational cycling between the active and inactive states of its catalytic triad. J Mol Biol 385:226–235. https://doi.org/10.1016/j.jmb.2008.10.050

Ong VZ, Wu TY, Chu KKL, Sun WY, Shak KPY (2021) A combined pretreatment with ultrasound-assisted alkaline solution and aqueous deep eutectic solvent for enhancing delignification and enzymatic hydrolysis from oil palm fronds. Ind Crop Prod 160:112974. https://doi.org/10.1016/j.indcrop.2020.112974

Oyetibo GO, Ilori MO, Adebusoye SA, Obayori OS, Amund OO (2010) Bacteria with dual resistance to elevated concentrations of heavy metals and antibiotics in Nigerian contaminated systems. Environ Monit Assess 168:305–314. https://doi.org/10.1007/s10661-009-1114-3

Pandey VP, Awasthi M, Singh S, Tiwari S, Dwivedi UN (2017) A Comprehensive Review on Function and Application of Plant Peroxidases. Biochem & Anal Biochem, 6:1. https://doi.org/10.4172/2161-1009.1000308

Paraschiv G, Ferdes M, Ionescu M, Moiceanu G, Zabava BS, Dinca MN (2011) Laccases—versatile enzymes used to reduce environmental pollution. J Appl Sci Res 7:1106–1110. https://doi.org/10.3329/bioethics.v1i1.9525

Pathak VM, Verma VK, Rawat BS et al (2022) Current status of pesticide effects on environment, human health and it's eco-friendly management as bioremediation: a comprehensive review. Front Microbiol 13:1–29. https://doi.org/10.3389/fmicb.2022.962619

Rahul K, Wilson DB (2000) Construction and characterization of an *Escherichia coli* strain genetically engineered for Ni(II) bioaccumulation. Appl Environ Microbiol 66:5383–5386. https://doi.org/10.1128/AEM.66.12.5383-5386.2000

Ramasamy K, Kamaludeen, Banu SP (2007) Bioremediation of metals: microbial processes and techniques. In: Singh SN, Tripathi RD (eds) Environmental bioremediation technologies. Springer Berlin Heidelberg, Berlin/Heidelberg, pp 173–187

Ramlee NN, Md Illias R, Toemen S, Abdul Manas NH, Wan Azelee NI (2021) Esterification of a waste cooking oil using metal-hybrid catalysts for chemoenzymatic reactions. Mater Today Proc 47:1280–1286. https://doi.org/10.1016/j.matpr.2021.02.795

Ramlee NN, Mahdi HI, Azelee NIW (2022a) Adsorption immobilization: mechanisms and interactions between support and lipase. In: Rahman RA, Shaarani SM (eds) Enzyme immobilization for bioprocessing. Penerbit UTM Press, UTM, Johor, pp 39–69

Ramlee NN, Mahdi HI, Azelee NIW (2022b) Biodiesel production using enzymatic catalyst. In: Biofuels and bioenergy. Elsevier, pp 133–169

Rashvanlou RB, Farzadkia M, Rezaee A, Gholami M, Kermani M, Pasalari H (2021) The influence of combined low-strength ultrasonics and micro-aerobic pretreatment process on methane generation and sludge digestion: lipase enzyme, microbial activation, and energy yield. Ultrason Sonochem 73:105531. https://doi.org/10.1016/j.ultsonch.2021.105531

Ravichandran A, Rao RG, Gopinath M, Sridhar M (2019) Purification and characterization of versatile peroxidase from *Lentinus squarrosulus* and its application in biodegradation of lignocellulosics. J Appl Biotechnol Bioeng 6:280–286. https://doi.org/10.15406/jabb.2019.06.00205

Saha P, Wagner TV, Ni J, Langenhoff AAM, Bruning H, Rijnaarts HHM (2020) Cooling tower water treatment using a combination of electrochemical oxidation and constructed wetlands. Process Saf Environ Prot 144:42–51. https://doi.org/10.1016/j.psep.2020.07.019

Sahu S, Kaur A, Khatri M, Singh G, Arya SK (2023) A review on cutinases enzyme in degradation of microplastics. J Environ Manag 347:119193. https://doi.org/10.1016/j.jenvman.2023.119193

Santillan JY, Muzlera A, Molina M, Lewkowicz ES, Iribarren AM (2020a) Microbial degradation of organophosphorus pesticides using whole cells and enzyme extracts. Biodegradation 31:423–433. https://doi.org/10.1007/s10532-020-09918-7

Santillan JY, Rojas NL, Ghiringhelli PD, Nóbile ML, Lewkowicz ES, Iribarren AM (2020b) Organophosphorus compounds biodegradation by novel bacterial isolates and their potential application in bioremediation of contaminated water. Bioresour Technol 317:124003. https://doi.org/10.1016/j.biortech.2020.124003

Santos-Beneit F, Chen LM, Bordel S et al (2023) Screening enzymes that can depolymerize commercial biodegradable polymers: heterologous expression of *Fusarium solani* cutinase in *Escherichia coli*. Microorganisms 11. https://doi.org/10.3390/microorganisms11020328

Saravanan A, Kumar PS, Vo DVN, Jeevanantham S, Karishma S, Yaashikaa PR (2021) A review on catalytic-enzyme degradation of toxic environmental pollutants: microbial enzymes. J Hazard Mater 419:126451. https://doi.org/10.1016/j.jhazmat.2021.126451

Selatnia A, Boukazoula A, Kechid N, Bakhti MZ, Chergui A, Kerchich Y (2004) Biosorption of lead (II) from aqueous solution by a bacterial dead *Streptomyces rimosus* biomass. Biochem Eng J 19:127–135. https://doi.org/10.1016/j.bej.2003.12.007

Seth CS (2012) A review on mechanisms of plant tolerance and role of transgenic plants in environmental clean-up. Bot Rev 78:32–62. https://doi.org/10.1007/s12229-011-9092-x

Shamim Hossain M, Kumar Paul G, Mahmud S et al (2022) Mixed dye degradation by *Bacillus pseudomycoides* and *Acinetobacter haemolyticus* isolated from industrial effluents: a combined affirmation with wetlab and in silico studies. Arab J Chem 15:104078. https://doi.org/10.1016/j.arabjc.2022.104078

Sharma B, Dangi AK, Shukla P (2018) Contemporary enzyme based technologies for bioremediation: a review. J Environ Manag 210:10–22. https://doi.org/10.1016/j.jenvman.2017.12.075

Shi T, Han P, You C, Zhang Y-HPJ (2018) An in vitro synthetic biology platform for emerging industrial biomanufacturing: bottom-up pathway design. Synth Syst Biotechnol 3:186–195. https://doi.org/10.1016/j.synbio.2018.05.002

Singh G, Dwivedi SK (2020) Decolorization and degradation of Direct Blue-1 (Azo dye) by newly isolated fungus *Aspergillus terreus* GS28, from sludge of carpet industry. Environ Technol Innov 18:100751. https://doi.org/10.1016/j.eti.2020.100751

Su CX-H, Low LW, Teng TT, Wong YS (2016) Combination and hybridisation of treatments in dye wastewater treatment: a review. J Environ Chem Eng 4:3618–3631. https://doi.org/10.1016/j.jece.2016.07.026

Su Y, Wang X, Dong S, Fu S, Zhou D, Rittmann BE (2020) Towards a simultaneous combination of ozonation and biodegradation for enhancing tetracycline decomposition and toxicity elimination. Bioresour Technol 304:123009. https://doi.org/10.1016/j.biortech.2020.123009

Subroto E, Indiarto R, Pangawikan AD, Huda S, Yarlina VP (2020) Characteristics, immobilization, and application of *Candida rugosa* lipase: a review. Food Res 4:1391–1401. https://doi.org/10.26656/fr.2017.4(5).060

Suhaimi S, Rizwana N, Jailani N et al (2023) In-situ fabrication of functionalized starch magnetic nanoparticles for immobilization of laccase. Jurnal Teknologi 85:169–179. https://doi.org/10.11113/jurnalteknologi.v85.20554

Sun Y, Guo M, Hu S, Fang X, Jin Z, Wu R (2022) Nanosurface-immobilized lipase and its degradation of phthalate wastewater. Mol Catal 529:112522. https://doi.org/10.1016/j.mcat.2022.112522

Tejo BA, Salleh AB, Pleiss J (2004) Structure and dynamics of *Candida rugosa* lipase: the role of organic solvent. J Mol Model 10:358–366. https://doi.org/10.1007/s00894-004-0203-z

Thamdrup B (2000) Bacterial manganese and iron reduction in aquatic sediments. Adv Microb Ecol 16:41–84. https://doi.org/10.1007/978-1-4615-4187-5_2

Thanavel M, Bankole PO, Selvam R, Govindwar SP, Sadasivam SK (2020) Synergistic effect of biological and advanced oxidation process treatment in the biodegradation of Remazol yellow RR dye. Sci Rep 10:1–9. https://doi.org/10.1038/s41598-020-77376-5

Theerachat M, Tanapong P, Chulalaksananukul W (2017) The culture or co-culture of *Candida rugosa* and *Yarrowia lipolytica* strain rM-4A, or incubation with their crude extracellular lipase and laccase preparations, for the biodegradation of palm oil mill wastewater. Int Biodeterior Biodegradation 121:11–18. https://doi.org/10.1016/j.ibiod.2017.03.002

Treviño-Reséndez J, Nacheva PM (2021) Removal of naphthalene and phenanthrene in synthetic solutions by electro-oxidation coupled with membrane bioreactor. Environ Sci Pollut Res 28:48543–48555. https://doi.org/10.1007/s11356-021-13787-9

Twala PP, Mitema A, Baburam C, Feto NA (2020) Breakthroughs in the discovery and use of different peroxidase isoforms of microbial origin. AIMS Microbiol 6:330–349. https://doi.org/10.3934/microbiol.2020020

Ueda H, Tabata J, Seshime Y, Masaki K, Sameshima-Yamashita Y, Kitamoto H (2021) Cutinase-like biodegradable plastic-degrading enzymes from phylloplane yeasts have cutinase activity. Biosci Biotechnol Biochem 85:1890–1898. https://doi.org/10.1093/bbb/zbab113

Ullah A, Hafsa M, Hazrat A, Muhammad FHM, Muhammad TJ, Hassan JC (2015) Diazotrophsassisted phytoremediation of heavy metals: a novel approach. Environ Sci Pollut Res, 22: 2505–2514. https://doi.org/10.1007/s11356-014-3699-5

Vaithyanathan VK, Vaidyanathan VK, Cabana H (2022) Laccase-driven transformation of high priority pesticides without redox mediators: towards bioremediation of contaminated wastewaters. Front Bioeng Biotechnol 9:1–13. https://doi.org/10.3389/fbioe.2021.770435

Vanleeuw E, Winderickx S, Thevissen K et al (2019) Substrate-specificity of *Candida rugosa* lipase and its industrial application. ACS Sustain Chem Eng 7:15828–15844. https://doi.org/10.1021/acssuschemeng.9b03257

Vasilachi IC, Asiminicesei DM, Fertu DI, Gavrilescu M (2021) Occurrence and fate of emerging pollutants in water environment and options for their removal. Water (Switzerland) 13:1–34. https://doi.org/10.3390/w13020181

Wang L, Sun Y (2021) Engineering organophosphate hydrolase for enhanced biocatalytic performance: a review. Biochem Eng J 168:107945. https://doi.org/10.1016/j.bej.2021.107945

Wang S-n, Chen Q-j, Zhu M-j et al (2018) An extracellular yellow laccase from white rot fungus *Trametes* sp. F1635 and its mediator systems for dye decolorization. Biochimie 148:46–54. https://doi.org/10.1016/j.biochi.2018.02.015

Wang Z, Su T, Zhao J (2021) Immobilization of *Fusarium solani*cutinase onto magnetic genipin-crosslinked chitosan beads. Catalysts. https://doi.org/10.3390/catal11101158

Wang J, Shu Z, Chen Z, Su J, Liu C (2022a) Iodide ions enhancing sulfamerazine degradation by horseradish peroxidase/H_2O_2: degradation products, degradation mechanism and toxicity assessment. J Clean Prod 337:130489. https://doi.org/10.1016/j.jclepro.2022.130489

Wang X, Wang Z, Yu L, Shi Q, Dong X, Sun Y (2022b) Zwitterionic polymer-mediated immobilization of organophosphorus hydrolase enhances hydrolysis of methyl parathion by substrate enrichment. Biochem Eng J 184:108491. https://doi.org/10.1016/j.bej.2022.108491

Wei R, von Haugwitz G, Pfaff L et al (2022) Mechanism-based design of efficient PET hydrolases. ACS Catal 12:3382–3396. https://doi.org/10.1021/acscatal.1c05856

Weralupitiya C, Wanigatunge R, Joseph S et al (2021) Anammox bacteria in treating ammonium rich wastewater: recent perspective and appraisal. Bioresour Technol 334:125240. https://doi.org/10.1016/j.biortech.2021.125240

Wu L, Liu Y, Chi B et al (2015) An innovative method for immobilizing sucrose isomerase on e-poly-L-lysine modified mesoporous TiO_2. Food Chem 187:182–188. https://doi.org/10.1016/j.foodchem.2015.04.072

Wu P, Zhang Y, Chen Z et al (2019) The organophosphorus pesticides in soil wasdegradated by *Rhodobacter sphaeroides* after wastewater treatment. Biochem Eng J 141:247–251. https://doi.org/10.1016/j.bej.2018.07.019

Xu M, Liu D, Sun P et al (2016) Degradation of 2,4,6-trinitrotoluene (TNT): involvement of Protocatechuate 3,4-dioxygenase (P34O) in Buttiauxella sp. S19-1. Toxics 9:1–23

Yadav A, Yadav P, Kumar Singh A et al (2021) Decolourisation of textile dye by laccase: process evaluation and assessment of its degradation bioproducts. Bioresour Technol 340:125591. https://doi.org/10.1016/j.biortech.2021.125591

Yan Z-F, Wang L, Xia W, Liu Z-Z, Gu L-T, Wu J (2021) Synergistic biodegradation of poly(ethylene terephthalate) using *Microbacterium oleivorans* and *Thermobifida fusca* cutinase. Appl Microbiol Biotechnol 105:4551–4560. https://doi.org/10.1007/s00253-020-11067-z

Yan Y, Wu J, Hu G et al (2022) Current state and future perspectives of cytochrome P450 enzymes for C—H and C = C oxygenation. Synth Syst Biotechnol 7:887–899. https://doi.org/10.1016/j.synbio.2022.04.009

Yang S, Xu H, Yan Q, Liu Y, Zhou P, Jiang Z (2013) A low molecular mass cutinase of *Thielavia terrestris* efficiently hydrolyzes poly(esters). J Ind Microbiol Biotechnol 40:217–226. https://doi.org/10.1007/s10295-012-1222-x

Yoshida S, Hiraga K, Takehana T et al (2016) A bacterium that degrades and assimilates poly(ethylene terephthalate). Science 351:1196–1199. https://doi.org/10.1126/science.aad6359

Yu X, Mao C, Wang W et al (2023) Reduction of metronidazole in municipal wastewater and protection of activated sludge system using a novel immobilized *Aspergillus tabacinus* LZ-M. Bioresour Technol 369:128509. https://doi.org/10.1016/j.biortech.2022.128509

Zhang H, Zhang X, Geng A (2020a) Expression of a novel manganese peroxidase from *Cerrena unicolor* BBP6 in *Pichia pastoris* and its application in dye decolorization and PAH degradation. Biochem Eng J 153:107402. https://doi.org/10.1016/j.bej.2019.107402

Zhang X, Wang X, Li C et al (2020b) Ligninolytic enzyme involved in removal of high molecular weight polycyclic aromatic hydrocarbons by *Fusarium* strain ZH-H2. Environ Sci Pollut Res 27:42969–42978. https://doi.org/10.1007/s11356-020-10192-6

Zhang Z, Wang W, Li D et al (2021) Decolorization of molasses alcohol wastewater by thermophilic hydrolase with practical application value. Bioresour Technol 323:124609. https://doi.org/10.1016/j.biortech.2020.124609

Zhu B, Wei N (2019) Biocatalytic degradation of parabens mediated by cell surface displayed cutinase. Environ Sci Technol 53:354–364. https://doi.org/10.1021/acs.est.8b05275

Zhu B, Ye Q, Seo Y, Wei N (2022) Enzymatic degradation of polyethylene terephthalate plastics by bacterial curli display PETase. Environ Sci Technol Lett 9:650–657. https://doi.org/10.1021/acs.estlett.2c00332

Zurier HS, Goddard JM (2022) Directed immobilization of PETase on mesoporous silica enables sustained depolymerase activity in synthetic wastewater conditions. ACS Appl Bio Mater 5:4981–4992. https://doi.org/10.1021/acsabm.2c00700

Solid Wastes to Renewable Energy: An Approach of the Processing Steps

16

Neeraj Kumar, Piyus Kumar Pathak, and Jahnvi Vashishth

Abstract

Solid waste encompasses any discarded or intended-for-disposal material that is not in liquid form. It includes a diverse array of materials originating from various sources, such as residential garbage, manufacturing by-products, building rubble, farming refuse, among others. The primary objective of solid waste management is to facilitate energy generation through effective treatment processes. Treatment methods not only decrease the amount of waste but also prepare it for subsequent processes such as incineration, verification, and disposal. Notably, disposal practices can yield valuable by-products, such as farmyard manure enriched with essential nutrients like nitrogen, phosphorus, and potassium (NPK). Solid waste can be categorized based on its source and characteristics, and it poses environmental and health challenges if not managed properly. Effective management includes strategies for reduction, reuse, recycling, and safe disposal. The use of municipal solid waste (MSW) as a means to generate free energy has become increasingly popular as part of the shift toward waste-to-energy solutions. The type of waste generated depends on several factors, such as the climate and the eating habits of people in a particular area. Consequently, identifying an appropriate technique for converting solid waste into energy,

N. Kumar (✉)
Department of Chemistry, Applied Science (NIET), NIMS University, Jaipur, Rajasthan, India
e-mail: neeraj.kumar1@nimsuniversity.org

P. K. Pathak
Department of Applied Science and Humanity, ITS Engineering College, Knowledge Park III, Greater Noida, Uttar Pradesh, India

J. Vashishth
NIMS Institute of Allied Medical Science and Technology, NIMS University, Jaipur, Rajasthan, India

N. Dwivedi et al. (eds.), *Wastewater Treatment Through Nature-Based Solutions*,
https://doi.org/10.1007/978-981-96-6699-7_16

along with exploring alternative methods, represents a crucial step in this process. This chapter explores the current state of solid waste management in India, focusing on energy generation through various technologies. It aims to contribute to the Clean India Mission (CIM) and serve as a valuable resource for new readers and researchers in the field, promoting sustainable practices and innovations in waste management.

Keywords

Biogas · Methane · Solid waste · Ultimate treatment · Anaerobic gas generation · Renewable energy · Solid waste management · Clean energy

16.1 Introduction

Solid waste refers to any non-liquid waste material that is discarded, including household trash, industrial waste, and agricultural waste. It can be categorized into various types. Municipal solid waste (MSW) includes everyday items like food scraps, paper, plastics, metals, and textiles. Hazardous waste includes batteries, chemicals, and medical waste. Construction and demolition waste includes waste from construction, renovation, demolition activities (concrete, wood, and metal). Industrial waste includes anything from scrap materials to chemical by-products. Agricultural waste is generated from farming activities, and includes crop residues and animal manure. This chapter covers the main types of agricultural waste: paddy straw, cow dung, and rice husk. Domestic and hotel waste is taken as kitchen waste for energy production. Distillery waste is taken as spent wash in the Indian context. Addressing contemporary environmental challenges necessitates the implementation of viable long-term solutions to achieve sustainable development (Deepanraj et al. 2017). The escalating global energy demand, coupled with the Ongoing depletion of fossil fuels, has heightened the urgency to develop renewable energy resources (Jabłonski et al. 2017). Among these, biofuels such as biogas have emerged as some of the most efficient and effective alternatives (Deepanraj et al. 2017). In the last 10 years, global solid waste production has risen sharply from 0.68 billion tons to 1.3 billion tons and is expected to hit 2.2 billion tons by 2025. This growth is fueled by factors like population increase, changes in consumption patterns, industrialization, fast economic growth, and rapid urbanization, especially in developing nations. These trends collectively contribute to the rising volumes of waste production. In India, 78.2 million tons of food is wasted annually, which is enough to feed almost 377 million people. This is equivalent to 55 kg of food waste per person per year. India is second in the world in household food waste, after China. Food waste in India is a complex issue with social, cultural, and economic factors (Food waste index report 2024). Some reasons include miscalculation at the time of meal preparation, Undervaluing the importance of food, and weakening traditional Indian value systems. Kitchen waste, classified as organic waste, consists of easily biodegradable materials rich in moisture, proteins, carbohydrates, and lipids. However, a major challenge in the anaerobic digestion of kitchen waste is the swift buildup of volatile fatty acids, causing a decrease in pH inside the reactor. This drop in pH inhibits the activity of methanogenic bacteria (Bouallagui et al. 2005; Lokshina et al. 2003). In mitigating this issue, strategies such as incorporating waste-activated sludge to balance the carbon-to-nitrogen (C/N) ratio (Lokshina et al. 2003) and implementing multi-stage systems, such as two-stage or three-stage processes (Matheri et al. 2017), have proven effective. India is the world's second-largest producer of paddy, after China. The country produces about 98 million tons of paddy each year, along with roughly 130 million tons of straw. About half of this straw is used as animal feed, while the rest is mostly burned in the fields. A small fraction is also consumed by industries such as brick kilns, paper manufacturing, and packaging (Badhan et al. 2007). The burning of paddy straw releases pollutants into the air, and the conversion of straw into biogas is one of the tools that will be beneficiary for our environment. India produces about 112 million tons of milled rice each year, along with roughly 140 million tons of paddy straw (Singh and Arya 2021). Use of biogas generated in this manner is one way to meet the increasing energy demand of people living in rural and semi-urban India and other rice-growing parts of the world, and Address the greenhouse effect through cost-effective measures (Pathak et al. 2005). The current focus involves analyzing and advocating for various solid waste management strategies centered on biogas production, which farmers can adopt. This approach not only addresses waste management but also has the potential to create local employment opportunities in rural regions. Consequently, it can lead to an enhancement in the socio-economic conditions of the farming community.

16.2 Significant Terminology

1. *Municipal Solid Waste (MSW)*: Often termed as trash or garbage, this category encompasses everyday items that are disposed of by households and other businesses.
2. *Hazardous Waste*: Hazardous waste refers to materials that endanger human health and the environment due to their chemical characteristics, including items like batteries, pesticides, and medical waste.
3. *Industrial Waste*: This refers to waste produced during manufacturing and industrial activities, which may consist of chemicals, metals, and other by-products.
4. *Construction and Demolition Waste (C&D)*: Waste produced from building, repairing, or tearing down structures, such as concrete, wood, and drywall.
5. *E-waste*: Discarded electrical and electronic gadgets like computers, phones, and TVs that can have harmful substances in them.
6. *Organic Waste*: Biodegradable waste that comes from living organisms, including food scraps, yard waste, and paper products.
7. *Landfill*: A specified location intended for the disposal of solid waste, where waste is buried and managed to minimize the environmental impact.

8. *Recycling*: Turning old or used items into new things to create less trash and save natural materials.
9. *Composting*: A method of recycling organic waste by decomposing it into nutrient-rich soil.
10. *Waste-to-Energy (WTE)*: A method that transforms non-recyclable waste into usable energy, such as electricity or heat, typically through combustion.
11. *Source Reduction*: Strategies aimed at reducing waste generation at the source, promoting practices that minimize waste before it is created.
12. *Litter*: Improperly discarded waste that is found in public places, which can harm the environment and wildlife.

16.3 Advantage of Solid Waste Treatment (SWT)

Solid waste management covers major areas, namely, environmental protection, resource recovery, public health improvement, landfill space conservation, energy generation, compliance with regulations, and community aesthetics. The waste that is used for energy generation is agricultural, domestic, and industrial waste. Solid waste treatment provides advantages in terms of environmental protection because proper treatment reduces pollution by minimizing landfill use and preventing leachate and methane emissions, which are harmful for soil and groundwater. Effective treatment reduces the risk of diseases and pests associated with untreated waste, leading to healthier communities. Some solid waste treatment methods, like incineration, can generate energy, contributing to renewable energy sources. Reducing visible waste and litter through effective treatment can enhance the appearance and livability of communities (Annual Govt. Report on SWM 2020–21; Singh et al. 2024; Municipal Solid waste generation 2024; Abdel-Shafy and Mansour 2018).

16.4 Appropriate Material for Energy Generation

The country generates approximately 160,050 metric tons of solid waste per day (TPD). Of this, around 152,740 TPD is collected, reflecting a collection efficiency of 96%. Out of the collected waste, roughly 79,970 TPD (~49%) is treated, while about 29,430 TPD (~19%) is sent to landfills. Notably, 50,655 TPD (~32%) of waste generated remains unaccounted for (Annual Report on Solid Waste Management 2020–21, CPCB, Delhi). State-wise details of SWM are provided in Table 16.1.

The trends in per capita solid waste generation over the past 6 years are shown in Fig. 16.1. A slight decline has been observed in per capita waste generation during this period. The daily per capita waste generation ranges between 2 kg and 6 kg in cities with populations varying from 1 lakh to 50 lakh (as per guidelines from the Govt. MHUA, available at https://mohua.gov.in).

In 2021, the Central Pollution Control Board (CPCB) reported that India's MSW (municipal solid waste) production rate was approximately 160,050 TPD, averaging

Table 16.1 An overview of the daily solid waste management status for each state, measured in metric tons per day (TPD)

S.No.	India States (Abbreviations as per Wikipedia)	Generated Waste	Collected	Treated	Landfilled
1.	AP[a]	6899	6830	1130	207
2.	AR	237	203	–	28
3.	AS	1200	1094	42	–
4.	BR	4283	4014	–	–
5.	CG	1649	1650	1650	–
6.	GA	227	218	199	23
7.	GJ	10,374	10,332	6946	3386
8.	HR	5353	5292	3124	2168
9.	HP	347	333	223	112
10.	JK	1464	1439	546	379
11.	JH	2227	1853	759	1088
12.	KA	11,086	10,199	6816	1252
13.	KL	3545	965	2553	–
14.	MP	8025	7237	6473	765
15.	MH	22,632	22,585	15,057	1355 (Negligently discarded = 6223)
16.	MN	283	192	108	82
17.	ML	108	95	11	84
18.	MZ	346	275	271	–
19.	NL	331	285.50	124	8
20.	OR	2134	2098	1039	1036
21.	PB	4338	4278	1895	2383
22.	RJ	6896	6721	1212	5083
23.	SK	71	73	21	51
24.	TN	13,421	12,842	9431	2301
25.	TL	9966	9964	7532	994
26.	TR	335	318	215	13
27.	UK	1459	1379	780	–
28.	UP	14,711	14,282	5522	–
29.	WB	13,708	13,355	668	203
30.	AN	90	83	76	8
31.	CG	515	515	70	445
32.	DH	269	269	238	16
33.	DL	10,992	10,991	5195	5534
34.	LD	36	18	18	–
35.	PY	505	484	37	447
36.	**Total**	~160,050	~52,849	~79,860	~29,441

Annual Report on Solid Waste Management (2020–21), CPCB, Delhi-India, ± 10variation in data

0.120 kg per person per day. Some cities in India such as Surat, Alleppy, Bobbili, Panji, and Pune have shown a positive attitude toward solid waste management (Fig. 16.2) strategies. However, municipalities are mainly focused on collection,

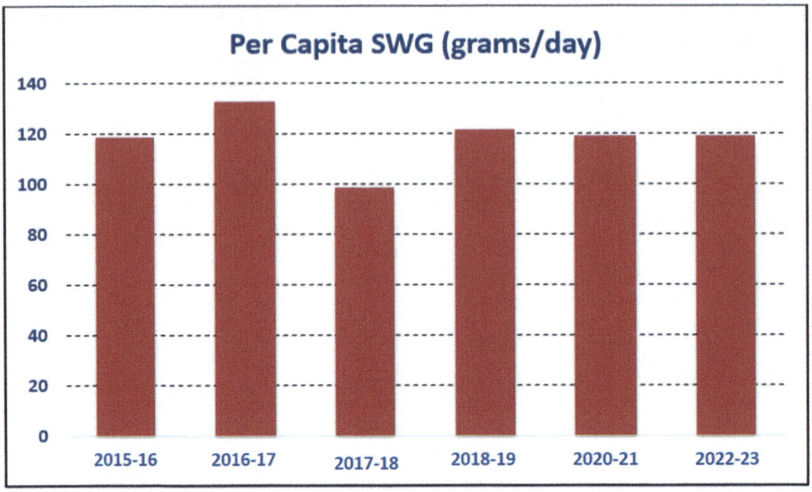

Fig. 16.1 Per capita SWG in gm/day (As per guidelines of Govt. MHUA, https://mohua.gov.in)

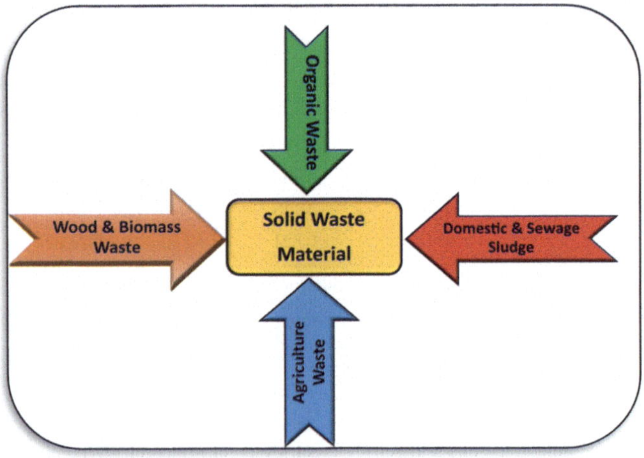

Fig. 16.2 Flow diagram of solid waste materials

whereas advanced treatment is lacking. In the financial year 2022, the total MSW generated in India exceeded 170,300 TPD. This marked an increase of approximately 6–7% compared to the previous year. To generate energy from waste, various materials such as paddy straw, cow dung, rice husk, kitchen waste, and spent wash can be used. This waste comes out of agricultural, domestic, and industrial waste. Each material used in energy generation from waste contributes differently to the energy recovery processes (MSW generation 2024).

The high-in-energy content can be converted into biogas through anaerobic digestion. Leftover crops and plant materials can be processed for bioenergy. The material is recyclable, but it can alternatively be incinerated or gasified to recover energy. The methane produced from decomposing organic waste in landfills can be captured and used for energy generation. By using these materials, waste-to-energy technologies offer a sustainable energy solution, decrease dependency on fossil fuels, and help manage waste effectively (Abdel-Shafy and Mansour 2018).

16.5 Methodology/Mechanism of Energy (Biogas) from Solid Waste

Household kitchen waste is decomposed by bacteria in an oxygen-free environment, a process known as anaerobic digestion. This produces biogas, which is mainly a mix of methane and carbon dioxide. Anaerobic digestion is a natural process that efficiently captures and uses biogas (about 65% methane and 15% carbon dioxide) to create bio-energy. The experimental setup plant competed in the SDGI Ghaziabad campus is shown in Fig. 16.3 and the related production outcome observations on different days like pH, temperature profile, and production of gas per day are given in Table 16.2.

Logan and Visvanathan (2019) suggested the general mechanism of biogas production (Fig. 16.4) in which organic waste was taken as raw material and the end product was biogas.

16.6 Description of Gas Digesters

Rajendran et al. (2012) also suggested in their study a large gas digester plant (Fig. 16.5) for solid waste materials like kitchen waste, agricultural waste (paddy straw, food waste, and cow dung) as methane culture which produces biogas after 20 days of batch digestion process.

To overcome the challenges of using and storing biogas, a modified version of the fixeddome, floating drum, and balloon-type digesters has been developed, designed to be free from technical faults. There are many advantages such as easy transportation and ease to use. This digester is easy to accept for local Indian

Fig. 16.3 Batch digestion setup biogas plant in laboratory

Table 16.2 Variation of biogas with factors, namely, pH and temperature

Day	pH	Temperature	Total Gas (ml)	Gas per day (ml)
1	6.5	40	60	60
5	6.5	39	115	55
10	6.3	43	440	325
15	6.8	42	850	410
20	6.2	45	1200	350

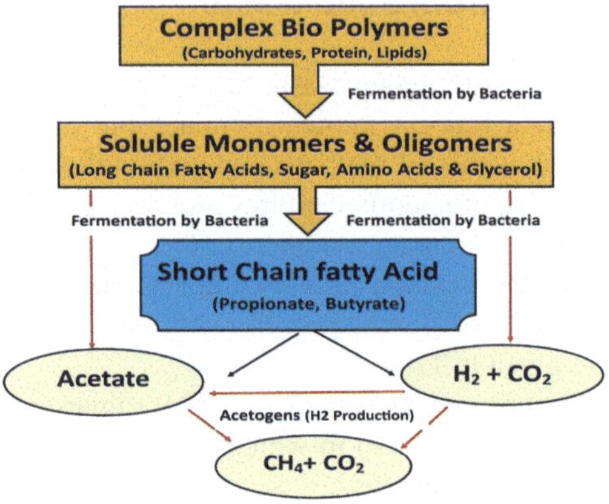

Fig. 16.4 Schematic anaerobic digestion of organic wastes. (The graph is directly taken from Logan and Visvanathan 2019)

families because of its low cost (Pathak et al. 2005). Singh et al. (2017) have designed another comfortable gas consumption assembly (Fig. 16.6) for direct kitchen use. Compactness, easy handling, and small size are the advantages of this assembly.

16.7 Factors Affecting Energy Generation from Solid Waste

Environmental factors like the type of waste, the balance between waste and bacteria, pH levels, temperature, the amount of waste added, how long it stays in the system, and the carbon-to-nitrogen ratio all play a big role in how well the anaerobic digestion process works and how much biogas is produced. Substrate: The nature of substrate depends on the sources of production like animal manure, animal dung, agricultural waste, municipal solid wastes, industrial waste, and algae biomass. The sources of substrate play a role as methane culture. Industrial wastes like digested slurry are used as methane culture for batch digestion (Nwokolo et al. 2020).

16.7.1 Inoculum-to-Substrate Ratio (I:S)

This is the crucial factor that affects stability and anaerobic digestion of food waste and the methanogenic capacity of the substrate. It also affects the anaerobic

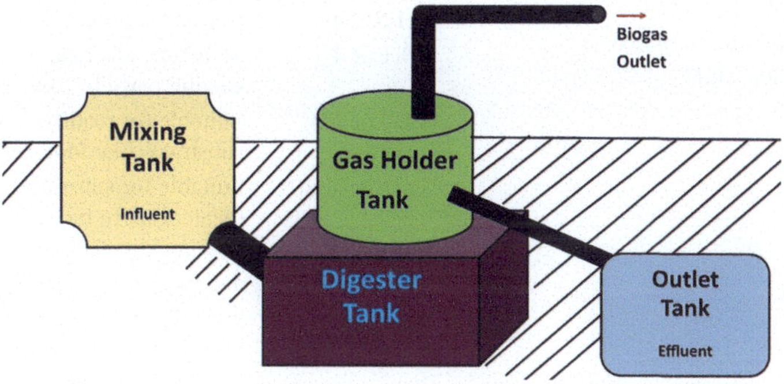

Fig. 16.5 Sketch diagram of floating type gas digestor. (The graph is directly obtained from Rajendran et al. 2012)

Fig. 16.6 Comfortable gas assembly. (Figure directly obtained from Singh et al. 2017)

digestion process, biogas composition, digestion period, reactor stability, and microbial activity. The optimal I/S ratio depends on various substrates as shown in Table 16.3 (Owamah et al. 2021).

Table 16.3 Impact of I:S ratio on production of biogas (Owamah et al. 2021)

(I/S) ratio	Impact on production
3:1 or 4:1	Stabilizes test reactors with smaller particle sizes
1:2 or 1:4	Suitable for readily degradable substrates like food waste
1:1 or higher	Suitable for substrates that are recalcitrant to biodegradation
1:3	Has the highest methane yield and the most abundant bacteria diversity
0.6	Has the maximum methane yield

16.7.2 pH

The optimum pH range for biogas production for agricultural solid wastes is 6.2–7.5. It leads to toxic effects on methane-producing bacteria when this value drops below 6.2. The pH of the digester depends on the degradable material. If it contains fatty acids, bicarbonate alkalinity, and carbon dioxide, the pH will fall below 5, which will have adverse effects on biogas production. The acidity level of the digester can affect biogas production (Abudi et al. 2020).

16.7.3 Temperature

Elevated temperatures can enhance biogas production; however, excessively high temperatures may adversely affect microbial activity and digester efficiency. Methanogenic bacteria exhibit reduced activity at both extremely high and low temperature thresholds. Consequently, the operational temperature of the reactor significantly influences the rate and yield of biogas production. The temperature within the digester directly impacts the retention time and the overall biogas output. The optimal temperature range for efficient biogas production is between 30 °C and 35 °C. Under winter conditions, when heating processes are neglected, temperatures can fall below 10 °C, leading to suboptimal digester performance (Krishania et al. 2012).

16.7.4 Organic Loading Rate (OLR)

It is a measure of how much organic matter is added to a system per unit of time and volume. It decides the performance of the system and is used in various applications including anaerobic digestion systems and trickling filters. It is an important factor for biogas production. An increase in Organic Loading Rate (OLR) can lead to higher biogas production, but if the organic matter is overloaded, it may actually reduce gas output. It also measures how much organic matter enters a trickling filter per unit of volume in a day. A higher OLR means a less efficient filter. The OLR is

Table 16.4 The chart of C/N ratio for different solid wastes (Krishania et al. 2012)

Solid wastes	Carbon-to-nitrogen ratio
Waterfowl (duck) and human	8
Poultry	10
Goat, pig, and sheep	12–19
Beef	24
Elephant	43
Harvested rice stalks	70
Timber residue	200

influenced by a number of factors, including the substrate and its concentration, temperature, and the type of digester being used (Ahmad et al. 2021).

16.7.5 Hydraulic Retention Time (HRT)

It is the average time required for the fermentation and production of biogas. It also determines how long the raw material remains in the digester or tank. It is an important factor for any digester because it influences treatment efficiency. The HRT for kitchen waste to biogas production is 20 days (Habashy et al. 2021).

16.7.6 Effect of C/N Ratio

Carbon is the fundamental structural component of any solid fuel, although nitrogen is central to the origination, growth, and progression of aerobic bacteria. The carbon-to-nitrogen (C/N) ratio should ideally be maintained within the range of 10/1 to 23/1. If the carbon-to-nitrogen ratio exceeds the upper limit, nitrogen will be rapidly depleted to meet the protein synthesis demands of microorganisms, leaving insufficient nitrogen to interact with the carbon compounds in the feedstock. This imbalance can hinder the metabolic processes necessary for biogas production. Conversely, a C/N ratio below the optimal range may also disrupt microbial activity. The optimal C/N ratio for efficient biogas production is approximately 23/1 (Krishania et al., 2012). Table 16.4 provides the C/N ratios for various types of solid waste materials.

The other factors that can affect biogas production include microbial communities, volatile fatty acids, ammonia, light/heavy metal ions, BOD, COD, among others.

16.8 Proposed Roadmap for India

The processing unit in Delhi covers the production and use of solid waste to energy. The details are given in Tables 16.5 and 16.6.

The implementation of systematic waste management practices is essential across all Indian cities. This means starting door-to-door garbage collection services. The current garbage trucks are being improved to have two separate covered

Table 16.5 Operational waste-to-energy plant in Delhi (Khanal et al. 2019)

S.No.	Name of the plant	Existing capacity (TPD)	Electricity generation (MW)
1.	Timarpur Okhla Waste Management Company Ltd.	1952	24
2.	East Delhi Waste Processing Company Ltd., Ghazipur	1301	12
3.	Delhi MSW Solutions Ltd.,	1301	24
	Cumulative	4551	60

Table 16.6 Waste treatment facilities and sanitary landfill sites in Delhi (Khanal et al. 2019)

S.No.	Treatment plant	Site	Local administrative zone	Capacity
1.	Waste to Energy Plant	Tehkhand	SDMC	2 K TPD
2.	Integrated Waste Management (IWM) and Energy Generation Facility	Ghonda	EDMC	2 K TPD
3.	Engineered Sanitary Landfill	Tehkhand	SDMC	---------
4.	IW Collection, Segregation, Transportation, Processing and Disposal plant	Ranikhera plant for Three zones of NDMC (City-SP, Karol Bagh and Narela zone)	NDMC	2500 TPD
5.	Bio-CNG Plants	Tehkhand and Keshopur	SDMC	560 TPD Each
6.	RDF Processing plant	At Tehkhand	SDMC	51 TPD
7.	Bio-methanation Plant	–	NDMC	30 TPD
8.	Biogas Plant	Dhogha Dairy	NDMC	200 TPD
9.	Biogas Plant	Ghogha and Nandi Dairies	SDMC	200 TPD each
10.	Biogas Plant	Ghazipur Dairy	EDMC	101 TPD

sections for collecting different types of waste. New trucks will also have two sections to keep wet and dry waste apart. There is also a special container or basket in the trucks or tricycles for safely collecting dangerous household waste. Local city offices are in charge of picking up biodegradable waste every day from homes, shops, and institutions, either from the doorstep or from shared bins. Big bins in fruit and vegetable markets are emptied during quiet times or at night by city workers. For the first step in waste collection, cities use covered handcarts, tricycles, tractor trolleys, compactors, or similar tools to move waste from different places. Existing waste collection vehicles are being retrofitted with dual compartments to facilitate segregated waste collection, while newly procured vehicles will come equipped with this feature. The next step in waste management is transporting solid waste from initial collection points to specific processing facilities called secondary treatment. Dumper placers with twin-bin containers are employed for the scientific disposal of waste in landfills. The leftover waste and materials that cannot be recycled are sent to approved landfill sites. There are also plans to treat this waste further to generate energy. Across India, there is a growing need for waste-to-energy plants and vermicomposting facilities. A public-private partnership (PPP) approach is suggested for these projects, where one organization or contractor will manage all aspects of solid waste handling, including door-to-door collection, transportation, processing, and final disposal. Furthermore, the establishment of waste-to-energy cluster projects is recommended for every city in India to ensure efficient and sustainable waste management.

16.9 Conclusion

The conclusion is related with environmental impacts of solid waste management and production of energy from solid waste. The energy conversation and reduction of volume is considering the specific technologies such as incineration, gasification, and pyrolysis. It also produces sustainability due to including emissions, energy recovery and resource recovery. A key determining factor is energy demand and environmental regulations. Functional unit assessments enable customized evaluations that take specific local conditions into account. Data-driven insights can guide policymakers in developing regulations and incentives that promote the most effective waste management strategies. Ultimately, clear comparisons can encourage dialogue among stakeholders including governments, industries, and communities promoting collaboration on sustainable waste management solutions. Understanding these system limitations helps assess the feasibility, sustainability, and environmental impact of each waste-to-energy technology. Decision-makers can use this information to select the most suitable technology for specific waste management and energy recovery needs. A functional unit is a quantified description of the service provided by a waste management system, often expressed in terms of mass (e.g., tons of waste processed) or energy produced (e.g., megawatt-hours). Biogas energy has several advantages, especially as a tool for solid waste management, though some challenges may still exist. The generation of biogas plays a significant role in mitigating ecological harm, as it is produced through the conversion of agricultural waste & effluents that would otherwise release harmful gases into the atmosphere. Each of these wastes-to-energy techniques has unique benefits and challenges. The choice among them depends on various factors, including local waste composition, energy needs, regulatory requirements, and public acceptance. An integrated approach that combines multiple methods may often yield the best results in terms of waste management and energy recovery. When assessing the operational boundaries of various waste-to-energy technologies, including incineration, conventional gasification, and two-stage plasma gasification, it is crucial to evaluate multiple factors that influence their efficiency, performance, and environmental implications.

References

Abdel-Shafy HI, Mansour MSM (2018) Solid waste issue: sources, composition, disposal, recycling, and valorization. Egypt J Pet 27(4):1275–1290

Abudi ZN, Hu Z, AboodA R, Gao A (2020) Effects of alkali pre-treatment, total solid content, substrate to inoculum ratio, and pH on biogas production from anaerobic digestion of mango leaves. Waste Biomass Valoriz 11(9):887. https://doi.org/10.1007/s12649-018-0437-0

Ahmad I, Abdullah N, Chelliapan S, Krishnan S, Koji I, Yuzir A (2021) Effect of organic loading rate on the performance of modified anaerobic baffled reactor treating landfill leachate containing heavy metals. Mater Today Proc 46(5):1913–1921. https://doi.org/10.1016/j.matpr.2021.02.027

Annual Report on Solid Waste Management (2020–21), CPCB, Delhi

Badhan A et al (2007) Production of multiple xylanolytic and cellulolytic enzymes by thermophilic fungus Myceliophthora sp. IMI 387099. Bioresour Technol 98(3):504–510

Bouallagui H, Touhami Y, Ben Cheikh R, Hamdi M (2005) Bioreactor performance in anaerobic-digestion of fruit and vegetable wastes. Process Biochem 40(3–4):989–995

Deepanraj B, Sivasubramanian V, Jayara S (2017) Effect of substrate pretreatment on biogasproduction through anaerobic digestion of food waste. Int J Hydrog Energy 42:26522–26528. https://doi.org/10.1016/j.ijhydene.2017.06.178

Food Waste Index Report (2024) https://www.unep.org

Guidelines, Ministry of Housing and Urban Affairs., https://mohua.gov.in

Habashy MM, Ong Ee S, Abdeldayem OM, Al-Sakkari EG, ReneE R (2021) Food waste: a promising source of sustainable biohydrogen fuel. Trends Biotechnol 39(12):1274–1288. https://doi.org/10.1016/j.tibtech.2021.04.001

Jabłonski SJ, Kułażyński M, Sikora I, Łukaszewicz M (2017) The influence of different pretreatment methods on biogas production from Jatropha curcas oil cake. J Environ Manag 203:714–719. https://doi.org/10.1016/j.jenvman.2016.06.001

Khanal SK, Nindhia TG, Nitayavardhana S (2019) Biogas from wastes: Processes and applications. In: Sustainable Resource Recovery and Zero Waste Approaches. Elsevier, Amsterdam, pp 165–174

Krishania M, Kumar V, Vijay VK, Malik A (2012) Analysis of different techniques used for improvement of biomethanation process: a review. Fuel 106:1–9

Logan MK, Visvanathan C (2019) Management strategies for anaerobic digestate of organic fraction of municipal solid waste: Current status and future prospects. Waste Manag Res 37(1):27–39. https://doi.org/10.1177/0734242X18816793

Lokshina L, Vavilin V, Salminen E, Rintala J (2003) Modeling of anaerobic degradation of solid slaughterhouse waste. Appl Biochem Biotechnol 109(1):15–32

Matheri AN, Ndiweni SN, Belaid M, Muzenda E, Hubert R (2017) Optimizing biogas production from anaerobic co-digestion of chicken manure and organic fraction of municipal solidwaste. Renew Sust Energ Rev 80:756–764. https://doi.org/10.1016/j.rser.2017.05.068

Municipal solid waste generation per day in India FY 2019–2022, Published by Bruna Alves, 2024

Nwokolo N, Mukumba P, Obileke Ke C, Enebe M (2020) Waste to energy: a focus on the impact of substrate type in biogas production energy systems. Processes 8(10):1224. https://doi.org/10.3390/pr8101224

Owamah HI, Ikpeseni SC, Alfa MI, Oyebisi SO, Gopikumar S, Samuel OD, Ilabor SC (2021) Influence of inoculum/substrate ratio on biogas yield and kinetics from the anaerobic co-digestion of food waste and maize husk. Environ Nanotechnol Monitor Manag 16:100558. https://doi.org/10.1016/j.enmm

Pathak P, Singh A, Rupainwar DC, Dhar P (2005) Production of Biogas from paddy straw- an environmental friendly process. Our Earth 2(2):5–8

Rajendran K, Aslanzadeh S, Taherzadeh MJ (2012) Household biogas digesters-a review. Energies 5(12). https://doi.org/10.3390/en5082911

Singh G, Arya SK (2021) A review on management of rice straw by use of cleaner technologies: Abundant opportunities and expectations for Indian farming. J Clean Prod 291(1):125278. https://doi.org/10.1016/j.jclepro.2020.125278
Singh AK, Singh A, Pathak PK (2017) Generation and Analysis of biogas by using kitchen waste. Int J Tech Res Sci 2(II):IIJTRS-V2-12-006
Singh M, Madhulika S, Singh SK (2024) Tackling municipal solid waste crisis in India: Insights into cutting-edge technologies and risk assessment. Sci Total Environ 917(20):170453

The manufacturer's authorised representative in the EU is Springer Nature Customer Service Centre GmbH, Europaplatz 3, 69115 Heidelberg, Germany. If you have any concerns regarding our products, please contact ProductSafety@springernature.com

Printed and bound by CPI Group (UK) Ltd, Croydon, CR0 4YY
01/10/2026
02225936-0013